Rolla C. Carpenter

Heating and Ventilating Buildings

A Manual for Heating Engineers and Architects

Rolla C. Carpenter

Heating and Ventilating Buildings
A Manual for Heating Engineers and Architects

ISBN/EAN: 9783743390072

Manufactured in Europe, USA, Canada, Australia, Japa

Cover: Foto ©Lupo / pixelio.de

Manufactured and distributed by brebook publishing software (www.brebook.com)

Rolla C. Carpenter

Heating and Ventilating Buildings

A MANUAL FOR HEATING ENGINEERS AND ARCHITECTS.

BY

ROLLA C. CARPENTER, M.S., C.E., M.M.E.,

PROFESSOR EXPERIMENTAL ENGINEERING, CORNELL UNIVERSITY.

*Past President American Society Heating and Ventilating Engineers;
Member American Society Mechanical Engineers.*

THIRD EDITION, REVISED.
FIRST THOUSAND.

NEW YORK:
JOHN WILEY & SONS.
LONDON: CHAPMAN & HALL, LIMITED.
1898

PREFACE.

THE subject of heating and ventilating buildings relates to a branch of engineering devoted to a practical application of the general physical laws of heat to the construction of heating and ventilating apparatus. A general discussion of this subject was given in treatises by Thomas Tredgold, in 1836, and by Charles Hood in 1855, in England, and by E. Péclet in 1850, in France, in which the condition of the art of heating and ventilating as it existed at that time was described. Since those early periods no treatise has been produced relating to the general principles and methods of construction in vogue, although many excellent works have been written relating to special systems or methods of heating, and one very complete and full treatise on ventilation has been published, to which reference is made in various places in the work.

The object of the present book is to present to the reader in as concise a form as possible a general idea of the principles which apply, and of the methods of construction which are in use at the present time in various systems of heating and ventilating. In writing the book the aim has been to present first the general principles which are well established, and later the methods of application to erection of systems of heating and ventilating. It has been the desire to render the reader familiar with general methods and important details of construction, also with methods of designing and estimating costs of apparatus. A full description of the various systems in use at the present time is given so that the reader may obtain an intelligent idea of the relative merits of different methods and the classes of buildings to which each is best adapted.

In preparing the present book, which is an elementary treatise on the subject, the writer has endeavored to present in as clear and concise a manner as possible, first, a statement of the general principles and laws of pure science which apply; second, a collection of important tests which give data and figures showing the relation of theoretical principles to practical construction; third, a description of the various practical methods which are in use in heating and ventilating buildings; fourth, a description of the methods of designing various systems of heating and ventilating; fifth, a collection of tables which will be useful in the practical application of the principles stated.

The writer has endeavored to arrange the matter so that it can be understood by any person possessing a practical knowledge of English and arithmetic. Algebraic demonstrations, when introduced, are printed in smaller type, and any conclusion deduced is stated in the form of a rule or general principle. Many valuable suggestions and much material aid have been given by J. J. Blackmore, J. G. Dudley, and W. S. Higgins, members of the Committee of Publication of the National Association of Manufacturers of Heating Apparatus, in adapting the book for practical use.

It has been the desire of the writer to arrange the work in a scientific manner, and to give no methods or rules of practice which were not based on the results of good, sound reasoning, modified by such coefficients as have been obtained by actual tests or experience. In the case of most systems of heating this has been possible, and it is believed in this respect that the book will be quite different from anything which has preceded it.

A great part of the material employed in writing the book has been used in a course of lectures on the subject of heating to the students in architecture in Cornell University, and one of the objects in preparing the work was to make it useful to the architect as giving a statement of principles and methods of practice applying to this branch of his profession. Professor Charles Babcock and C. F. Osborne of the Department of Architecture, Cornell University, have given material aid and service by suggesting the nature of the information needed in connection with building design.

The book generally presents such information as the writer has found in an extensive practice in the erection and operation of heating apparatus to be that which is required by contractors and by mechanics who have charge of erection of plants. The limited size of the book does not permit any extensive illustration of plants actually constructed, but a few examples are presented, selected from work done by our most noted engineers in this line.

For the literary part of the work obligation is due to nearly every writer who has preceded; in nearly every case special credit has been given; but in the back part of the book will be found a complete list of references. The writer has had the cordial assistance of many noted heating engineers, many manufacturers of heating apparatus, and all the publishers of current literature devoted to this subject.

The principal portion of the practical part of the book is devoted to construction of gravity heating systems with steam and hot water, but systems of heating with hot air, with or without a blower, with exhaust steam and with electricity, are considered, and practical directions for construction are given. The general character of the contents will be best seen by consulting the appended table.

ITHACA, N. Y., October 1, 1895.

TABLE OF CONTENTS.

CHAPTER I.

NATURE AND PROPERTIES OF HEAT.

ARTICLE	PAGE
1. Demand for Artificial Heat	1
2. Magnitude of the Industry of Manufacturing and Installing Heating Apparatus	1
3. Nature of Heat	2
4. Measure of Heat—Heat-unit	4
5. Relation to Mechanical and to Electrical Units	4
6. Temperature—Absolute Zero	6
7. Thermometer Scales	7
8. Special Forms of Thermometers	9
9. Pyrometers and Thermometers for High Temperatures	11
10. Maxima and Minima Thermometers	12
11. Use of Thermometers	13
12. Specific Heat	14
13. Latent Heat	15
14. Radiation	15
15. Reflection and Transmission of Radiant Heat	16
16. Diffusion of Heat	17
17. Conduction of Heat	17
18. Convection, or Heating by Contact	19
19. Systems of Warming	20

CHAPTER II.

PRINCIPLES OF VENTILATION.

20. Relation of Ventilation to Heating	21
21. Composition and Pressure of the Atmosphere	21
22. Diffusion of Gases	24
23. Oxygen	24
24. Carbonic Acid or Carbon Dioxide, CO_2, and Carbonic Oxide, CO	25

ARTICLE	PAGE
25. Nitrogen—Argon	27
26. Analysis of Air	27
27. Determination of Humidity of the Air	29
28. Amount of Air Required for Ventilation	31
29. Influence of the Size of the Room on Ventilation	34
30. Force for Moving the Air	35
31. Measurements of the Velocity of Air	37
32. The Flow of Air and Gases	40
33. The Effect of Heat in Producing Motion of Air	43
34. The Inlet for Air	44
35. The Outlet for Air	48
36. Ventilation-flues	49
37. Summary of Problems of Ventilation	50
38. Dimensions of Registers and Flues	52

CHAPTER III.

AMOUNT OF HEAT REQUIRED FOR WARMING.

39. Loss of Heat from Buildings	54
40. Loss of Heat from Windows	54
41. Loss of Heat from Walls of Buildings	55
42. Heat Required for Purposes of Ventilation—Total Heat Required	59

CHAPTER IV.

HEAT GIVEN OFF FROM RADIATING SURFACES.

43. The Heat Supplied by Radiating Surfaces	60
44. Heat Emitted by Radiation	61
45. Heat Removed by Convection (Indirect Heating)	63
46. Total Heat Emitted	64
47. Material of Radiators	67
48. Methods of Testing Radiators	69
49. Measurement of Radiating Surface	73
50. Effect of Painting Radiating Surfaces	74
51. Results of Tests of Radiating Surface	75
52. Tests of Indirect Heating Surfaces	79
53. Conclusions from Radiator Tests	83
54. Probable Efficiency of Indirect Radiators	84
55. Temperature Produced in a Room by a given Amount of Surface when Outside Temperature is High	84

CHAPTER V.

PIPE AND FITTINGS USED IN STEAM AND HOT-WATER HEATING.

ARTICLE	PAGE
56. General Remarks..	87
57. Cast-iron Pipes and Fittings.................................	87
58. Wrought-iron Pipe...	89
59. Pipe Fittings...	92
60. Valves and Cocks..	98
61. Air-valves..	102
62. Expansion Joints..	105

CHAPTER VI.

RADIATORS AND HEATING SURFACES.

63. Introduction...	107
64. Radiating Surface of Pipe..................................	107
65. Vertical Pipe Steam Radiators............................	109
66. Cast-iron Steam Radiator..................................	110
67. Hot-water Radiator...	112
68. Direct Indirect Radiator....................................	116
69. Indirect Radiators...	116
70. Proportion of Parts of a Radiator......................	119

CHAPTER VII.

STEAM-HEATING BOILERS AND HOT-WATER HEATERS.

71. General Properties of Steam—Explanation of Steam Tables...	120
72. General Requisites of Steam Boilers................	121
73. Boiler Horse-power...	122
74. Relative Proportions of Heating to Grate Surface............	123
75. Water Surface in Boiler—Steam and Water Space............	126
76. Requisites for Perfect Steam-boiler.................	127
77. Classification of Boilers...................................	128
78. Horizontal Tubular Boiler................................	130
79. Locomotive and Marine Boilers.......................	131
79a. Vertical Boilers...	132
80. Water-tube Boilers...	133
81. Hot-water Heaters..	135
82. Classes of Heating-boilers and Heaters...........	136
83. Heating-boilers with Magazines......................	141
84. Heating-boilers for Soft Coal..........................	142

CHAPTER VIII.

SETTINGS AND APPLIANCES, METHODS OF OPERATING.

ARTICLE	PAGE
85. Brick Settings for Boilers	143
86. Setting of Heating-boilers	147
87. The Safety-valve	149
88. Appliances for Showing the Level of the Water in Boiler	152
89. Methods of Measuring Pressure	153
90. Thermometers	156
91. Damper Regulators	156
92. Blow-off Cocks or Valves	157
93. Expansion Tank	158
94. Form of Chimneys	160
95. Size of Chimneys	161
96. Chimney-tops	162
97. Grates	163
98. Traps	164
99. Return Traps	167
100. General Directions for the Care of Steam-heating Boilers	169
101. Care of Hot-water Heaters	171
102. Boiler Explosions	171
103. Explosions of Hot-water Heaters	176
104. Prevention of Boiler Explosions	176

CHAPTER IX.

VARIOUS SYSTEMS OF PIPING.

105. Systems Employed in Steam-heating	176
106. Definitions of Terms Used	176
107. Systems of Piping	180
108. Systems of Piping Used in Hot-water Heating	185
109. Combination Systems of Heating	188
110. Pipe Connections, Steam-heating Systems	191
111. Pipe Connections, Hot-water Heating Systems	193
112. Position of Valves in Pipes	195
113. Piping for Indirect Heaters	196
114. Comparisons of Pipe Systems	197
115. Systems of Piping where Steam does not return to the Boiler	197
116. Protection of Main Pipe from Loss of Heat	198

CHAPTER X.

DESIGN OF STEAM AND HOT-WATER SYSTEMS.

117. General Principles	201
118. Amount of Heat and Radiating Surface Required for Warming	202

ARTICLE	PAGE
119. The Amount of Surface Required for Indirect Heating.......	209
120. Summary of Approximate Rules for Estimating Radiating Surface..	215
121. Flow of Water and Steam..	217
122. Size of Pipes to Supply Radiating Surfaces..................	222
123. Size of Return Pipes, Steam Heating........................	227
124. Size of Pipes for Hot-water Radiators......................	228
125. Size of Ducts and Ventilating Flue for Conveying Air........	232
126. Dimensions of Registers......................................	235
127. Summary of Various Methods of Computing Quantities Required for Heating	236
128. Heating of Greenhouses..	236
129. Heating of Workshops and Factories........................	245

CHAPTER XI.

HEATING WITH EXHAUST STEAM. NON-GRAVITY RETURN SYSTEMS.

130. General Remarks	247
131. Systems of Exhaust Heating..................................	247
132. Proportions of Radiating Surface and Main Pipes Required in Exhaust Heating...	249
133. Systems of Exhaust Heating with Less than Atmospheric Pressure..	251
134. Combined High- and Low-pressure Heating Systems..........	255
135. Pump Governors...	256
136. The Steam Loop...	257
137. Reducing Valves...	258
138. Transmission of Steam Long Distances......................	260

CHAPTER XII.

HEATING WITH HOT AIR.

139. General Principles...	268
140. General Form of a Furnace....................................	270
141. Proportions Required for Furnace Heating..................	272
142. Air-supply for the Furnace....................................	275
143. Pipes for Heated Air..	276
144. The Areas of Registers or Openings into Various Rooms.....	278
145. Circulating Systems of Hot Air..............................	280
146. Combination Heaters...	281
147. Heating with Stoves and Fireplaces.........................	281
148. General Directions for Operating a Furnace.................	282

CHAPTER XIII.

FORCED-BLAST SYSTEMS OF HEATING AND VENTILATING.

ARTICLE	PAGE
149. General Remarks	283
150. Form of Steam-heated Surface	283
151. Ducts or Flues—Registers	284
152. Blowers or Fans	289
153. Heating Surface Required	291
154. Size of Boiler Required	292
155. Practical Construction of Hot-blast System of Heating	292
156. Systems of Ventilation without Heating	298
157. Heating with Refrigerating Machines	299
158. Cooling of Rooms	300

CHAPTER XIV.

HEATING WITH ELECTRICITY.

159. Equivalents of Electrical and Heat Energy	301
160. Expense of Heating by Electricity	301
161. Formulæ and General Considerations	304
162. Construction of Electrical Heaters	306
163. Connections for Electrical Heaters	309

CHAPTER XV.

TEMPERATURE REGULATORS.

164. General Remarks	310
165. Regulators Acting by Change of Pressure	311
166. Regulators Operated by Direct Expansion	315
167. Regulators Operated with Motor—General Types	316
168. Pneumatic Motor System	318
169. Saving Due to Temperature Regulation	320

CHAPTER XVI.

SPECIFICATION PROPOSALS AND BUSINESS SUGGESTIONS.

170. General Business Methods	322
171. General Requirements	323
172. Form Proposed by the National Association of Manufacturers of Heating Apparatus	326
173. Form of Uniform Contract	336

ARTICLE	PAGE
174. Specifications for Plain Tabular and Water-tube Boilers	340
175. Protection from Fire—Hot Air and Steam Heating	344
176. Duty of the Architect	347
177. Methods of Estimating Cost of Construction	347
178. Suggestions for Pipe-fitting	348

APPENDIX.

LITERATURE AND REFERENCES	353
EXPLANATIONS OF TABLES	356
TABLES	359
INDEX	401

A TREATISE
ON
HEATING AND VENTILATING BUILDINGS.

CHAPTER I.

INTRODUCTION.

NATURE AND PROPERTIES OF HEAT.

1. Demand for Artificial Heat.—The necessity for artificial heat depends to a great extent upon the climate, but to a certain extent on the customs or habits of the people. In all the colder regions of the earth artificial heat is necessary for the preservation of life, yet there will be found a great difference in the temperature required by people of different nations or races living under the same circumstances. On the continent of Europe, 15 degrees centigrade, corresponding to about 59 degrees F., is considered a comfortable temperature; in America it is the general practice and custom to maintain a temperature of 70 degrees in dwellings, offices, stores, and most workshops, and a heating apparatus is considered inadequate which will not maintain this temperature under all conditions of weather.

2. Magnitude of the Industry of Manufacturing and Installing Heating Apparatus.—The industry connected with the manufacture and installation of the various systems for warming is a great one and gives employment to many thousand workmen. The manufacture of heating apparatus is not only of great magnitude, but it is varied in its nature; all kinds of apparatus for heating—as, for instance, the open fireplace built at the base of a brick chimney, the cast-iron stove with its unsightly piping, the furnace and appliances for warming air, apparatus for heating by steam and also by hot water—can be readily bought on the market in almost every form, from that of the simplest to that of the most complicated design.

The exact amount of capital invested in this industry could not be ascertained by the author, but in twenty cities, selected in alphabetical order from a list of one hundred and sixty-five cities of the United States containing over twenty thousand inhabitants, the total amount invested in the business of erecting and installing heating apparatus as given in the Census Report by the U. S. Government for 1890 was $12,910,250, and the yearly receipts for 1890 from this business in the same cities was $5,592,148. The aggregate population of these cities was 1,573,508 people. This would indicate an investment of $8.20 and a yearly expenditure of $3.52 for each inhabitant. Reckoning on the same basis for the cities of the United States which contain over 25,000 inhabitants each, we should have an invested capital of over $106,000,000 and a yearly expenditure of over $46,000,000. These numbers are probably less than the amount actually invested, but they serve to give an idea of the magnitude of the industry connected with the supply of apparatus for artificial warming.

3. Nature of Heat.—Before consideration of the methods of utilizing heat in warming buildings a short discussion of the nature and scientific properties of heat seems necessary.

Heat is recognized by a bodily sensation, that of feeling, by means of which we are able to determine roughly by comparison that one body is warmer or colder than another. From a scientific standpoint heat is a peculiar form of energy, similar in many respects to electricity or light, and is capable, under favorable conditions, of being reduced into either of the above or into mechanical work. We shall have little to do with the theoretical discussion of its nature, but, as it is well to have a distinct understanding of its various forms and equivalents, we will consider briefly some of its important properties.

Heat was at one time considered a material substance which might enter into or depart from a body by some kind of conduction, and the terms which are in use to-day were largely founded on that early idea of its material existence. The theory that heat is a form of energy and is capable of transformation into work or electricity is thoroughly established by fact and experiment. It probably produces a molecular motion among the particles of bodies into which it enters, the rate of such motion being proportional to the intensity of the heat.

Heat has two qualities which correspond in a general way to *intensity* on the one hand and *quantity* on the other. The intensity of heat is termed *temperature*—this can be measured by a thermometer; but, except in scientific discussion, no name has been applied to designate the unit-quantity of heat,* and there is no method of measuring it directly, although it is of as much importance as temperature.

It is a fact which will appear from later statements that the amount of heat contained in two bodies of different kinds, but of the same weight and temperature, may be essentially different. A familiar analogy might perhaps be seen in the case of the dimensions and weight of men. The weight would depend on the general dimensions, height, breadth, etc., and it would probably be the case that two men having equal heights would have quite different weights. In a similar manner the amount of heat depends upon the temperature and also upon the property of the body to absorb heat without showing any effects which may be measured on a thermometer. This latter property in itself depends upon the nature of the body and also upon that peculiar quality of heat to which reference has been made. Under every condition heat must be quite different in nature from temperature.

Note that heat is equivalent, not to mechanical force, but to mechanical work. Work, defined scientifically, is the application of force in overcoming some resistance; it is the result of a force acting through a certain distance; the distance moved through having as much effect on the result as the force acting. The work done is proportional to the product of the force exerted, multiplied by the space passed through. In English measures the unit of this product is a *foot-pound*, which signifies one pound raised to a height equal to one foot; it is itself a complex quantity resembling heat in this respect. Heat can be transformed into work.

---

* The term *entropy* is now applied in scientific discussions to this property.

4. Measure of Heat—Heat-unit.—As explained heat cannot be measured by the thermometer; it can, however, be measured by the amount that some standard is raised in temperature. The standard adopted is water, and heat is universally measured by its power to raise the temperature of a given weight of water. In English-speaking countries the *heat-unit* is that required to raise one pound of water from a temperature of 62 to 63 degrees, and this quantity is termed a British thermal unit; this will be referred to in this work, by its initial letters B. T. U., or simply as a heat-unit. The amount of heat required to change the temperature of one pound of water one degree is not the same at all temperatures; the variation, however, is slight and for practical purposes can be entirely disregarded. The unit of heat used by the French and Germans, and for scientific purposes generally, is called the *calorie;* it is equal to one kilogramme (2.20 pounds) of water raised one degree centigrade (1.8 degrees Fahrenheit) and is equal to 3.9672 B. T. U. The *calorie* is referred to water at a temperature of 15–16° Centigrade (60 degrees Fahrenheit).

5. Relation to Mechanical Work and to Electrical Units.—The relation of heat to mechanical work was accurately measured by Joule in 1838 by noting the heating effects produced in revolving a paddle-wheel immersed in water. The wheel being revolved by a weight falling a given distance, the mechanical work was known; this compared with the rise in temperature of the water enabled him to determine that the value of one heat-unit estimated from 39° to 40° F. was equivalent to 772 foot-pounds. Later investigation has slightly increased this result, so that when reduced to a temperature of 62 degrees F., and for this latitude, it is 6 foot-pounds greater, so that at present the work equivalent of one heat-unit is generally regarded as 778 foot-pounds. This signifies that the work of raising 1 lb. 778 feet is equivalent to the energy required to change the temperature of 1 lb. of water, at 62° F. in temperature, 1 degree.

The equivalent value of heat and mechanical work is now thoroughly established, and under favorable conditions the one can always be transformed into the other. As illustrations of the transformation of heat into work we have only to consider the numerous forms of steam-engines, gas-engines, and the like. A transformation from mechanical work into heat is shown in the rise of temperature accompanying friction in the use of machines of all classes. The heat produced in the performance of any mechanical work is exactly equivalent to the work accomplished, 778 foot-pounds of mechanical work being performed in order to produce a heating effect equivalent to raising 1 lb. of water 1° Fahr.

The term horse-power has been used as the measure of the amount of work. It has been fixed as 33,000 foot-pounds per minute. This is equivalent to 42.42 B. T. U. per minute, or to 746 watts in electrical measures. For the work done in one second the above numbers should be divided by 60; for that done in one hour they should be multiplied by 60. In all English-speaking countries the capacity of engines and machinery in general is expressed in horse-power, so that it is necessary to become familiar with this term and its equivalents in heat and electrical units.

The electrical units are all based on French measures, the centimetre (0.3937 inch) being the standard of length, the gramme (15.432 grains) the standard of mass, and the second the unit of time; the system being generally denominated the C. G. S. system. In this system the unit of force, the *dyne*, is 1 gramme moved so as to acquire a velocity of one centimetre per second. As the force of gravity in latitude of Paris is 32.2 feet = 981 cm., the *dyne* is equal to the weight moved, expressed in grammes divided by 981, for latitude of Paris.

The unit of work and of energy is called an *erg* and is equal to the force of one *dyne* acting through one centimetre, or to a gramme-centimetre divided by 981.

One million ergs is equal to 0.0738 foot-pound.

One watt is equal to 10 million ergs per second, or 738 foot-pounds per second.

One calorie is 42,000 million ergs, one minor calorie 42 million ergs.

One B. T. U. is 10,550 million ergs.

Expressed in work we have the following equivalents:

One horse-power = 746 watts = 550 foot-pounds per second
                = 0.707 B. T. U. per second.
                = 0.1767 calories per second
                = 176.7 minor calories per second
                = 7460 millions of ergs per second.

(See full table of equivalents in back of book.)

6. Temperature—Absolute Zero.—One of the properties of heat is called temperature; this property can be measured by a thermometer and is proportional to the intensity of the heat. All our knowledge of heat, as obtained by the sensation of feeling, deals only with the temperature, and the terms in common use by means of which bodies are compared and denominated hot, hotter, hottest, have reference, not to the heat actually in the different bodies, but to the temperature.

There is always a tendency for heat to flow through intervening mediums from a hotter to a colder body, and there is no tendency for heat to flow from a cold to a hot body, although the relative amounts of heat in the two bodies might be different from that indicated by the thermometer. Thus, as an illustration, a pound of water requires about eight times as much heat to raise it one degree in temperature as a pound of iron, and hence when equal weights of both of these materials are at the same temperature the water contains eight times as much heat as the iron, although in common parlance the two bodies would be equally hot.

The tendency for the hotter body to cool off and give up its heat to surrounding objects is characteristic of all materials, and if no other heat were supplied all bodies would come sooner or later to one common temperature. This temperature, when finally reached by all bodies in the universe, will represent the ultimate limit of all cooling and almost the entire absence of heat. It will be near absolute zero for all thermometric scales, and no greater cold will be possible or even conceivable. The inter-planetary space is believed by many to be very nearly at this limit, at the present time. Scientific men have made very careful determinations to ascertain what such a temperature must be, compared with the ordinary thermometric scales.

A perfect gas which remains under constant pressure will contract in volume an amount directly proportional to the change of temperature when reckoned from the point of greatest cold, which point is known as the *absolute zero*. By experiment it is found that when air is at a temperature of 32 degrees its volume is reduced one part in 492 each time that the temperature is lowered one degree. From this fact it has been concluded that the absolute zero is 492 degrees on the Fahrenheit scale or 273 degrees on the Centigrade scale, below the freezing-point of water. Strictly speaking there is no perfect gas, yet the results obtained with different gases by different observers are so nearly in accord that there is no question but that the results as given above are for all practical purposes correct.

7. Thermometer Scales.—The thermometer is an instrument used to measure temperature. The effect of heat is to expand or to increase the volume of most bodies. For perfect gases the amount of this expansion is strictly proportional to the change of temperature; for liquids and solids the expansion, while not exactly proportional to the increase of temperature, is very nearly proportional to it, and these bodies can be used for an approximate and even a close measure of difference of temperature. In nearly all thermometers the temperature is measured by the expansion of some body, mercury, alcohol, or air being commonly used as the thermometric substance.

The first thermometer was probably made by Galileo before 1597. It consisted of a glass bulb containing air, terminated below in a long glass tube which dipped into a vessel containing a colored fluid. The variations of volume of the enclosed air caused the fluid to rise or fall in the tube, the temperature being read by an arbitrary scale. Alcohol thermometers were in use as early as 1647, being made by connecting a spherical bulb with a long glass stem, on which graduations were made by beads of blue enamel placed in positions corresponding to one thousandths of the volume.

Fig. 1. Ordinary form of Mercurial Thermometer.

Fahrenheit, a German merchant, in 1721 was the first to make a mercurial thermometer, and the instrument which he designed, with certain modifications, has been retained in use by the English-speaking people up to the present time. Fahrenheit took as fixed points the temperature of the human body, which he called 24 degrees, and a mixture of salt and sal-ammoniac, which he supposed the greatest cold possible, as zero. On this scale the freezing-point is 8 degrees. These degrees were afterwards divided into quarters, and later these subdivisions themselves, termed degrees. On this modified scale the freezing-point of water becomes 32 degrees, blood-heat 96 * degrees, and the point of boiling water at atmospheric pressure 212 degrees. Unscientific as this thermometer is, it has been retained by two of the principal nations of the world, the English and the American; it is awkward to use, it was borrowed from a foreign nation which had itself adopted a more scientific instrument, and except for the fact that it has been long in use it has not a single feature to recommend it.

In 1724 Delisle introduced a scale in which the boiling-point of water was called zero and the temperature of a cellar in the Paris Observatory was called 100 degrees. This thermometer was used for many years in Russia, but is now obsolete. In 1730 Réaumur made alcohol thermometers in which the boiling-point of water was marked 80 degrees. This thermometer is still in use in Russia.

Celsius adopted a centesimal scale in 1742 on which the boiling-point was marked zero and the freezing-point of water 100 degrees. This instrument is not now in use, although the centigrade scale is often called after Celsius. The botanist Linnæus introduced the centigrade thermometer, in which the freezing-point of water is marked zero and the boiling-point of water 100 degrees. This themometer is now adopted for ordinary use by the nations of continential Europe and for scientific use by every nation in the world.

The relative values of the degrees on the different thermometers used by various nations are given in the following table:

THERMOMETRIC SCALES.

	Fahrenheit.	Centigrade.	Réaumur.	Celsius.
Degrees between freezing and boiling..	180	100	80	100
Temperature at freezing-point........	32	0	0	100
Temperature at boiling-point.........	212	100	80	0
Comparative length of degree	1	9/5	9/4	9/5
" " " "	5/9	1	5/4	1
Countries where used	England and America	France and Germany	Russia	Not in use

* As determined later, this should be 98°.

In all thermometric scales as given above, fixed points are determined by reference to the freezing and boiling points of water, with barometer at 29.92 inches, and all thermometers are constructed by marking these two points and then subdividing into the required number of degrees. The boiling-point of water changes with the atmospheric pressure and with the purity of the water. The greater the pressure the higher the boiling-temperature. A table in the Appendix of this book shows the relation between the barometer' pressure and the temperature of boiling water at atmospheric pressure. Mercury, alcohol, liquids and solids generally do not expand uniformly for each degree of temperature, or, in other words, they are not perfect thermometric substances. The error, however, is slight and is of more scientific than practical importance. Any perfect gas, however, does expand uniformly and is a perfect thermometric substance, but gas varies in volume with slight change in barometric pressure, and, while of great value as material for a scientific thermometer, is too bulky and awkward for ordinary use. It is at the present time considered doubtful if there is any perfect gas in existence, or one which cannot be liquefied by intense cold or great pressure. Air, hydrogen, and nitrogen act like perfect gases at ordinary temperatures; the same is true in a slightly less degree of oxygen. Yet oxygen is a liquid whose boiling-point is 119 degrees centigrade (182 degrees Fahrenheit) below zero. Nitrogen and air are liquids boiling at a temperature of 193 degrees centigrade (315 degrees Fahrenheit) below zero. Pictet and Cailletet have reduced the temperature to 200 degrees C. below zero, finding air at that temperature to be a liquid as limpid as water and, like water, having a decided blue tint when seen by transmitted light.

8. Special Forms of Thermometers.—The mercurial thermometers, as ordinarily constructed (Fig. 1), consist of a bulb of glass joined to a capillary glass tube filled so as to leave a vacuum in the upper part of the glass stem, above the mercury; they cannot be used for any temperature higher than that of the boiling-point of mercury, which is about 575° F. More recently these thermometers have been filled with nitrogen or carbonic dioxide in the upper part of the glass stem, which by pressure prevents the mercury boiling. Thermometers constructed in this way can be used safely in temperatures as high as the melting-point of ordinary glass, say to 1000° F.

Mercurial thermometers are made in various ways; the cheaper ones have graduations on an attached frame of wood or metal, Fig. 1, but the more accurate and better grades have the graduations cut directly on to the glass stem, Fig. 2. It has been found that the glass from which these thermometers are made changes volume slowly for many months after construction, so that it is necessary to fill the thermometer with mercury a long time before graduation. In the better grade of thermometers the graduations are obtained by comparing point by point with an accurate standard; in the cheaper ones by simply subdividing into equal parts between freezing and boiling points. At very low temperatures ($-38°$ F.) mercury solidifies and its rate of expansion changes; alcohol or spirits of similar nature are not so affected, and hence are better suited for use in thermometers for measuring extremely low temperatures. Air thermometers, while rather difficult to use and of somewhat clumsy construction, are accurate through any range of temperature. These are made either by confining the air in a constant volume and measuring the increase in pressure (Fig. 3), or else by maintaining the pressure constant and noting the increase in volume. If the volume be maintained constant, the pressure will increase directly proportional to the increase in absolute temperature. In the air thermometer (Fig. 3) the volume is kept constant and the increase in pressure is measured by the rise of mercury in the tube OC above the line AB. That is, in passing from the freezing to the boiling point of water, the barometer being at 29.92, the pressure will increase 180/492, as expressed on the Fahr. scale, or 100/273 on the Cen. scale.

FIG. 2

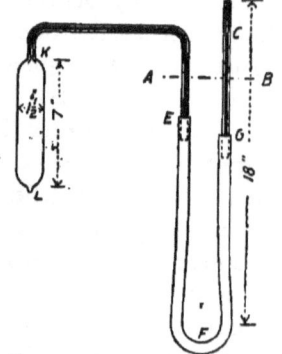

FIG. 3.— AIR THERMOMETER.

The determination of temperature with the air thermometer, even if the instrument is calibrated to read in degrees, needs a correction for barometer-reading, since the height to which the mercury will rise in the tube will depend on the pressure of the air. The directions for using the instrument would be: 1st. Find the constant of the instrument by putting the bulb in melting ice, and dividing the absolute temperature, 492, by the sum of barometer-reading and reading of tube of the thermometer; 2d. To find any temperature, multiply the *constant* as found above by the sum of barometer-reading and reading of thermometer, and subtract from this product 460°.

NOTE.—In using the instrument always keep the mercury at or near point A, so as to keep volume of air constant.

9. Pyrometers and Thermometers for High Temperatures.

—Most metals have rates of expansion which differ sensibly from each other, and this fact has been utilized in the construction of thermometers.

Metallic thermometers are frequently used for high temperatures and have often been called pyrometers. If two bars of metal with unequal rates of expansion be fastened together at one end and heated, the difference of extension of the two ends can be utilized in moving a hand over a dial graduated to show change of temperature (Fig. 4). The metal may also be bent into the form of a helix, in which case the heating will tend to change the curvature and thus move a hand which can be used to measure the temperature.

A thermometer consisting of an iron bulb and a dial, very much like the metallic pyrometer in appearance, is made by filling the bulb with ether or hydrocarbon vapor, and constructing it on the same principle as gauges used to register pressure on boilers. The vapor has a temperature corresponding to a given pressure, so that the dial can be calibrated to read in degrees of temperature instead of pounds of pressure.

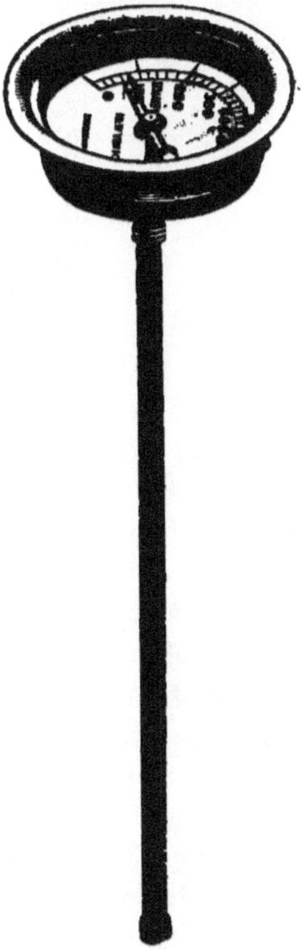

FIG. 4.
METALLIC PYROMETER.

12 HEATING AND VENTILATING BUILDINGS.

These instruments are extremely convenient and answer admirably for temperatures not exceeding 1000° F.

Calorimetric Pyrometers.—The principle of operation used in determining specific heat, Art. 13, can, if the specific heat is known, be employed to ascertain the temperature of any hot body.

Temperature by the Color of Incandescent Bodies and by Melting-points.—Pouillet, as the result of a large number of experiments, concluded that all incandescent bodies have a definite and fixed color corresponding to each temperature.

This color and temperature scale was given as follows:

Color.	Temp. C.	Temp. F.
Faint red	525	977
Dark red	700	1295
Faint cherry	800	1652
Cherry	900	1652
Bright cherry	1000	1932
Dark orange	1160	1850
Bright orange	1200	2192
White heat	1300	2372
Bright white	1400	2552
Dazzling white	1500	2732

This scale applies only to bodies that shine by incandescent light and not from actual combustion. A pyrometer making practical application of this scale has been invented by *Noel*, and consists of a telescope with polarizing attachment and a scale so fixed as to read the angle through which a part of the instrument turns while a sudden transition of color takes place.

Temperature by the Melting-points of Bodies.—The melting-points of bodies often provide an excellent means of determining temperature. The temperature is obtained by using metallic alloys having known melting-points, it being higher than those which have melted, but lower than those which remain unmelted. A table of temperature of melting-points is given in the Appendix. In Germany a carefully prepared set of alloys can be purchased for temperature determinations in this manner.

10. Maxima and Minima Thermometers.—The ordinary method of making a thermometer for recording the highest temperature is by introducing a small piece of steel wire about half an inch in length and finer than the bore of the thermometer into the tube above the mercury, in a mercurial thermometer. The thermometer is placed with its stem in a horizontal position, and the steel index is brought into contact with the extremity of the column of mercury. Now when the heat increases and the mercury expands the steel wire will be thrust forward; but when the temperature falls and the mercury contracts the index will be left behind, showing the maximum temperature. For showing minimum temperature a spirit thermometer prepared in a similar manner is used, as the spirits in contracting draw the index with the alcohol because of the capillary adhesion between the alcohol and the glass; but when the alcohol expands it passes by the index, without displacing it, so that its position shows the lowest temperature to which the instrument has been subjected.

11. **Use of Thermometers.**—In the use of thermometers for determining the temperature of the air, they should be exposed to unobstructed circulation in a dry place and in the shade. Any drops of moisture on the bulb

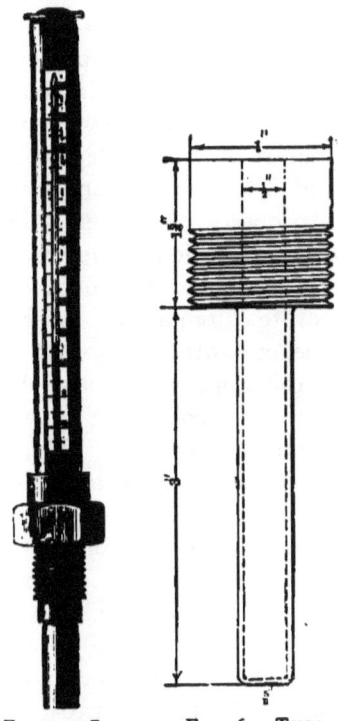

FIG. 5.—STEAM-THERMOMETER. FIG. 6.—THERMOMETER-CUP.

of the thermometer tend to evaporate and lower the temperature. For determining the temperature of steam or water under pressure thermometers are set into a brass frame so that they will screw directly into the liquid (Fig. 5) without permitting leakage. In other cases the thermometer can be inserted into a cup made as shown in Fig. 6. Cylinder-oil or mercury is put into the cup, and the reading of the thermometer will then indicate the temperature of the surrounding fluid. When the thermometer is inserted into a cup some time will be required to obtain the correct temperature. The temperature of steam-pipes or hot-water pipes cannot be obtained accurately by any system of applying the thermometers externally to the pipes, and in case thermometers are used they should be set deep into the current of flowing steam or water, not placed in a pocket where air can gather.

12. Specific Heat.—The capacity which bodies have of absorbing heat when changing temperature varies greatly; for instance, the same amount of heat which would raise one pound of water one degree in temperature would raise about 8 pounds of iron 1 degree in temperature or would raise 1 pound 8 degrees in temperature. The term used to express this property of bodies is *specific heat*, which is defined as follows: Specific heat is the quantity of heat required to raise the temperature of a body one degree, expressed in percentage of that required to raise the same amount of water one degree, or in other words with water considered as one. Specific heat can always be found by heating the body to a given temperature, cooling it in water, and noting the increase in temperature of water. Thus if 1 pound of iron in cooling 8 degrees heats one pound of water one degree, its specific heat is $\frac{1}{8} = 0.125$. A table of *specific heats* of the principal materials is given in the back of the book, from which it will be seen that the specific heat of water is greater than that of any other known substance.

A knowledge of the specific heat of various materials is of considerable importance in the design of heating apparatus, since it indicates the capacity for absorbing heat without increase of temperature. The heat which is absorbed in raising the temperature of a body is all given out when the body cools, so that although there is a difference in the amount absorbed, there is no difference in the final result due to heating and cooling.

The total heat which a body contains is equivalent to the product obtained by multiplying difference of temperature, specific heat and weight. The results will be expressed in heat-units or in capacity of heating one pound of water.

The specific heat of bodies in general increases slightly with the temperature, the values in the table being true from 32° to 212°.

13. Latent Heat.—When heat is applied to any liquid the temperature will rise until the boiling-point is reached, after which heat will be absorbed; but the temperature will not change until the entire process of evaporation is complete, or until the liquid is all converted into vapor. The heat absorbed during evaporation has been termed *latent*, since it does not change the temperature and its effects cannot be measured by a thermometer. In the evaporation of water about five and one-half times as much heat is required to evaporate the water when at 212 degrees, into steam at the same temperature, as to heat the water from the freezing to the boiling point. Heat stored during evaporation is given out when the vapor condenses, so that there is no loss or gain in the total operation of evaporating and condensing. A similar storage of heat takes place when bodies pass from the solid to the liquid state, but in a less degree. Although similar in some respects, latent heat differs in nature from specific heat. In both cases, heat not measured by the thermometer is stored; when the temperature is lowered the stored heat is given up in both cases: in the first it represents a change in the physical condition, as from a solid to a liquid or a liquid to a gas; in the second the condition remains unchanged.

14. Radiation.—Heat passes from a warmer body to a colder by three general methods, each of which is of considerable importance in connection with the methods of heating. These methods are *radiation, conduction,* and *convection.* The heat which leaves a body by radiation travels directly and in a straight line until it is intercepted or absorbed by some other body. Radiant heat obeys the same laws as light, its amount varying inversely as the square of the distance, and with the sine of the angle of inclination. The amount of radiant heat which is emitted or which is absorbed depends largely, if not altogether, upon the character of the surface of the hot and cold body; it is found by experiment that the power of absorbing radiant heat is exactly the same as that of emitting it. The relative amount of heat emitted or absorbed by different surfaces is given in the following table.

RELATIVE EMISSIVE POWERS AT THE BOILING TEMPERATURE.

Lamp-black	100	Steel	17
White-lead	100	Platinum	17
Paper	98	Polished brass	7
Glass	90	Copper	7
India ink	85	Polished gold	3
Shellac	72	Polished silver	3

Radiant heat passes through gases without affecting their temperature or being absorbed to any appreciable extent. It is probably true that a very large body of air, especially air containing watery vapor, does absorb radiant heat, for it is known that the earth's atmosphere intercepts a sensible proportion of the heat radiated from the sun.

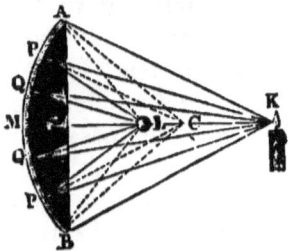

FIG. 7.—REFLECTION OF HEAT.

15. Reflection and Transmission of Radiant Heat.—Radiant heat, like light, may be reflected and sent in various directions by materials of various kinds. Thus in Fig. 7 heat radiated from K is reflected to L, and *vice versa*. The following table shows the proportion of radiant heat which would be reflected by various substances:

REFLECTING POWER.

Silver-plate	97	Polished platinum	80
Gold	95	Steel	83
Brass	93	Zinc	81
Speculum-metal	86	Iron	77
Tin	85		

Radiant heat also possesses the property of passing through certain substances in very much the same manner that light will pass through glass. This property is called diathermancy. The following table gives the diathermanous value of various substances, the heat being obtained from a lamp. The transmission power varies with the source of heat.

PER CENT OF HEAT TRANSMITTED THROUGH DIFFERENT SUBSTANCES.

WHEN RECEIVED FROM AN ARGAND LAMP (DESCHAUD'S PHYSICS).

SOLIDS.

Colorless Glass 1.88 mm. thick.

Flint-glass............from	67 to 64%
Plate-glass	62 to 59
Crown-glass (French)	58
Crown-glass (English)	49
Window-glass................	54 to 50

Colored Glass 1.85 mm. thick.

Deep violet.......................	53
Pale violet........................	45
Very deep blue...................	19
Deep blue........................	33
Light blue........................	42
Mineral-green....................	23
Apple-green.....................	26
Deep yellow.....................	40
Orange..........................	44
Yellowish red...................	53
Crimson.........................	51

LIQUIDS 9.21 MM. THICK.

Colorless Liquids.

Distilled water...................	11%
Absolute alcohol.................	15
Sulphuric ether..................	21
Sulphide of carbon...............	63
Spirits of turpentine.............	31
Pure sulphuric acid	17
Pure nitric acid..................	15
Solution of sea-salt..............	12
Solution of alum.................	12
Solution of sugar................	12
Solution of potash...............	13
Solution of ammonia.............	15

Colored Liquids.

Nut-oil (yellow)	31
Colza-oil (yellow)................	30
Olive-oil (greenish yellow)	30
Oil-carnations (yellowish)........	26
Chloride sulphur (reddish brown)..	63
Pyroligneous acid (brown).......	12
White of egg (slightly yellow)....	11

CRYSTALLIZED BODIES 63.62 MM. THICK.

COLORLESS.

Rock-salt.......................	92%
Iceland spar	12
Rock-crystal....................	57
Brazilian topaz..................	54
Carbonate of lead...............	52
Borate of soda..................	28
Sulphate of lime................	20
Citric acid.....................	15
Rock-alum.....................	12

COLORED.

Smoky quartz (brown)..........	57%
Aqua-marina (light blue)........	29
Yellow agate...................	29
Green tourmaline...............	27
Sulphate of copper (blue).......	0

16. Diffusion of Heat.— Various materials possess the property of reflecting the radiant heat in such a manner as to diffuse it in all directions, instead of concentrating the heat in any one direction. If the heat were all returned, the temperature of the body would not rise, but would remain constant. The diffusive power as determined by Laprovostaye and Desains was found to be as follows for the following substances, the heat received being 100:

White-lead..................................	.82
Powdered silver	.76
Chromate of lead...........................	.66

17. Conduction of Heat.—When heat is applied to one end of a bar of metal it is propagated through the substance of the bar, producing a rise of temperature which gradually travels to the remote portions. This transmission of heat is called conduction. It differs from radiation, first, in being gradual instead of instantaneous; second, in exhibiting no preference for travelling in straight lines, the propagation being as rapid through a crooked as a straight bar. In heating a body the heat is at first largely absorbed by the body without changing its temperature, then for a time it is applied in raising the temperature; the time required for this operation will depend upon its specific heat. After a certain time the temperature of the body will remain constant, the heat being removed as rapidly as it reaches a given position, and in this case we have an illustration of the transmission of heat by conduction. The amount of heat which passes is directly proportional to the area of cross-section, to the difference of temperature divided by the thickness, and to a coefficient which depends upon the character of the material. The *coefficient* is the quantity of heat which flows, in unit time, through a cross-section of unit area, when the thickness of the plate is unity and the difference of temperature is one degree.*

The conducting power of materials varies greatly. The metals are in general good conductors of heat, but differ greatly among themselves. The following table gives the relative values of the conducting powers for different metals:

RELATIVE CONDUCTING POWERS.

Silver	100	Steel	12
Copper	77.6	Iron	17
Gold	53.2	Lead	8.5
Brass	33	Platinum	8.2
Zinc	19.9	Palladium	6.3
Tin	14.5	Bismuth	1.9

Rocks and earthy materials have very much less power of conducting heat than the metals. Table XIV in the back part

---

* This can be expressed in a formula as follows:

$$Q = kA \frac{t_2 - t_1}{x},$$

in which Q = quantity of heat, k = coefficient, A = area, x = thickness, $t_2 - t_1$ = difference of temperature on the two sides of the plate.

of the book gives the value of the coefficient of various materials in terms of the absolute amount of heat conveyed. The relative conductive powers of stone is about 4 per cent of that of iron and ⅔ of one per cent of that of copper. The conducting power of woods does not differ greatly from that of water, and is about 1½ per cent of that of iron. The conducting power of the air and gases is very small, and for practical purposes may be considered as zero. As compared with iron the conducting power is about as 1 to 3500. A knowledge of the conductive powers of bodies is of very great importance in connection with the loss of heat in buildings of various classes.

The bodily sensation of heat or cold is affected to a great extent by the conducting power of the material with which the body comes in contact. Thus if the hand were placed upon a metal plate at a temperature of 40 degrees, or plunged into mercury of the same temperature, a very marked sensation of cold is experienced. This sensation is less intense with a plate of marble of the same temperature, and still less with a piece of wood. The reason is that the heat is more rapidly conducted away in the case of the metals, and this causes a more marked sensation of cold.

Where heat is applied to one surface of a metallic body, it passes through the body by conduction and is given off on the opposite side, usually to the air or to bodies in the surrounding room, by radiation and convection. It will be found that the rate of conduction through the metallic body is many times greater than the rate of passage of the heat from the metallic substance. The knowledge of the conductive power is of little practical importance, as regards heating surface, because of this fact, but it is of great value in the selection of materials which will prevent the escape of heat from dwellings. This subject will be taken up in Chapter III, and applications given showing the loss of heat from different constructions of building.

18. Convection or Heating by Contact.—When bodies are in motion there is more or less rubbing contact of their particles with each other and against stationary objects. When the particles rub against hot bodies they will themselves become warm; it is only by such motion that liquids or gases can be heated any appreciable amount. The heating of the air of a room is practically all accomplished by currents, which brings the particles into contact with radiators, heated pipes, or even the walls of a room. If the air enters a room at a higher temperature, then by the reverse process the heat is given up to the colder objects, and the air is lowered in temperature. The heating of water in steam-boilers is largely due to a circulation which brings the particles of water in direct contact with highly heated surfaces, so that the heating in that case is accomplished largely by convection.

19. Systems of Warming.—Any general consideration of a system of warming must include, first, the combustion of fuel which may take place in a fireplace, stove, steam or hot-water boiler; second, a system of transmission by means of which the heat shall be conveyed with as little loss as possible to the position where it can be utilized for heating; third, a system of diffusion of heat so that it shall be conveyed from any reservoir, radiator, etc., which is heated to objects, persons, or to the air of a room, in the most economical way possible.

In case stoves are used the heat is directly applied by radiation and convection to heating the objects and air in the room in which the stove is placed. There is in this case no special system for the transmission of heat. In the case of hot-air heating, the air is drawn over a heated surface and then transmitted by pipes while at a high temperature to the rooms where heat is required. In the case of steam-heating, steam is formed in a boiler, transmitted through pipes to radiators which are placed either directly in the room or in passages leading to the rooms, and the condensed steam is returned either directly or by means of a pump to the boiler. In the case of hot-water heating the general system is much the same —water instead of steam circulates from the heater to the rooms where heat is required and back to the heater; the motive force which produces the circulation being the difference in weight between the hot and cold water.

CHAPTER II.

PRINCIPLES OF VENTILATION.

20. Relation of Ventilation to Heating.—Intimately connected with the subject of heating is the problem of maintaining air of a certain standard of purity in the various buildings occupied. The introduction of pure air can only be done properly in connection with the system of heating, and any system of heating is incomplete and imperfect which does not provide a proper supply of air.

The general principles relating to ventilation are considered in this chapter, but the practical methods of securing ventilation are considered in connection with systems of indirect heating.

The subject of ventilation often receives very little consideration in connection with the erection of apparatus for heating.

21. Composition and Pressure of the Atmosphere.—Atmospheric air is not a simple substance, but consists of a mechanical mixture of nitrogen and oxygen, together with more or less vapor of water, and almost always a little carbonic acid and a peculiarly active form of oxygen, known as ozone. The nitrogen and oxygen are combined in the ratio of 79.1 to 20.9 by volume, and these proportions are generally the same in all parts of the globe, and at all accessible elevations above the earth's surface.

The amount of carbonic acid in the air varies in the open country from 4 to 6 parts in 10,000 by volume. The amount of moisture in the atmosphere sometimes forms 4 per cent of its entire weight, and sometimes is less than one tenth of one per cent.

The weight of the atmosphere is measured by the height in inches at which it will maintain a column of mercury in an instrument called a barometer. The pressure of the atmosphere is less as the distance from the centre of the earth becomes greater. For that reason points of different elevation give different average readings of the barometer. The normal reading of the barometer at sea-level, which corresponds to a boiling-point for pure water of 212° F., is 29.905 inches.

The weight of the atmosphere, even at the same place, is constantly fluctuating with various conditions of the weather. The variation in barometer-reading from the mean may be 1.5 inches in either direction.

The fall of the barometer due to different elevations from the sea-level would be approximately as follows:

At 917 feet the barometer sinks 1 inch.
" 1860 "   "   "   2 inches.
" 2830 "   "   "   3   "
" 3830 "   "   "   4   "
" 4861 "   "   "   5   "

The atmospheric pressure has great effect upon the boiling-temperature of water; thus pure water will boil at the temperatures corresponding to the various barometric pressures, as shown in the following table:*

Boiling-temperature F.	Barometer, Inches	Boiling-temperature F.	Barometer, Inches
212	29.905	205	25.990
211	29.331	204	25.465
210	28.751	203	24.949
209	28.180	202	24.442
208	27.618	201	23.943
207	27.066	200	23.453
206	26.523		

The weight of a cubic foot of air is inversely proportional to the absolute temperature; if freed from aqueous vapor and under a pressure of 30 inches of mercury, it weighs, according to Regnault, 536.29 grains or 0.076613 pound. The rate of expansion in volume or decrease in density is $\dfrac{1}{460 + t}$ for each degree Fahrenheit, t being temperature above 32°.

Table VIII in the Appendix gives the weights of air for different temperatures. For the temperature of 60° air is

---

* Encyc. Brit., vol. III. p. 387.

813.67 times lighter than water. Various other units are sometimes used to measure the head or pressure, and for convenience of reference these equivalents can be arranged as follows:

30 inches of mercury = 14.7304 lbs.
= 407.07 in. water = 33.92 ft. water
= 11985.4 ft. air at 60° Fahr.

1 inch water = 0.57902 oz.

Air contains more or less impurities which are to be found only in places where the ventilation is not perfect. These impurities consist of carbon monoxide, CO, ammoniacal compounds, sulphuretted hydrogen, and sulphuric and sulphurous and nitric and nitrous acids. It also contains some ozone, which is a peculiarly active form of oxygen, and is believed by many to have an important influence in the preservation of the purity of the atmosphere. Authorities, however, differ very widely as to its distribution and action. Lately a new constituent called *argon* has been discovered.

Air contains more or less solid matter in the form of minute particles of dust. The dust particles are thought to bear an important part in the propagation and distribution of the bacteria of various diseases, and also in the production of storms.

Air contains microbe organisms, or bacteria, in greater or less numbers. The number of bacteria may be determined by slowly passing* a given volume of the air through a glass tube coated inside with beef jelly; the germs are deposited on the nutrient jelly, and each becomes in a few days the centre of a very visible colony. In outside air the number of microbe organisms varies greatly, being often less than one per litre (61 cubic inches); in well-ventilated rooms they vary from 1 to 20, while in close schoolrooms as many as 600 per litre have been found. Carnelley, Haldane, and Anderson found in their researches in mechanically ventilated schoolrooms an average number of 17 microbe organisms per litre. The results of stopping the mechanical ventilation was to increase the carbonic acid without changing the number of microbe organisms.

---

* Encyc. Britannica, article "Ventilation."

22. Diffusion of Gases.—Gases which have no chemical action on each other will, regardless of weights or densities, mingle with each other so as to form a perfectly uniform mixture. This peculiar property is called *diffusion*, and is of great importance in connection with ventilation, since it indicates the impossibility of separating gases of different densities.

Liquids of different densities do not make uniform mixtures, unless they have a special affinity for each other; the heavier invariably settles to the bottom.

Perfect diffusion is a process which requires some time, so that the composition of samples from the same room may in some instances be sensibly different. The time required for the diffusion of gases is inversely proportional to the density, and directly proportional to the square root of the absolute temperature. Diffusion is a molecular action, and can be calculated from the kinetic theory of gases. One computation of this character indicates that the time required for the equal diffusion of carbonic acid throughout the atmosphere was 2,220,000 years.

Dr. Angus Smith found the following percentages of oxygen present in the air, in samples collected in various places, which serve to show the variation which may exist under different conditions:*

Seashore of Scotland, on the Atlantic	20.99%
Top of Scottish hills	20.98
Sitting-room, feeling close, but not excessively so	20.89
Backs of houses and closets	20.70
Under shafts in metal mines	20.424
When candles go out	18.50
When difficult to remain in air many minutes	17.20

The variation in amount of carbonic acid is equally great, the quantity being as follows:

London parks	0.0301%	In workshops	0.3%
On the Thames	0.0343	In theatres	0.32
London streets	0.0380	Cornwall mines	2.5
Manchester fogs	0.0679		

23. Oxygen.—Oxygen is one of the most important elements of the atmosphere, so far as both heating and ventilation

---

* Encyc. Brit., vol. XVI. p. 617; also vol. II. p. 35.

is concerned. It is the active element in the chemical process of combustion, and also of a somewhat similar physiological process which takes place in the respiration of human beings. It exists in a free state mixed with about four parts of nitrogen in the air, and is essential not only for the support of any combustion, but for the support of life. It is not to be considered as having any properties as a food, but is rather the necessary element which makes it possible to assimilate and utilize the food. Taken into the lungs it acts upon the excess of carbon of the blood, and possibly also upon other ingredients, forming chemical compounds which are thrown off in the act of respiration. The chemical action of oxygen with the other elements can generally be considered as a sanitary one. In many respects the process of respiration resembles that of combustion; for in both cases oxygen is derived from the air, carbon or other impurities are oxidized, and the products of this oxidization are rejected. In both cases heat is given off as the result of this process. Its weight is sixteen times that of hydrogen. It is sometimes found in a peculiarly active form called *ozone*.

24. Carbonic Acid or Carbon Dioxide, CO_2, and Carbonic Oxide, CO.—The first is a product resulting from the perfect combustion of carbon; it is always found in small quantities,—3 to 5 parts in 10,000 in the atmosphere of the country.

This gas, although very heavy as compared with that of pure air (22 times that of hydrogen), will, if sufficient time be given, mix uniformly with the air. It is not a poisonous gas, although in an atmosphere containing large quantities of carbonic dioxide a person might die from suffocation or for want of oxygen.

While carbonic dioxide is not of itself injurious, yet as it is a product of combustion and respiration, and is usually accompanied with other injurious products, it is regarded as an index of the quality of the air, and the amount of it present in the air is taken as the standard by which we can judge of the ventilation.* In such a case pure air, containing 4 parts of car-

---

* J. S. Billings, in his work on Ventilation and Heating, cites an experiment by Carnelley and Mackie, showing that the ordinary theory of increase

bonic dioxide in 10,000 would be the standard of absolute purity. Authorities differ as to the greatest amount of carbon dioxide which might be permitted. It is quite certain that any unpleasant sensation is not experienced until the amount is increased to 10 or 12 parts in 10,000; yet authorities are generally agreed that the maximum amount should not exceed 10 parts in 10,000, at least for sleeping-rooms. The standard of good ventilation usually adopted at present would permit about 8 parts in 10,000 in the air. There has been a tendency to make the standard of ventilation higher and higher during the last few years, thus requiring the introduction of greater quantities of air.

Carbonic acid is continually increased by the processes of combustion and respiration, yet for the past thirty years the amount in the air has not sensibly changed.

Plant-growth and vegetable life assimilate carbonic acid and give off oxygen.* There exists in the air about 28 tons of carbonic acid to each acre of ground, yet an acre of beech-forest annually absorbs about one ton, according to Chevandier; and no doubt the total vegetation growing is sufficient to absorb the excess due to combustion and respiration, so that the total does not experience much change.

Carbonic Oxide, CO.—This compound is not found in the air except under unusual circumstances. It is distinctly a poison, and has a characteristic reaction on the blood. Hempel,† the German chemist, experimented on its poisonous

of organic matter with increase of carbon dioxide is a reasonable on The results of the experiment were as follows:

Proportion of Organic Matter. Oxygen required to Oxidize 1,000,000 Volumes.	Average Carbonic Acid in 10,000 Volumes of Air.	Number of Trials.
0 to 2.5	2.8	20
2.5 to 9.5	3.0	20
4.5 to 1.0	3.2	20
7.0 to 15.8	3.7	20

* "How Crops Feed," by Johnson, page 47.
† Hempel's Gas Analysis. Macmillan & Co.

effects with a mouse. No symptoms of poisoning were detected until there were 6 parts CO in 10,000 of air, in which case after 3 hours' time respiration was difficult; in another case the mouse could scarcely breathe in 47 minutes. With 12 parts in 10,000 the mouse showed symptoms of poisoning in 7 minutes; with 29 parts in 10,000 the mouse died in convulsions in about two minutes.

25. Nitrogen—Argon.—The principal bulk of the earth's atmosphere is nitrogen, which exists uniformly diffused with oxygen and carbonic acid. This element is practically inert in all the processes of combustion or respiration. It is not affected in composition either by passing through a furnace during combustion or in passing through the lungs in process of respiration. Its action is to render the oxygen less active, and to absorb some part of the heat produced by the process of oxidation. It is an element very difficult to measure directly, as it can be made to enter into combination with only a few other elements, and then under peculiarly favorable circumstances.

A very small amount of ammonia, which is a compound of nitrogen and hydrogen, is found in the atmosphere.

Argon.—A constituent of the atmosphere recently discovered, which amounts to about one per cent of the total, was announced at the meeting of the Royal Society, January 31, 1895. This element is very soluble in water, and liquefies at a temperature 232° below zero F., under a pressure of 50.6 atmospheres. It is even more inert in action than nitrogen, and practically may be considered the same.

26. Analysis of Air.—The accurate analysis of air requires the determination of aqueous vapor, carbon dioxide, carbon monoxide, oxygen and ozone, but for sanitary purposes the determination of carbon dioxide and water is the most frequently called for. For a complete discussion of these various methods the reader is referred to Hempel's Gas Analysis, translated by Dennis and published by Macmillan & Co. The nitrogen of the atmosphere cannot be determined by any known method of analysis; it is obtained by deducting the sum of all the other elements from the total. The approximate determination of the oxygen is done very readily by drawing a certain volume of the air into a measuring-vessel and then passing it over a mixture of pyrogallic acid and caustic potash; the oxygen is absorbed, reducing the volume in amount proportional to the quantity of oxygen. This process is, however, not of extreme accuracy, and for minute quantities very much more complicated methods must be resorted to.

Method of Finding Carbon Dioxide (CO_2).—The amount of this material present in the atmosphere is so small that the most delicate methods are required in order to measure it. The writer gives here the only simple method which can be rapidly applied, and which is said to be accurate to one part in one hundred thousand. This system of finding CO_2 was devised by Otto Pettersson and A. Palmqvist, two European chemists. The instrument used for this determination is shown in Fig. 8, and can be had from any dealer in physical apparatus. It consists of a measuring-vessel, A, connected with a U-shaped burette B, from which communication can be made by a small stop-cock, b; a manometer, fg, containing a graduated scale nearly horizontal; and two stop-cocks, f and g, by means of which communication can be made with the air. One side of this manometer, f, is in communication with the closed vessel C; the other side can be put in communication with the measuring-vessel A. The burette B contains a saturated solution of caustic potash (KOH). The flask E contains mercury, and by raising it, when the stop-cock c is open, the mercury will rise in the flask A, and the air will be driven out. If the flask E be lowered the mercury will flow from the measuring-tube, and the amount of air entering A can be measured by the gradua-

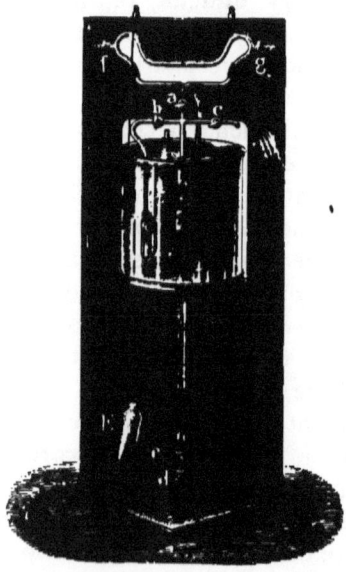

FIG. 8.—PETTERSSON'S APPARATUS FOR DETERMINING CO_2 IN AIR.

tions. When the measuring-tube A is full of air, the stop-cocks c, b, f, and g being open, the position of the drop of liquid in the horizontal tube of the manometer is accurately read. The stop-cocks c, a, f, and g are then closed, that at b opened, and the vessel E raised, driving the air out of the measuring-tube A into the absorption burette B. This operation of raising and lowering the flask E is repeated several times; it is then lowered, and the air is drawn over into the measuring burette; the cock a is then opened and the vessel E manipulated until the reading of the manometer on the horizontal scale agrees with that in the beginning of the test. The reading of the graduated tube A gives directly the amount of CO_2. The determinations are made with air of ordinary humidity, and there is a very slight correction due to this fact, which is not likely to equal, in any case, one part of CO_2 in one million parts of air.*

27. **Determination of Humidity of the Air.**—The humidity of the air is determined by gradually cooling a body and observing at what temperature the vapor of the air condenses on the body as dew. When dew is deposited the air is saturated for the given temperature, and if the temperature of the air be known, at which dew will be deposited, and also the temperature of the air in its normal condition, we can compute the amount of moisture contained in the air. The instrument generally employed for this purpose consists of two thermometers, the bulb of one of which is exposed in its ordinary condition to the air; the bulb of the other is kept constantly wet by means of a bit of cloth extending to a vessel filled with water. If the air were saturated with moisture these two thermometers would give the same reading, but if the air is not saturated the readings will differ an amount depending upon the humidity. The table following, and a more complete one in the Appendix, give the amount of moisture expressed as percentage of saturation for different readings of the wet and dry bulb thermometer.

---

* For approximate methods of determining the purity of air see Appendix to book.

Moisture in Grains per Cubic Foot Absorbed by Saturated Air.				Per cent of Saturation for Difference in Readings, Wet and Dry Bulb Thermometer.			
Temp. Air, degs.	Grains per cu. ft.	Temp. Air, degs.	Grains per cu. ft.	Difference in Reading.	Temp. Air, 32° F.	Temp. Air, 70° F.	Temp. Air, 95° F.
20	1.56	70	7.94	0	100	100	100
32	2.35	80	10.73	1	96	97	97
40	3.06	90	14.38	2	92	93	94
50	4.24	100	19.12	3	88	90	91
60	5.82	110	25.5	5	81	84	86
				7.5	72	77	79
				10	65	71	73
				15	52	59	62
				20	41	49	53

The first table gives the weight of moisture contained in one cubic foot of saturated air; the second shows the percentage of saturation for any difference in reading of the wet and dry bulb thermometer. The weight of moisture is the product of the results. Thus, saturated air at 70° F. contains 7.94 grains per cubic foot, and if at the same time the difference between the wet and dry bulb thermometers was 10, this air would be 71 per cent saturated, and would contain 71 per cent of 7.94 grains, or 5.62 grains. Since there are 7000 grains in one pound, this weight may, if desired, be reduced to pounds.

Moisture in air can also be determined approximately, but with sufficient accuracy for practical purposes, by the hair hygrometer. This instrument is illustrated in Fig. 9. It is constructed by fastening a hair, from which the oil has been removed, in the top part of a suitable frame, and winding the lower part on a cylinder which is free to revolve, and which carries a balanced pointer. The hair increases or diminishes in length, quite exactly, in proportion to the amount of moisture in the air, and this acquired property seems to be a permanent one. A scale graduated by comparison with determinations made with a wet and dry bulb thermometer serves to show the amount of moisture present, as a percentage of saturated air.

The degree of moisture in the air has an important influence on ventilation. When air is saturated with moisture water is deposited on all bodies which conduct heat readily and have a lower temperature than the air. On the other hand, if the air is entirely deprived of watery vapor it evaporates moisture from the body, and thus causes an unpleasant sensation. It also takes up a great deal of heat. When the air is saturated no evaporation can take place from the body. When the air is very dry, very rapid evaporation will take place. A mean condition between these two extremes is required in every case. The air should be from 50 to 70 per cent saturated in order to feel pleasant, and be of the most value for ventilating purposes.

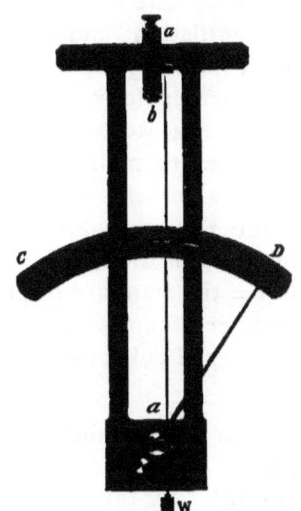

FIG. 9.—THE HAIR HYGROMETER

28. Amount of Air required for Ventilation. — The amount of air required in order to maintain the standard of purity below a certain given amount can be very readily determined, provided we know the amount of carbon dioxide which is given off in the process of respiration.

It is estimated that at each respiration of an adult person 20 cubic inches of air on the average are required, and that 16 to 24 respirations take place per minute; so that from 320 to 480 cubic inches, or about one fourth of a cubic foot, are required per minute.* The air ejected from the lungs is delivered at a temperature from 70 to 90 degrees, and very nearly saturated with watery vapor; hence it is about 2.3 per cent lighter than pure air.

The following table shows the approximate effect of respiration on the composition of air : †

---

* This is estimated by Box as 800 cubic inches, but is given by recent physiologists as above. See works of Dalton, Dr. Carpenter, Art. Respiration in Ency. Brit., etc. This is increased by violent exercise, and to make the allowance liberal 576 cubic inches or ⅓ cubic foot is taken as the amount to be supplied.

† Ency. Britannica, Art. Respiration.

32 HEATING AND VENTILATING BUILDINGS.

	Entering Air.	Respired Gases.
Oxygen, per cent of volume	20.26	16
Nitrogen, " " " "	78.00	75
Watery vapor " " "	1.70	5
Carbonic acid " " "	0.04	4

If we take the carbon dioxide as an index of the character of ventilation, and consider that each person uses one third cubic foot of gas per minute, and that the respired gas contains 400 parts in 10,000 of carbonic acid, while the entering air contains but 4, we can calculate the amount of air which must be provided to maintain any standard of purity desired. The formula for this operation would be as follows: If $a =$ the number of parts of CO_2 in 10,000, thrown out in respiration or other impurities; if $b =$ the cubic feet of air used per minute; if $n =$ the standard of purity to be preserved, expressed as the number of units of CO_2 permissible in 10,000, and $C =$ the number of cubic feet of air required,—we shall have

$$C = ab/(n-4).$$

For the condition we have just considered, for each adult person $a = 400$, $b = \frac{1}{3}$, so that the formula becomes $C = 133/(n-4)$. By taking n as 8, $C = 33$, and n as 10, $C = 22$.

The following table shows the amount of air which must be introduced for each person in order to maintain various standards of purity:

AMOUNT OF AIR REQUIRED PER PERSON FOR VARIOUS STANDARDS OF PURITY.

Standard Parts of CO_2 in 10,000 of Air in the Room.	Cubic Feet of Air required per Person.	
	Per Minute.	Per Hour.
5	133.3	8000
6	67	4000
7	44	2667
8	33	2000
9	27	1600
10	22	1333
11	19	1151
12	17	1000
13	15	889
14	13	800
15	12	727
16	11	667
18	9.5	571
20	8.3	500

The combustion of one cubic foot of gas per hour contaminates about the same amount of air as one person, so that an allowance, equivalent to that required for four or five people, should be made for each gas-burner.

Authorities differ greatly as to the amount of air to be provided per person, but at the present time they seem well united in considering the admission of <u>30 cubic feet</u> of air per minute for each person as giving good ventilation, and this amount is required by law for school buildings in Massachusetts.*

Some authorities insist that a higher standard should be required, but there is little doubt that present conditions would be very much improved could the above amount be obtained in every case.

The amount advised by various authorities has been as follows: Parkes advises 2000 cubic feet of air per hour for persons in health and 3000 to 4000 for sick persons. The English Barracks Improvement Commissioners require that the supply be not less than 1200 cubic feet per man per hour. Pettenkofer recommends 2100 cubic feet; and Morin considers that the following allowances are not too high:†

```
Hospitals (ordinary).................  2000 to 2400 cubic feet per hour.
    "     (epidemic)..................  5000          "    "    "    "
Workshops (ordinary)................   2000          "    "    "    "
    "     (unhealthy trades)..........  3500          "    "    "    "
Prisons..............................   1700          "    "    "    "
Theatres.............................  1400 to 1700  "    "    "    "
Meeting-halls........................  1000 "  2000  "    "    "    "
Schools (per child)..................   400 "   500  "    "    "    "
    "   (per adult)..................   800 "  1000  "    "    "    "
```

Tredgold ‡ in 1836 considered 4 cubic feet of air per minute as good ventilation for healthy people, and 6 cubic feet per minute for the sick in hospitals.

---

* Dr. Billings states that this amount could be increased 25 to 50 per cent, with good results. Ventilation and Heating.
† Études sur la Ventilation.
‡ Warming and Ventilating Buildings; third edition.

29. Influence of the Size of the Room on Ventilation.—The purity of the air of a room depends to some extent on the proportion of its cubic capacity to the number of inmates. This influence is often overestimated, and even in a large room if no fresh air be supplied the atmosphere will quickly fall below the standard of purity. It must be considered that no room is hermetically sealed. Ventilation takes place through every crack and cranny, and even by diffusion through the walls of the room. Such ventilation is generally, however, uncertain and inadequate. Large rooms have the advantage over small ones that they act as reservoirs of air, and also because there is chance for intermittent ventilation such as occurs when doors or windows are opened, and for the casual ventilation which takes place through the walls and around the windows. They are also advantageous, because a larger volume of air may be introduced with less danger of producing disagreeable air-currents or draughts. The following table, taken in part from article "Ventilation," Encyc. Britannica, gives a general idea of the cubic capacity per person usually allowed in certain cases, and the time which would be required to reduce the air inclosed to the lowest admissible standard of purity (12 parts of CO_2 in 10,000 of air), provided no fresh air was admitted.

Class of Building.	Cubic Contents.	Time required for contaminating the Air.
Hospitals.	1200 cu. ft. and above	70 min.
Middle-class houses.	1000 " " " "	59 "
Barracks.	600 " " " "	35 "
Good secondary schools.	500 " " " "	29 "
London Board schools.	130 " " " "	8 "
Workhouse dormitories.	300 " " " "	18 "
London lodging-houses.	240 " " " "	14 "
One-roomed houses.	212 " " " "	13 "

It is seen from the above table that in the ordinary grade of middle-class houses it would require about one hour to render the air unfit for breathing, while for the lowest grade of houses the time required would be only 13 minutes. It may be said, however, respecting the cheaper grade of houses, that while the amount of space allowed per person is small, the character of construction is such that air can usually enter or leave the room without very great retardation, and consequently this table does not fairly represent the character of ventilation actually secured.

Pettenkofer found that, by diffusion through the walls, the air of a room in his house containing 2650 cubic feet was changed once every hour when the difference of exterior and interior temperatures was 34 degrees. With the same difference of temperature, but with the addition of a good fire in a stove, the change rose to 3320 cubic feet per hour. With all the crevices and openings about doors and windows pasted up air tight the change amounted to 1060 cubic feet per hour; with a difference of 40 degrees the ventilation through the walls amounted to 7 cubic feet per hour for each square yard of wall surface. The effect of diffusion in changing the air of a room should generally be neglected in practical ventilation, because it is very uncertain in amount and character.

30. Force for Moving the Air.—No ventilation can be secured unless provision is made for (1) power for moving the air, (2) passages and inlet for admitting the air, (3) passages and outlet for escape of air. Air is moved for ventilating purposes in two ways: first, by expansion due to heating; and second, by mechanical means.

The effect of heat on the air is to increase its volume and lessen its density directly in proportion to the increase in absolute temperature. The lighter air simply because of its less density (tends to rise,) and is replaced by the colder air below. The head which induces the flow is a column of air corresponding in weight to the difference in heights of columns of equal weight of cold and heated air. The velocity can be computed, since theoretically it will be equal to the square root of twice the force of gravity into this difference of height. The result so computed will apply only when there is unrestricted openings at both ends. It is scarcely ever applicable to chimneys, for the reason that the flow of air is retarded by passing through the fuel.

The amount of air which may be made to pass through a ventilating flue of ordinary construction and of different heights is given in a table on page 45.

The available force for moving the air which is obtained by heating is very feeble, and quite likely to be overcome by the wind or external causes. Thus to produce the slight pressure equivalent to one tenth inch of water in a flue 50 feet in height would require a difference in temperature of 50 degrees. In a flue of the same height a difference of temperature of 150 degrees would produce the same velocity as that caused by a pressure of 0.5 inch of water. To produce the same velocity as that due to a pressure sufficient to balance 0.1 inch of water will require that the product of height of chimney and difference of temperature should be 1760.

It will in general be found that the heat used for producing velocity, when transformed into work in a steam-engine is considerably in excess of that required to produce draught by mechanical means. In a rough way, an increase in temperature of one degree increases the head producing the velocity only about one part in 500.

Ventilation by Mechanical Means is performed either by pressure or by suction. In the first case the air is increased in density and discharged by mechanical force into the flue, the flow being produced by an excess of pressure over that of the atmosphere, so that the air tends to move in the direction of least resistance, which is outward to the atmosphere. In the second case, pressure in the flue is less than that of the atmosphere, and the velocity is produced by the flowing in of the outside air. By both processes of mechanical ventilation the air is supposed to be moved without change in temperature, and the force for moving it must be sufficient to overcome effects of wind or change of temperature, otherwise the introduction of air will not be positive and certain. The velocity in feet per second for various differences of pressure is computed as explained in Article 32, and tables are given on pages 42 and 45 for use in computing the amount discharged per square foot of the area of the cross-section of the flue.

31. Measurements of the Velocity of Air.—The velocity of air or other gases is measured directly by an instrument

Fig. 10.—Biram's Portable Anemometer.

called an anemometer, or it is measured indirectly by difference of pressure. The anemometer which is ordinarily employed for this purpose consists of a series of flat vanes attached to an axis and a series of dials. The revolution of the axis causes motion of the hands in proportion to the velocity of the air. In the forms shown in Figs. 10 and 11 the dial mechanism can be started or stopped by a trip arranged conveniently to the operator. In some instances the dial mechanism is operated by an electric current, in which case

Fig. 11.—Portable Anemometer.

it can be located at a distance from the vanes. For measuring the velocity of the wind an anemometer, which consists of hemispherical cups mounted on a vertical axis, is much used. The anemometers are all calibrated by moving them in still air at a constant velocity and noting the readings of the dials. This is usually done by mounting the anemometer rigidly on a long horizontal arm which can be rotated about a vertical axis at a constant speed.

When the pressure is light it can be measured by using a U-tube partly filled with water. Such an instrument is shown in Fig. 12, attached to a flue. There being less than atmospheric pressure in the flue K, the water rises in the leg FE and sinks in the leg DE. The difference of level in the two legs is ab, which is usually measured in inches. If the flue is under pressure the water will stand higher in the leg DE than in FE, but the method of use is essentially the same in all cases.

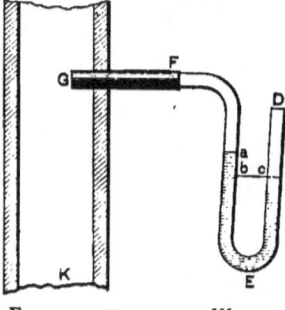

FIG. 12.—U-SHAPED WATER GAUGE.

In case the pressure and velocity are great, considerable error will be made by using the open tube as above, and for such a case a Pitot's tube arranged as shown in Fig. 13 should be used.

This tube consists of two parts, one of which is straight and enters at right angles to the current dB; the other is curved so as to face the current at right angles, cA. These are connected to a U-shaped manometer containing water or some light liquid. The pressure in the two tubes will be the same except for the velocity of the current. This will tend to make the liquid stand higher in the arm fm than in the arm en. The difference in elevations of these two arms will be the velocity-head producing the flow. Call this difference in height h, and the ratio of specific gravity of the liquid in the tube and of the gas in the flue r; then will $v = \sqrt{2ghr}$. *That is, the velocity is equal to 8.03 multiplied by the square root of the difference in height multiplied by ratio of weight.*

In case water is used in the manometer and the gas is **air** at a temperature of 60 degrees, r will equal 813. Hence v will equal 228 $\sqrt{h}$, in which h is in feet, and will equal 65.7 $\sqrt{h}$ when h is in inches of water. For any other temperature than 60 degrees this quantity must be multiplied by the square root of 460 + the temperature, and then divided by $\sqrt{520}$. Practically for air the velocity will equal 228 times the square root of the difference in the heights of the columns.

The velocity of air may also be computed by the heating effects, provided the amount of heat is accurately measured

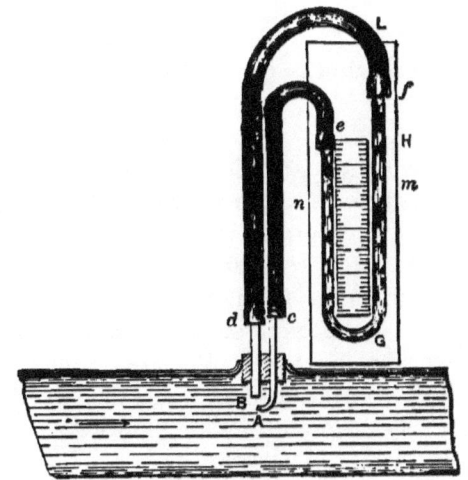

FIG. 13.—SKETCH OF PITOT'S TUBE FOR GREAT PRESSURES.

and the increase in temperature of the air be known. The specific heat of air is 0.238, hence the heat sufficient to warm one pound of water would heat (1/.238) = 4.2 pounds of air. This at 60 degrees would correspond to about 231 cubic feet. By consulting Table VIII the volume heated 1 degree by 1 heat-unit at any other temperature can be found.

The total number of cubic feet of air heated would be equal to the total number of heat-units absorbed divided by the number of degrees the air is heated, and this result multiplied by the volume of one pound divided by the specific heat (the latter number can be taken directly from Table VIII). Having the total amount of air in a given time, the velocity can be obtained by dividing by the area of the passage.

Note.—In the shape of a formula these results are as follows: Let T equal temperature of discharged air, t that of entering air; H equal the total number of heat-units given off per unit of time; V equal the number of cubic feet of air heated 1 degree by 1 heat-unit (see Table VIII); A equal area of passage in square feet; v equal velocity for the same time that the total number of heat-units are taken. Then we shall have

$$C = \text{Total amount of air in cu. ft.} = \frac{HV}{T-t}; \quad v = \frac{C}{A}.$$

32. The Flow of Air and Gases.—The flow of air obeys the same general laws as those which apply to liquids. The gases are, however, compressible, and the volume is affected very much by change of temperature, so that the actual results differ considerably from those obtained for liquids. These laws can only be expressed in mathematical formulæ, from which, however, practical tables are derived.

The flow of air from an orifice takes place under the same general conditions as those of liquids, and we have the general formula $v = \sqrt{2gh}$ as applicable. In this case h is the head which is equal to the height of a column of air of sufficient weight to produce the pressure. Air under a barometric pressure of 30 inches and at 60 degrees in temperature is 813 times lighter than water. The pressure of air is usually measured by its capacity of balancing a column of water in a U-shaped tube (see Article 31), and this pressure is expressed in inches of water. One inch of water-pressure is equivalent to 65.7 feet of air at 60°, and increases $\frac{1}{480}$ part for each degree of increase in temperature. The above formula is only approximate, and does not account for the change in temperatures and of pressures due to expansion, although sufficiently accurate for the designing of ventilating apparatus. Prof. Unwin gives in the article "Hydromechanics," Encyc. Brit., the following formula for computing the velocity of flow of air:

$$\frac{v^2}{2g} = 183.6 T \left\{ 1 - \left(\frac{P_2}{P_1}\right)^{0.29} \right\}.$$

T = absolute temperature;
P_1 = absolute pressure in vessel from which flow takes place;
P_2 = absolute pressure in surrounding space.

To find the volume discharged the velocity must be multiplied by the area and that result by a coefficient which Prof. Unwin gives as follows:

	$c =$
Conoidal mouthpieces of the form of the contracted vein, with effective pressures of .23 to 1.1 atmosphere..................................	.097 to 0.99
Circular sharp-edged orifices....................	0.563 " 0.788
Short cylindrical mouthpieces.................	0.81 " 0.84
The same, rounded at the inner end...........	0.92 " 0.93
Conical converging mouthpieces..............	0.90 " 0.99

In the flow of air or gases through pipes the same considerations hold that have been stated for water. There is the same condition respecting the head which produces pressure and that which produces velocity, and in addition we have those changes due to the compressible nature of the fluid moved.

Taking into account all these conditions, Prof. Unwin gives as a formula for the flow of air in a circular pipe

$$u_0 = \sqrt{\left\{\frac{gctd}{4\zeta l}\right\}\frac{p_0{}^2 - p_1{}^2}{p_0{}^2}},$$

in which u_0 = velocity in feet per second;
    $c = 53.15$;
    t = absolute temperature;
    $g = 32.16$;
    d = diameter in feet;

42   HEATING AND VENTILATING BUILDINGS.

l = length in feet;
ζ = coefficient of friction = $0.005(1 + 3/10d)$;
p_0 = greatest absolute pressure;
p_1 = least absolute pressure.

For a velocity of 100 feet per second ζ varies from 0.00484 to 0.01212 for a diameter varying from 1.64 ft. to 0.164 ft.

For a temperature of 60° F. and for a pipe one foot in diameter and 100 feet long, $\zeta = 0.006$. For barometer reading of 30 inches, pressure being expressed in inches of water, $p_0 = 407$, we have

$$u_0 = 1.512 \sqrt{(p_0 - p_1)(p_0 + p_1)},$$

from which the third column of the following table is calculated.

VOLUME OF AIR DISCHARGED AT VARIOUS PRESSURES.

Difference of Pressure.		Velocity in Feet per Second.	
Inches of Water.	Ounces per Square Inch.	By Accurate Formula Pipe 1 Ft. in Diam, 100 Ft. Long.	By Approximate Formula. (Coefficient 0.7.)
0.01	0.006	4.3	4.6
0.05	0.030	9.6	9.5
0.1	0.058	14.5	14.5
0.2	0.116	19.4	20.5
0.3	0.174	23.6	25.1
0.4	0.232	27.4	29.1
0.5	0.289	30.5	32.5
0.6	0.347	34.0	35.2
0.7	0.405	36.0	38.3
0.8	0.463	39.2	40.7
0.9	0.512	41.0	43.7
1.0	0.579	43.0	45.7
2.0	1.158	61.1	65.2
3.0	1.303	78.0	78.2
4.0	2.316	85.3	91.1
5.0	2.895	86.2	103.3
6.0	3.474	104.0	113.3
7.0	4.053	114.0	122.1
8.0	4.622	121.0	130.6
9.0	5.221	128.0	138.8
10.0	5.790	136.0	145.7
11.0	6.369	142.0	153.0
12.0	6.948	148.0	159.6

The preceding table gives the velocity of air in feet per second as calculated from the accurate formula of Prof. Unwin, and also from the approximate formula $v = \sqrt{2gh}$, using a coefficient of 0.7. The table is calculated for a barometric pressure of 30 inches and for a temperature of 60° F. For any other temperature the results must be multiplied by factors which are calculated as explained below.

For the discharge at any other temperature divide the above results by the square root of 520 multiplied by 460 plus the temperature. For temperature of 32 degrees multiply by .972, 40 degrees .981, 50 degrees .987, 70 degrees 1.01, 80 degrees 1.018, 90 degrees 1.03, 100 degrees 1.04, 110 degrees 1.05, 120 degrees 1.06, 130 degrees 1.07.

33. The Effect of Heat in producing Motion of Air.— The effect of heat is to expand air in proportion to its absolute temperature for each degree of increase. If a column of air be heated it will expand and occupy more space. In other words, a given bulk will have less weight as its temperature is increased; which has the effect of producing lack of equilibrium, and the warmer air will be replaced by colder air, causing a velocity which is in proportion to the change in temperature. The case is analogous to the action of two fluids in the branches of a U-tube, Fig. 14, $DABC$,—the heavier fluid in DA and the lighter fluid in BC. The action of gravity causes the heavier fluid to flow downward and displace the lighter fluid, causing an upward motion in BC. If a volume of the lighter fluid with height greater than BC balances the weight of the heavier fluid DA, the

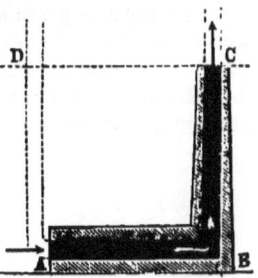

FIG. 14.

flow which is produced will take place with a head equal to the difference in height of AD, and an equal weight of the lighter fluid. The flow will take place in the same manner whether the heavier fluid be confined in a tube arranged as in the dotted lines, Fig. 14, or whether it be drawn from a large vessel, or from the surrounding air. Let the head which produced the draught be equal to h', the height of the flue BC as h; let t be the temperature of the outside air or heavier fluid and t' that of the lighter fluid; and let a be the coefficient of expansion, which for one degree of temperature of air will be

$\frac{1}{460}$. Since the expansion is directly proportional to the increase in temperature, we shall have in general:

$$\frac{h}{1+at} = \frac{h+h'}{1+at'}, \text{ from which } h' = \frac{ha(t'-t)}{1+at}.$$

By substituting for a its value $\frac{1}{460}$, we shall have the following for the head producing the flow in case air is the moving fluid:

$$h' = \frac{h(t'-t)}{460(1+\frac{1}{460}t)} = \frac{h(t'-t)}{460+t}.$$

$460 + t$ is the absolute temperature of the air.

The velocity is equal to the square root of twice the force of gravity, 32.16, into the head which produces the flow, as follows:

$$V = \sqrt{2gh'} = \sqrt{\frac{2gha(t'-t)}{1+at}} = \sqrt{\frac{2gh(t'-t)}{460+t}} = 8\sqrt{\frac{h(t'-t)}{460+t}}, \text{ nearly.}$$

The velocities given above, multiplied by 60 and by the area of cross-section, will give the discharge in cubic feet per minute. Mr. Alfred R. Wolff takes the actual discharge as 0.5 of that given by the formula, so that the actual discharge in cubic feet per minute would be, with 50 per cent allowance for friction,

$$Q = 240F\sqrt{\frac{h(t'-t)}{460+t}},$$

in which F equals the area of cross-section of the flue in square feet. The table on next page gives the discharge per square foot of area of flue for various temperatures and heights computed from the formula.

The above formulæ are for the discharge of air from a flue. The volume, and consequently the velocity, for the entering air will be proportional to its absolute temperature; and hence to obtain the quantity of air entering when T' is the temperature at entering and T that at discharging multiply the preceding formula by $\frac{460+T'}{460+T}$.

34. The Inlet for Air.—The air for ventilation is usually warmed and a portion or all of the heat required for warming is introduced at the same time.

TABLE SHOWING THE QUANTITY OF AIR, IN CUBIC FEET, DISCHARGED PER MINUTE THROUGH A FLUE, OF WHICH THE CROSS-SECTIONAL AREA IS ONE SQUARE FOOT.
(External Temperature of the Air, 32° Fahr; allowance for Friction, 50 per Cent.)

Height of Flue in Feet.	Excess of Temperature of Air in Flue above that of External Air.								
	5°	10°	15°	20°	25°	30°	50°	100°	150°
1	24	34	42	48	54	59	76	108	133
5	55	76	94	109	121	134	167	242	298
10	77	108	133	153	171	188	242	342	419
15	94	133	162	188	210	230	297	419	514
20	108	153	188	217	242	265	342	484	593
25	121	171	210	242	271	297	383	541	663
30	133	188	230	265	297	325	419	593	726
35	143	203	248	286	320	351	453	640	784
40	153	217	265	306	342	375	484	684	838
45	162	230	282	325	363	398	514	724	889
50	171	242	297	342	383	419	541	765	937
60	188	264	325	373	420	461	594	835	1006
70	203	286	351	405	465	497	643	900	1115
80	217	306	375	453	485	530	688	965	1185
90	220	324	398	460	516	564	727	1027	1225
100	243	342	420	485	534	594	768	1080	1325
125	273	383	468	542	604	662	855	1210	1480
150	298	420	515	596	665	730	942	1330	1630

It is found from experience that if the velocity of the entering air is very great it produces a disagreeable current, which is generally known as a draught, and is more or less dangerous to health. The following table from Loomis' Meteorology gives the relation between the velocity and force of air:

RELATION BETWEEN VELOCITY AND FORCE OF AIR.

Sensation.	Velocity.		Pressure, Lbs. per Sq. Foot.
	Miles per Hour.	Feet per Second.	
Just perceptible............	2	2.92	0.02
Gently pleasant............	4	5.85	0.08
Pleasant brisk.............	12.5	18.3	0.750
Very brisk................	25	36.6	3.0
High wind	35	51.5	6
Very high wind............	45	66	10
Strong gale...............	60	88	18
Violent gale...............	70	105	24
Hurricane.................	80	117	31
Most violent hurricane....	100	146	49

It is quite generally agreed that the velocity of the entering air should not exceed four to six feet per second unless it can be introduced in such a position as to make an insensible current. The table which has just been given, while only approximately correct, gives a very fair idea of the sensations produced by air-currents of different velocities and pressures, and is useful in fixing limiting values.

The most effective location for the air-inlet is probably in or near the ceiling of a room, although authorities differ much in this respect. The advantages of introducing warm air at or near the top of the room are: first, the warmer air tends to rise and hence spreads uniformly under the ceiling; second, it gradually displaces other air, and the room becomes filled with pure air without sensible currents or draughts; third, the cooler air sinks to the bottom and can be taken off by a ventilating-shaft. So far as the system introduces air at the top of a room it is a forced distribution, and produces better results than other methods. When the inlet is placed in the floor or near the bottom part of the walls it is a receptacle for dust from the room, and a lodging- and breeding-place for microbe organisms. In the ventilation of large buildings the inlets can usually be located in the ceiling, especially if the lighting be done by electricity or in some manner not affected by air-currents.

Some experiments were made by Mr. Warren R. Briggs, of Bridgeport, Conn., on the subject of the proper method of introducing pure air into rooms and the best location for the inlet and outlet. The experiments were conducted with a model having about one sixth of the capacity of a schoolroom to which the perfected system was to be applied. The movements of the air in the model of the building were made visible by mingling the inflowing air stream with smoke, which rendered all the changes undergone by it in its passage apparent to the eye.

The results of the experiments are shown graphically in the six sketches, Figs. 15 to 20. In each case the distribution of the fresh air is indicated by the curved lines of shading. A study of these sketches is very suggestive, as it indicates the best results when the inlet is on the side near the top, and the outlet is in the bottom and near the centre of the room. The tendency of the entering air to form air-currents or draughts, which in some instances tend to pass out without perfect diffusion, is well shown. This tendency is less as the velocity of the entering air is reduced, and we probably get nearly perfect diffusion in every case where the outlet is well below that of the inlet, provided the velocity of the entering air is small —less than 4 feet per second.

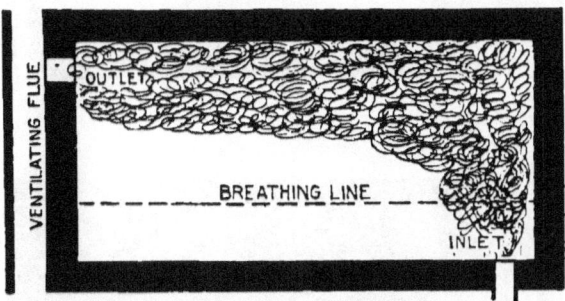

FIG. 15.—AIR INTRODUCED AT BOTTOM, DISCHARGED AT TOP.

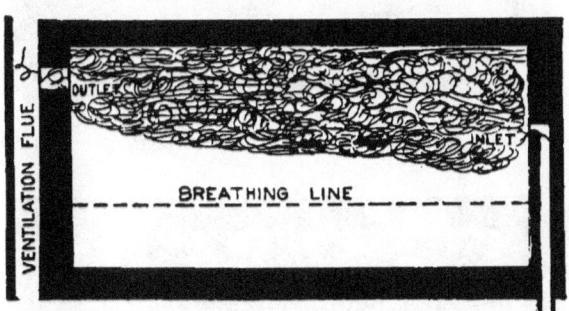

FIG. 16.—AIR INTRODUCED ON SIDE, DISCHARGED AT TOP.

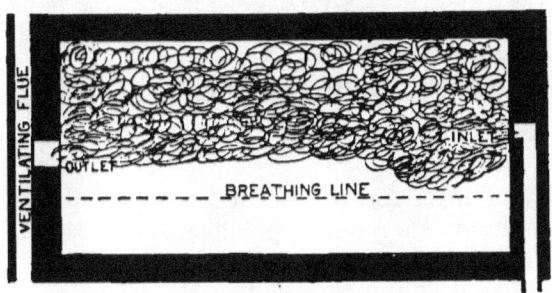

FIG. 17.—AIR INTRODUCED ON SIDE, DISCHARGED ON OPPOSITE SIDE.

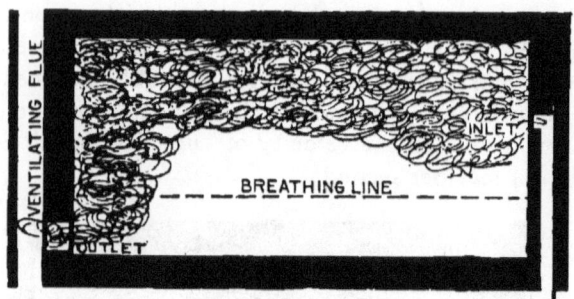

Fig. 18.—Air Admitted on Side, Discharged near Bottom.

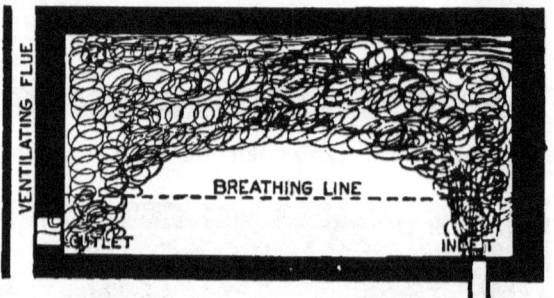

Fig. 19.—Air Admitted at Bottom, Discharged near Bottom.

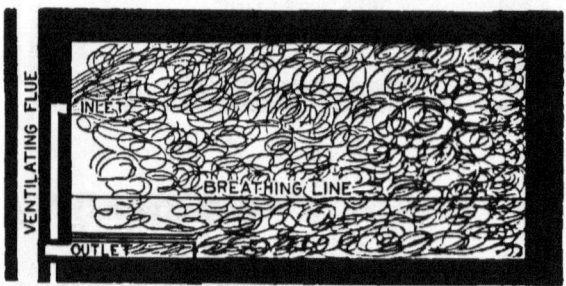

Fig. 20.—Inlet near Top, Discharge near Bottom.

35. The Outlet for Air.—The outlet for air should be as near the bottom of a room as possible, and it should be connected with a flue of ample size maintained at a temperature higher than that of the surrounding air, unless forced circulation is in use, in which case the excess of pressure in a room will produce the required circulation. If the temperature in a room is higher than that of the surrounding air, and if the flue leading to the outside air can be kept from cooling and is of ample size and well proportioned, the amount of air which will be discharged will be given quite accurately by the tables referred to. These conditions should lead us to locate vent-flues on the inside walls of a house or building, and where they will be kept as warm as possible by the surrounding bodies. If for any reason the temperature in the flue becomes lower than that of the surrounding air the current will move in a reverse direction, and the ventilation system will be obstructed.

The conditions as to size of the outlet register are the same as those for the inlet; the register should be of ample size, the opening should be gradually contracted into the flue, and every precaution should be taken to prevent friction losses.

36. Ventilation-flues.—The size of ventilation-flue will depend to a great extent upon the character of system adopted, but will in all cases be computed as previously explained. A practical system of ventilation generally, is intimately connected with a system of heating, and the various problems relating to the size and construction of ventilating ducts will be considered later. In general the ducts should be of such an area as not to require a high velocity, since friction and eddies are to a great extent due to this cause.

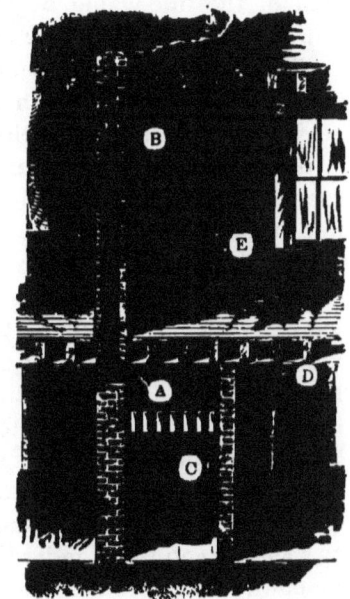

FIG. 21.—VENTILATION-FLUE.

The size of the ventilating duct can be computed, knowing its rise, length, and the difference of temperature by dividing the total amount to be discharged by the amount flowing through one square foot of area of the flue under the same conditions.

In introducing heated air into a room, it is very much better to bring in a large volume heated but slightly above the required temperature of the room rather than a small volume at an excessively high temperature. If the temperature of the air be brought in 25 degrees above that of the air in the room, the discharge in a flue one square foot in area would be in cubic feet per minute, 171 for a height of 10 feet, 271 for a height of 25 feet, 342 for a height of 40 feet. By referring to the table, Article 33, the discharge for any condition can be readily determined.

As the difference of temperature of the air in the room and outside may usually be taken as 20°, the velocity in feet per minute for heights corresponding to the distance of floor to roof in a building of 3 stories would be about as follows: 1st floor, 306; 2d floor, 242; attic or top floor, 188,—or about 5, 4, and 3 feet per second. For air discharged, the order of the velocities would be reversed on the particular floors. The area of the flue would be found by dividing the total air required per second by these numbers.

The general arrangement for heating the air and introducing it into a room is shown in Fig. 21. In this case the cold air is drawn in at D and delivered into the chamber C, whence it passes through the heater, thence into the flue, entering the room at the register B. The vitiated air enters the ventilating flue at E.

37. Summary of Problems of Ventilation.—From the foregoing considerations it is to be noted that the practical problems of ventilation require the introduction, first, of thirty or more cubic feet of air per minute for each occupant of the room, and in addition sufficient air to provide perfect combustion for gas-jets, candles, etc., which are discharging the products of combustion directly into the room. Second, the problem requires the fresh air to be introduced in such a manner as to make no sensible air-currents, and to be in such quantities as to keep the standard of contamination below a certain amount. This problem can be solved by either, first, moving the air by heat, in which case the motive force is very feeble and likely to be counteracted by winds and adverse conditions; second, by moving the air by fans or blowers, in which case the circulation is more positive, and less influenced by other conditions.

The methods for meeting these conditions will be given under appropriate heads in later articles.

It will generally be found much more convenient to estimate the air required, not in cubic feet per minute for each person, but by the number of times the air in the room will need to be changed per hour. If the number of people who occupy a room be known, and each one requires 30 cubic feet of air per minute or 1800 cubic feet per hour, one can easily compute the number of times the air in a room must be changed to meet this requirement. Thus a room containing 1800 cubic feet, in which five people might be expected to stay, would need to have the air changed five times per hour in order to supply the required amount for ventilation purposes.

By consulting the table Properties of Air, No. VIII, it will be seen that one heat-unit contains sufficient heat to warm 55 cubic feet of air, at average pressures and temperatures, one degree; so that practically to find the number of heat-units required for warming the air one degree we must simply divide by 55 the number of cubic feet to be supplied. If the cubic contents of the room is to be changed from five to ten times per hour, we can very readily make the necessary computations by knowing the volume of the room.

Even in the case of direct heating, where no air is purposely supplied for ventilation, there will be a change by diffusion of the air in a room which the writer has found practically met by an allowance equal to one to three changes in the cubic contents per hour, which serves to supply heat for ventilation purposes in addition to that transmitted by the walls.

The number of times that air will need to be changed per minute in a given room will depend upon its size as compared with the number of occupants. If we take the smallest size of rooms, in which we allow only 400 cubic feet of space per occupant, a supply of 30 cubic feet per minute would change the air in this space in $13\frac{1}{3}$ minutes, or at the rate of $4\frac{1}{2}$ times per hour. If 600 cubic feet are supplied per occupant, the air of the room would be changed once in 20 minutes, or at the rate of 3 times per hour. The following table may be of practical value, as it shows the number of changes per hour required to supply each person with 30 cubic feet per minute when the space supplied is as given in the table:

Space to each Person. Cubic Feet.	Number of Times Air to be Changed per Hour.
100	18
200	9
300	6
400	4.5
500	3.6
600	3
700	2.6
800	2.25
900	2

38. Dimensions of Registers and Flues.—The approximate dimensions of registers and flues can be computed from considerations of the limiting velocity of entering air.

For residence heating the velocity in flues is likely to be as follows, in feet per second:

	Warm-air Duct.	Ventilating Duct.	Entering Air at Register.	Discharge Air at Register.
First story	2.5 to 4	6	3	4
Second story	5	5	3	4
Third story	6	4	3	3
Attic floor	7	3	3	$2\frac{1}{2}$

The velocity per hour is 3600 times that per second. The area of the duct can be found by dividing the cubic feet of air needed per hour by 3600 times that in the above columns. If the air required is taken as a certain number of times the cubic contents of the room the following method is applicable:

If we denote the cubic contents of a room by C, the number of times the air is to be changed per hour by n, the velocity in feet per second by V, then will the area in square feet $A = \dfrac{nC}{3600 V}$. In square inches $a = \dfrac{nC}{25 V}$.

The following table gives the net area in square inches for

each 1000 cubic feet of space, of either the hot air or ventilating register, for any required velocity of the air. The net area is about 0.7 the nominal area. (See Table of Registers, Article 144.)

AREA IN SQUARE INCHES FOR EACH 1000 CUBIC FEET OF SPACE.

Velocity, Feet per Second.	Number of Times Air changed per Hour.							
	1	2	3	4	5	6	8	10
1.....	40	80	120	160	200	240	320	400
2.....	20	40	60	80	100	120	160	200
3.....	13.3	26	40	53	67	80	107	133
4.....	10	20	30	40	50	60	80	100
5.....	8	16	24	34	40	48	64	80
6.....	6.7	13	20	27	33	40	53	67
8.....	5	4	15	20	25	30	20	50
10.....	4	8	12	17	20	24	32	40
15.....	2.7	5.3	8	11.3	13.3	16	21	26.6
20.....	2	4	6	8.5	10	12	16	20
25.....	1.6	3.4	4.8	6.8	8	9.6	12.8	16
30.....	1.3	2.7	4	5.7	6.7	8	10.5	13.3

CHAPTER III.

AMOUNT OF HEAT REQUIRED FOR WARMING.

39. Loss of Heat from Buildings.—Heat is required to warm the air of a room to a given temperature, to supply the loss due to the radiation and conduction of heat from windows and walls, and to supply the heat for the air required for ventilation. The amount of heat required for these various purposes will depend largely upon the construction of the building and the supply needed for ventilation purposes.

This question was investigated experimentally by Péclet, and it also received attention by Tredgold at about the same time, and has been more recently investigated by the German Government. Péclet's investigations were carried out with extreme care, and reduced to general laws. He divides the loss into two parts: first, that from the windows; second, that lost by conduction through the walls. He considers the loss in each case from the exterior of the wall as due in part to radiation and in part to convection.

40. Loss of Heat from Windows.—The values which Péclet found for glass, reduced to English measures, were as follows :*

LOSS PER SQUARE FOOT PER DEGREE DIFFERENCE OF TEMPERATURE FAHR. PER HOUR FOR WINDOWS.

Height of Window.	3 ft. 3 in.	6 ft. 7 in.	10 ft.	13 ft. 3 in.	16 ft. 3 in.
Loss in B. T. U. per square foot per degree difference of temperature,	0.98	0.945	0.93	0.92	0.91

* The general formula which Péclet gives as expressing this loss is as follows: $M = \frac{1}{4}(T-\theta)(K+K')$, in which T equals temperature of the room, θ = temperature of the air, K = coefficient loss for radiation, K' = coefficient loss for convection. K' varies with the height. K is constant, and in all cases equal to 291 when the temperature is measured by a centigrade thermometer. The values of the coefficients K and K' were determined by experiment.

For multiple glass the above numbers are to be multiplied by the following coefficients:

Double $\frac{2}{3}$, Triple $\frac{1}{2}$, Quadruple $\frac{2}{5}$, n layers $\frac{2}{1+n}$.

The coefficients given above do not differ greatly from unity for each square foot of single glass and two thirds as much for each square foot of double glass per degree difference of temperature.

Tredgold, in his work on "Warming and Ventilation," states that one square foot of glass will cool 90 cubic feet of air one degree per hour. This is about equivalent to 1.7 B. T. U. per degree difference of temperature per hour. This number was used in computation by both Tredgold and Hood, neglecting the cooling effect of the walls. Hood, in his work "Warming of Buildings," third edition, page 213, gives various other experiments of the same nature.

Mr. Alfred R. Wolff, M.E., in a recent pamphlet gives coefficients adopted by the German Government, as follows:

Heat transmission in B.T.U. per square foot per hour, per degree difference of temperature: Single window, 1.09; single skylight, 1.118; double window, 0.518; double skylight, 0.621. These coefficients are to be increased, as explained in the next article, for exposed buildings.

41. Loss of Heat from Walls of Buildings.—The loss of heat depends upon the material used, its thickness, the number of layers, the difference of temperature between outside and inside surfaces, and air exposure.

The problem is one very difficult of theoretical solution, and we depend principally for our knowledge on the results of experiments.

The following tables were computed from formulæ given by Péclet and reduced to English measures by the writer: *

---

* $M = CQ(T - \theta) + (2C + Qe)$, in which $Q = K + K'$, $e =$ thickness, and $C =$ coefficient of conduction. See Table XIV. Other values as on page 54.

AMOUNT OF HEAT IN BRITISH THERMAL UNITS PASSING THROUGH WALLS PER SQUARE FOOT OF AREA PER DEGREE DIFFERENCE OF TEMPERATURE PER HOUR.

Thickness, inches.	Single Wall.		Wall with Air-space.
	Brick or Stone.	Wood.*	Brick or Stone.
4	0.43	0.12	0.36
8	0.37	0.065	0.30
12	0.32	0.045	0.25
16	0.28	0.033	0.21
18	0.26	0.031	0.19
20	0.25	0.03	0.18
24	0.24	0.029	0.17
28	0.22	0.027	0.15
32	0.21	0.025	0.13
36	0.20	0.020	0.12
40	0.18	0.018	0.10

Mr. Alfred R. Wolff, in a lecture before the Franklin Institute,† gives coefficients for loss of heat from walls of various thicknesses, which he translated from and transformed into American units from tables prescribed by the German Government as follows:

FOR EACH SQUARE FOOT OF BRICK WALL.

Thickness of wall =	4″	8″	12″	16″	20″	24″	28″	32″	36″	40″
Loss of heat per square foot per hour per degree difference of temperature	0.68	0.46	0.32	0.26	0.23	0.20	0.174	0.15	0.129	0.115

1 square foot, wooden beam, planked over or ceiled, { as flooring.... $K = 0.083$; as ceiling..... $K = 0.104$
1 square foot, fireproof construction, floored over, { as flooring.... $K = 0.124$; as ceiling..... $K = 0.145$
1 square foot, single window .... $K = 1.09$
1 square foot, single skylight............................. $K = 1.115$
1 square foot, double window .... $K = 0.518$
1 square foot, double skylight........................... $K = 0.621$
1 square foot, door...................................... $K = 0.414$

---

* This experiment applies to solid wood; it is evidently of little use when applied to wooden buildings, since these buildings generally present so many opportunities for loss of heat through crevices.

† Lecture on Heating of Large Buildings, published in pamphlet form.

These coefficients are to be increased respectively as follows:

Ten per cent where the exposure is a northerly one and the winds are to be counted on as important factors.

Ten per cent when the building is heated during the daytime only, and the location of the building is not an exposed one.

Thirty per cent when the building is heated during the daytime only, and the location of the building is exposed.

Fifty per cent when the building is heated during the winter months intermittently, with long intervals (say days or weeks) of non-heating.

Mr. Wolff has arranged the results in a graphical form (Fig. 22), so that the values for heat losses can be obtained by inspection.

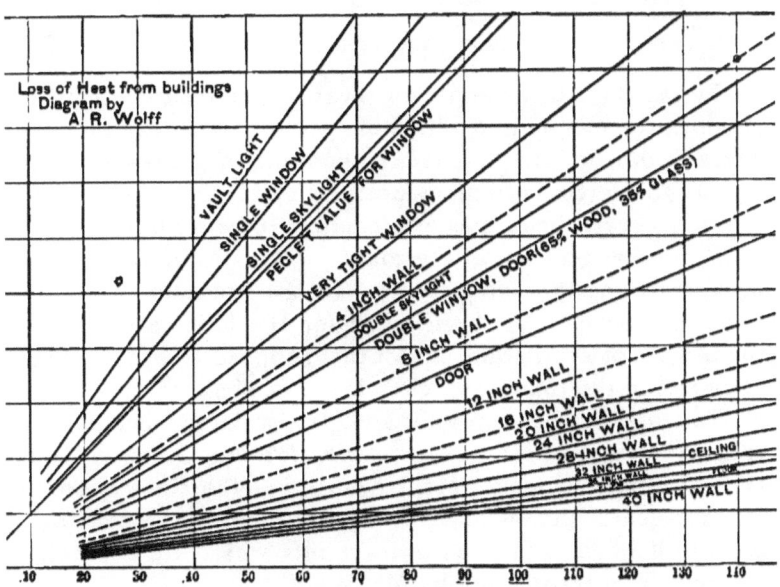

FIG. 22.—WOLFF'S DIAGRAM OF LOSS OF HEAT FROM WALLS.

In this diagram distance in horizontal direction is the required difference in temperature between that of the room and the outside air; the various diagonal lines correspond to the different radiating surfaces of the building, floors, ceiling, doors, windows, etc. The heat transmitted per square foot of surface per hour is given by the numbers in the vertical column.

The German Government require computations to be made on the following assumed lowest temperatures: *

External temperature	4° Fahr.
Assumed lowest temperature of non-heated cellar and other portions of building permanently non-heated	32
Vestibules, corridors, etc., non-heated, and at frequent intervals in direct contact with external air	23
Air-spaces between roof and ceiling of rooms, { Metal and slate roofs	14
{ Denser methods of roofing, such as brick, concrete, etc.	23°

As the temperature to be attained in rooms of various kinds, the German Government prescribes for—

Stores and dwellings	68° Fahr.
Halls, auditoriums, etc.	64°
Corridors, staircase-halls, etc.	54°
Prisons, occupied by day and night	64°

In making calculations for heat losses for buildings in America the minimum external temperature is usually assumed as zero Fahr., and the required temperature in stores and dwellings as 70 degrees. In many portions of the country the corridors, staircase, halls, etc., are required to be from 65° to 68°; while in other portions of the country the halls are required to be as warm as the living-rooms. In the preceding computations no allowance has been made for the heat carried off in the process of ventilation, nor for that supplied from the bodies of people in the room, gas, electric lights, etc.

The loss of heat from walls and glass surfaces has also been considered by Leicester Allen, *Metal Worker*, October, 1892; and by John J. Hogan. Mr. Hogan gives the cooling power of one square foot of glass as 1.57 heat-units, and that of a brick wall 4 inches thick as .231—results which are somewhat different from those given by Mr. Wolff.

42. Heat required for Purposes of Ventilation.—In addition to the loss of heat through walls of buildings, more or less heat will be carried off by the air which escapes from various cracks and crevices.

By consulting Table VIII it will be seen that, for ordinary

---

* Lecture by Alfred Wolff before Franklin Institute.

AMOUNT OF HEAT REQUIRED FOR WARMING. 59

temperatures and pressures, 55 cubic * feet of air will absorb one heat-unit in being warmed one degree F., and hence can be considered the equivalent of one pound of water.

The heat-units required for ventilation can then be found by multiplying the number of cubic feet of air by the difference of temperature between warm and outside air, and dividing by 55.†

Total Heat Required.—By referring to the values for heat losses given by Wolff and Péclet, it will be noted that a fair average value would be 1 heat-unit for glass and 0.25 heat-unit for walls per degree difference of temperature per square foot per hour. Usually we can neglect all inside walls, floors, and ceilings, and consider only the exposed or outside walls with sufficient accuracy.

For direct heating of residences it seems necessary to consider the air of halls changed 3 times per hour, that of rooms on first floor 2 times per hour, and that of rooms on the upper floors once per hour, to account for changes taking place by diffusion.

If C represent cubic contents of room, W the area of exposed wall surface, G the area of glass, n the number of times air is changed per hour, t the difference of temperature between air in room and outside, we have, as a general formula for heat required, in heat-units per hour,

$$h = \left(\frac{nC}{55} + G + \tfrac{1}{4}W\right)t.$$

Very elaborate methods of computing the loss of heat through the walls of a building are given by Box in his Treatise on Heat as a translation from the experiments by Péclet.‡ These methods have in some instances been employed by amateurs in this art in computing the loss of heat through the walls. It seems necessary to remark here that the coefficients obtained by Péclet are accurate only under the conditions gov-

---

* This quantity varies somewhat with barometric pressure and temperature.

† If $C=$ cubic contents of room, n the number of times air is changed, t the difference of temperature, h the heat-units for ventilation, $h = \dfrac{nC}{55}t$.

‡ Traité de la Chaleur, Paris.

erning his experiments, and there is little or no proof that the loss of heat from the walls of a building was ever actually measured by him. Recent writers* on heat have found that Péclet made an error in the position of the decimal point in reporting the coefficient of conductivity, and that his values in consequence were ten times too small at least for metals of high conductibility and were probably in error for all cases. Not only are the coefficients given by Péclet doubtful, but his method or rule for computing the heat lost through the walls is erroneous. For computing the loss of heat he employs formulæ of the same general nature as those given on page 63 for loss of heat from a heated body in still air. For such cases there is a decrease in the loss of heat per unit of area with increase in height, but different conditions apply to the the side of a building freely exposed to air-currents. Actually there is in many cases an increase in heat transmission, due to stronger air-currents near the top of a building. The application of the formulæ quoted by Box † shows that the loss of heat from a building with one side exposed is greater per unit of area than from a building with all sides exposed, which is rarely ever true. The principal objection to the methods referred to lies in the fact that, while the loss of heat through the walls is computed with great elaborateness of detail, no consideration is given to the heat required to warm the air, which in spite of all precautions will constantly enter and leave an apartment and for which considerable heat is in all cases required.

Practically there is little or no difference in the amount of heat required to warm a wooden or a brick building, which is due to the fact that air-spaces lined with heavy building-paper make the heat losses in the one practically as small as in the other. There is, however, a great difference in the amount of heat transmitted through the walls of different buildings, due to good or bad construction or to use of inferior or superior materials; this fact renders any elaborate formula for this purpose abortive. The best that can be expected of any rule is agreement with the average condition.

---

* Theory of Heat, Preston, London.
† A Practical Treatise on Heat, p. 218.

The author in two cases measured the loss of heat, with the following results:* In the first case a room on the second floor with exposed side and end had 246 sq. ft. of wall surface and 96 sq. ft. of window surface. When the air in the room was 28 degrees above that outside the loss was 4247 B. T. U. per hour, and when 27 degrees above, was 4240 B. T. U. per hour. To supply loss of heat by the rule stated would require respectively 4410 and 4253 B. T. U. per hour, the error varying from a fraction of one per cent to nearly five per cent. In the second case a test was made in the N. Y. State Veterinary College; this showed that to maintain the room 31 degrees warmer than the outside air 16,000 B. T. U. were required per minute, of which 39 per cent escaped in the ventilation-flues, and 61 per cent passed by conduction through the walls and windows. The building was exposed on all sides, was 3 stories in height, had 9281 sq. ft. of glass and 31,644 sq. ft. of exposed wall surface. By the rule quoted the building loss should be 532,952 B. T. U. per hour. The actual loss by experiment was 9120 B. T. U. per minute or 547,200 B. T. U. per hour, which is within two per cent of that called for by the rule. In this case the building was of brick, the thickness of walls varied from 24 to 16 inches, the windows had single glass.

The above experiments, which were made on a large scale and on actual buildings, indicate the substantial accuracy of the rule quoted.

Data regarding the number of changes of air which take place per hour under different conditions of direct heating in buildings are still very deficient. The following seems to be reliable:

Number of Changes of Air per Hour.

Residence heating .... Halls, 3; sitting-room, etc., 2; sleeping-rooms, 1.
Stores .... First floor, 2 to 3; second floor, 1½ to 2.
Offices .... First floor, 2 to 2½; second floor, 1½ to 2.
Churches and public assembly-rooms, 2 to 2½.

---

* Transactions of American Society of Heating and Ventilating Engineers, vols. III. and IV.

CHAPTER IV.

HEAT GIVEN OFF FROM RADIATING SURFACES.

43. The Heat Supplied by Radiating Surfaces.—The heat used in warming is obtained either by directly placing a heated surface in the apartment, in which case the heat is said to be obtained by *direct radiation*, or else by heating the air which is to be used for ventilating purposes while on passage to the room, in which case the heating is said to be by *indirect radiation*. As air is not heated appreciably by radiant heat, this latter term is very clearly one which is used in a wrong sense. In this treatise we shall use the terms *direct heating* or *radiation* and *indirect heating*.

Direct heating is performed by locating the heated surface directly in the apartment: this surface may be heated by fire directly, as is the case with stoves and fireplaces; or it may receive its heat from steam or from hot water warmed in some other portion of the premises and conveyed in pipes. The general principles of warming are the same in all cases, but for the case of stoves the temperature is greatly in excess of that for steam or hot-water heating surfaces. The heat is carried away from the heated surface partly by radiation, in which case the heat passes directly in straight lines and is absorbed by people, furniture, and objects in the room, without warming up the intervening air directly, and also by particles of air coming in contact with the heated surface, which may be the radiating surface, or the people and objects in the room which have been warmed by radiant heat.

The sensation caused by radiant and convected heat is quite different: the radiant heat has the effect of intensely heating a person on the side towards the source of heat, and of producing no warming effect whatever on the opposite side. The heat which has passed off by convection is first utilized in warming the air, and the sensation produced on any person is that of lower temperature-heat equably distributed. Radiant and convected heat are essentially of the same nature: in the one case it is received by the person directly from the source of heat, and at a high temperature; in the other case it is received from the air, which is at a comparatively low temperature.

The heat in passing through any metallic surface raises its temperature an amount which depends upon the facility with which heat is conducted by the body and discharged from the outer surface. The phenomena of the flow of heat through any metallic substance can be illustrated by the sketch in Fig. 23. If E represents the source of heat, and $ABCD$ a section of a metallic wall surrounding, the flow of heat takes place into the metallic surface, then through the solid metal, and finally through the outer surface.

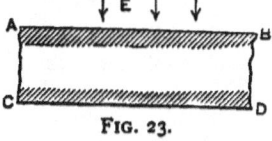

FIG. 23.

It is noted that the heat meets with three distinct classes of resistances: first, that due to the inner surface; second, that due to the thickness of the material; and third, that due to the outer surface. The first and third resistances are due to change of media, and when the material under consideration is a good conductor, constitute the principal portion of the resistance to the passage of heat.

If the resistance on the inner surface AB is small and that on the outer surface CD is great, the temperature of the metallic body will approach that of the source of heat, for the reason that the heat will be delivered to the surface CD faster than it is discharged. In this case the thickness of the material is of little or no importance, and the rate at which heat will pass will depend entirely upon the rapidity with which it can be discharged from the outer surface.

44. Heat Emitted by Radiation.—Heat emitted by radiation, per unit of surface and per unit of time, is independent of the form and extent of the heated body, provided there are no re-entrant surfaces which intercept the rays of radiant heat. The amount of heat projected from a surface of such form as to radiate heat equally in all directions, depends only on the nature of its surface, the excess of its temperature over that of the surrounding air, and the absolute value of its temperature.

Radiation of heat was stated by Sir Isaac Newton to be in exact proportion to the difference of temperature of the heated surface and the surrounding media, but this law was found to be inaccurate by Dulong and Petit. They found that the radiation increased at a greater rate than the difference in temperature, and for high temperature, was much in excess of that given by the law of Newton. From a large number of experiments on the cooling of bodies they were able to determine the following law: "The rate of cooling due to radiation is the same for all bodies, but its absolute value varies with the nature of the surface." It is represented by the formula

$$V = ma^\theta(a^t - 1),$$

in which m represents a number depending on the nature of the surface of the body, a represents a constant number, which for the centigrade thermometer is equal to 1.0077 and for the Fahrenheit above 32° to 1.00196, θ the temperature of the surrounding air, and t the excess of temperature of the body over that of the surrounding space.

Péclet found that if the radiant heat be received by a dull surface the value of m becomes equal to a constant 124.72 multiplied by K, a coefficient which depends on the nature of the surface. A table giving the rapidity of cooling for different values of difference of temperature in both Fahrenheit and metric units is given on page 64, and the value of the coefficient K for different surfaces, which is to be multiplied by the numbers which express the relative rates of cooling, is given in a subsequent table.

The results of the experiments by Péclet accord very well with recent experiments made in testing radiators for steam and hot-water heating. For these cases either wrought or cast iron is used, and the difference in radiating power is immaterial. The construction of the ordinary form of radiator is such as to present very little free radiating surface, as all the heat which impinges from one tube on another is reradiated back, and consequently not of use in heating the apartment. The greater portion of the heat removed is no doubt absorbed by the air which comes in contact with the radiator, or, in other words, it is removed by convection.

45. Heat Removed by Convection (Indirect Heating).—The heat removed by convection is independent of the nature of the surface of the body and of the surrounding absolute temperature. It depends on the velocity of the moving air, and is thought to vary with the square root of the velocity. It also depends on the form and dimensions of the body and of the excess of temperature over that of the surrounding air. We are indebted to Péclet for exact experiments giving us the value of the loss from this cause. Péclet's experiments were, however, made in ordinary still air, and if the velocity is increased should be multiplied by factors which will be given later. The formulæ which Péclet found as applying to bodies of different form were as follows, the results below being given in heat-units per square foot per hour.

The general formula for loss by convection is, in metric units,

$$A = 0.552 K' t^{1.233}.$$

The values of K' depend upon the form and surface of the body and are as follows:

For a sphere, radius r,

$$K' = 1.778 + 0.13/r.$$

For a vertical cylinder, circular base, radius r, height h,

$$K' = (0.726 + 0.0345/\sqrt{r})(2.43 + 0.8758 \sqrt{h}).$$

For horizontal cylinder, radius r,

$$K' = 2.058 + 0.0382/r.$$

For vertical planes, height h,

$$K' = 1.764 + 0.636/\sqrt{h}.$$

Numerical values of these various quantities are given in tables, Art. 46.

46. Total Heat Emitted.—The amount of heat given off by radiation and convection for various differences of tempera-

HEAT-UNITS PER HOUR.

Excess of Temperature.		Radiation.				Convection.			
		Total Radiation.		Per Degree Difference.		Total.		Per Degree Difference.	
Deg. Cent.	Deg. Fahr.	Calories per Sq. Metre.	B. T. U. per Sq. Ft.	Calories per Sq. Metre.	B. T. U. per Sq. Ft.	Calories per Sq. Metre	B. T. U. per Sq. Ft.	Calories per Sq. Metre.	B. T. U. per Sq. Ft.
10	18	11.2 K	4.1 K	1.12 K	.228 K	9.4 K'	3.4 K'	0.94 K'	.189 K'
20	36	23.2 "	8.6 "	1.16 "	.239 "	22.2 "	8.2 "	1.11 "	.228 "
30	54	36.1 "	13.2 "	1.20 "	.243 "	36.6 "	13.5 "	1.22 "	.025 "
40	72	50.1 "	18.5 "	1.25 "	.257 "	52.2 K	19.2 "	1.30 "	.265 "
50	90	65.3 "	24.2 "	1.31 "	.269 "	68.6 "	25.3 "	1.37 "	.284 "
60	108	81.7 "	30.2 "	1.36 "	.281 "	86.0 "	31.8 "	1.43 "	.295 "
70	126	99.3 "	36.6 "	1.42 "	.291 "	104.0 "	38.4 "	1.49 "	.306 "
80	144	118.5 "	43.7 "	1.48 "	.304 "	122.6 "	45.0 "	1.53 "	.311 "
90	162	138.7 "	51.2 "	1.54 "	.317 "	141.7 "	52.2 "	1.57 "	.32 "
100	180	161.3 "	59.5 "	1.61 "	.33 "	161.5 "	59.5 "	1.61 "	.33 "
110	198	185.3 "	68.5 "	1.69 "	.035 "	181.5 "	67.0 "	1.64 "	.334 "
120	216	211.3 "	78.0 "	1.76 "	.361 "	202.1 "	75.5 "	1.68 "	.345 "
130	234	239.3 "	88.3 "	1.83 "	.377 "	223.1 "	82.2 "	1.72 "	.35 "
140	252	269.5 "	99.0 "	1.92 "	.395 "	244.4 "	90.0 "	1.74 "	.355 "
150	270	302.1 "	112 "	2.01 "	.416 "	266.1 "	98.0 "	1.76 "	.36 "
160	288	339.0 "	125 "	2.12 "	.435 "	288.1 "	106 "	1.79 "	.365 "
170	306	371.4 "	139 "	2.22 "	.454 "	310.5 "	115 "	1.82 "	.372 "
180	324	418.5 "	155 "	2.32 "	.478 "	333.2 "	123 "	1.85 "	.38 "
190	342	463.2 "	172 "	2.43 "	.503 "	356.1 "	132 "	1.87 "	.384 "
200	360	511.2 "	188 "	2.55 "	.523 "	379.4 "	140 "	1.89 "	.39 "
210	378	563.1 "	208 "	2.68 "	.553 "	402.9 "	149 "	1.91 "	.394 "
220	396	619.0 "	229 "	2.81 "	.573 "	426.7 "	157 "	1.93 "	.40 "
230	414	679.5 "	255 "	2.95 "	.617 "	450.4 "	166 "	1.95 "	.403 "
240	432	744.8 "	275 "	3.10 "	.665 "	475.0 "	175 "	1.97 "	.406 "
250	450	848.7 "	314 "	3.39 "	.700 "	498.6 "	184 "	1.99 "	.408 "

FACTOR TO DETERMINE RADIATION LOSS FROM VARIOUS SURFACES.

VALUE OF COEFFICIENT K.

Polished silver	0.43	Powdered wood	3.53
Silvered paper	0.42	" charcoal	3.42
Polished brass	0.258	Fine sand	3.62
Gilded paper	0.23	Oil painting	3.71
Red copper	0.16	Paper	3.71
Zinc	0.24	Soot	4.01
Tin	0.215	Building stone	3.60
Polished sheet iron	0.45	Plaster	3.60
Sheet lead	0.65	Wood	3.60
Ordinary sheet iron	2.77	Calico	3.65
Rusty sheet iron	3.36	Woollens	3.68
Cast iron, new	3.17	Silk	3.71
Rusty cast iron	3.36	Water	5.31
Glass	2.91	Oil	7.24
Powdered chalk	3.32		

NOTE.—To find the total heat emitted by radiation, multiply the value of K as given in the above table by the numbers corresponding to radiation due to difference of temperature as in the preceding table.

HEAT GIVEN OFF FROM RADIATING SURFACES.

ture and from any surface when K or K' is unity is given in the first table on p. 64, as computed from Péclet's experiments. The total heat emitted by any surface will be obtained by multiplying the results given in the first table by the factor of radiation and convection for the required conditions. This table is exact for the surrounding air at 15° Centigrade or 59° Fahrenheit.

FACTOR TO DETERMINE CONVECTION LOSS FROM BODIES OF VARIOUS DIMENSIONS.
Value of Coefficient K'.

Diameter.		Sphere.	Horizontal Cylinder.	Vertical Cylinder, Height in Metres and Feet.						
Metres.	Inches.			0.5 m. 1.64 ft.	1 m. 3.28 ft.	h 2 m. 6.56 ft.	h 3 m. 9.84 ft.	h 4 m. 13.12 ft.	h 5 m. 16.4 ft.	h 10 m. 32.8 ft.
0.025	0.984	......	5.114	......	......	......	......	......	......	......
0.05	1.968	6.9	3.59	3.55	3.2	2.95	2.84	2.79	2.73	2.62
0.10	3.94	4.38	2.82	3.22	2.9	2.68	2.57	2.52	2.48	2.38
0.20	7.88	3.08	2.44	3.05	2.75	2.54	2.44	2.39	2.35	2.26
0.40	15.74	2.43	2.25	2.93	2.65	2.45	2.35	2.30	2.26	2.17
0.60	23.62	......	2.18	2.88	2.60	2.40	2.31	2.26	2.22	2.13
0.8	31.50	2.10	2.15	2.85	2.57	2.37	2.28	2.23	2.20	2.11
0.10	39.38	......	......	2.83	2.55	2.36	2.26	2.22	2.18	2.09
0.16	63.0	1.94	......	......	......	......	......	......	......	......
			ratio $\frac{l}{d}$	20	20	20	15	13¼	12.5	20

The table on p. 66 gives the total loss from various forms of direct radiating surfaces in still air, calculated by Péclet's coefficients, slightly modified by recent experiments.

The loss of effective surface due to rays of radiant heat impinging on hot surfaces can be calculated as follows:

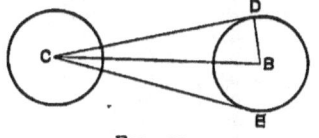

Fig. 24.

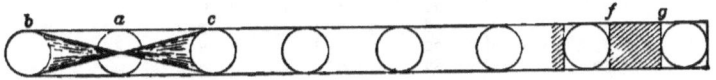

Fig. 25.

Thus in Fig. 24, supposing pipes equally hot, occupying the relative positions of *C* and *B*, the effective radiating sur-

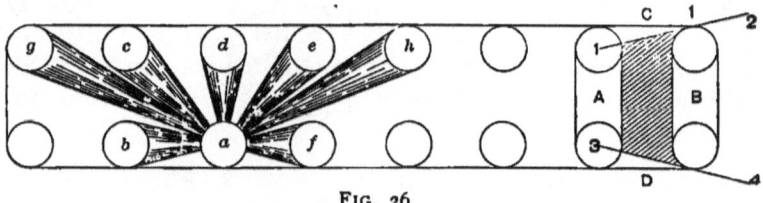

FIG. 26.

face of *C* will be diminished by that portion of the circumference intercepted by the lines *CD* and *CE*. The angle *DCB*

HEAT-UNITS EMITTED PER HOUR PER SQUARE FOOT FROM VARIOUS SURFACES, DIRECT RADIATION, STILL AIR.

Difference of Temperature. Deg. F.	Coefficient or Amount per Degree Difference of Temperature.				Total per Square Foot per Hour.*			
	Horizontal Pipe, Diameter.				Horizontal Pipe, Diameter.			
	6 in.	4 in.	2 in.	1 in.	6 in.	4 in.	2 in.	1 in.
	Radiator, Height.				Radiator, Height.			
	40 in. Massed Surface.	40 in. Thin.	24 in. Massed.	12 in. Thin.	40 in. Massed Surface.	40 in. Thin.	24 in. Massed.	12 in. Thin.
10	0.55	0.62	0.66	0.85	5.50	6.7	6.6	8.5
20	1.11	1.25	1.32	1.72	20.2	24.9	26.4	34.4
30	1.18	1.34	1.42	1.84	35	39.7	42.7	55.2
40	1.24	1.40	1.48	1.92	49.6	56.2	59.0	77
50	1.29	1.46	1.54	2.01	64.5	73.0	77	100
60	1.33	1.50	1.58	2.06	79.8	90	95	124
70	1.36	1.54	1.63	2.12	95.2	108	113	148
80	1.40	1.58	1.67	2.18	112	127	133	173
90	1.43	1.63	1.72	2.24	128	147	153	199
100	1.47	1.66	1.76	2.28	147	167	175	228
110	1.51	1.71	1.80	2.34	166	188	198	257
120	1.54	1.74	1.84	2.39	184	208	219	287
130	1.57	1.78	1.88	2.44	203	230	242	318
140	1.61	1.81	1.91	2.48	223	252	266	346
150	1.64	1.84	1.94	2.53	244	276	291	378
160	1.66	1.87	1.97	2.57	265	300	316	410
170	1.69	1.91	2.02	2.62	286	324	341	443
180	1.72	1.94	2.05	2.65	307	348	367	475
190	1.75	1.98	2.09	2.71	330	375	393	512
200	1.78	2.01	2.12	2.76	356	403	415	552
225	1.87	2.12	2.24	2.91	420	477	500	650
250	1.97	2.23	2.35	3.06	493	557	587	762
275	2.07	2.34	2.47	3.22	563	637	670	872
300	2.17	2.45	2.58	3.37	654	742	780	1020
325	2.27	2.55	2.70	3.50	740	840	882	1150
350	2.37	2.67	2.82	3.66	835	945	995	1295

* Results divided by 1000 give approximate weight of steam condensed per hour.

has for its sine DB/BC. DB is the external radius of the pipes, BC the distance between the centres, which is usually not far from two diameters. In Figs. 25, 26, and 27 the shaded

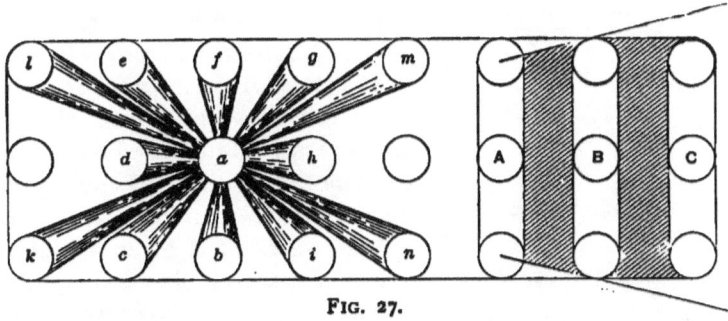

FIG. 27.

areas show the position of surface, by which the radiant heat coming from a single pipe or a single section is intercepted and reradiated to its source.

Supposing the distance apart to be as given above, the following table gives the percentage of reduction in amount of heat transmitted due to this cause:

Number of Rows of Tubes.	Amount of Surface from which no Radiation takes place.	Probable Reduction in Heat transmitted.
	Per cent.	Per cent.
1	16	8
2	42.7	21.3
3	55	27.5
4	66	33
5	73	36.5
6	79	39.5

47. Material of Radiators.*—As bearing directly upon the above subjects, the writer planned a series of experiments which were conducted by E. T. Adams and M. H. Gerry in Sibley College during the winter of 1893-4. The results of these experiments show that the amount of heat transmitted does not depend so much upon the kind of metal as upon the

---

* Transactions of American Society Heating and Ventilating Engineers, vol. I.

media in contact with the metal on both sides. The experiments performed were quite elaborate ones, and every precaution was taken to secure accuracy.

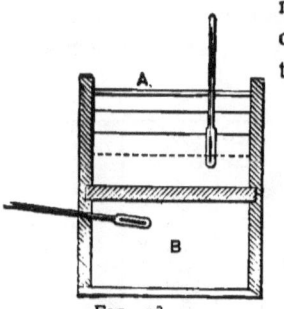

FIG. 28.

The heat measurements were made with an apparatus arranged as in Fig. 28. The box A was fitted so that plates of cast or wrought iron of various thickness could be used as a bottom. The box B directly beneath the box A was so constructed that steam, air, oil, or water of a given temperature could be supplied, and would transmit its heat upward through the bottom of the box A. The heat thus transmitted was measured by its effect on the temperature of the water in box A. The results were reduced to thermal units passing through one square foot of metal per hour, for each degree difference in temperature in box B and box A.

We see from the table that when steam and water are in contact with the plates considerable difference in the results were obtained by varying the thickness of the plate in the bottom of the vessel A. But when air is on one side, the results are little affected by varying the nature and thickness of the plate. The experiments were made with a clean cast-iron plate, also with a wrought-iron plate, and then with each of these plates covered with a thick layer of boiler-scale, neatly fitted. In the latter case the heat had to pass through both the scale and the plate.

The table (page 69) is of interest, since it shows that with the same material for heat transmission, and with the same difference of temperature, there is a great difference in the results. These depend more upon the material which receives or takes up heat than upon the material which conducts it. It can be readily seen, however, that such would not have been the case with poor conductors.

Thus the heat transmission through iron, from steam to water, varies from 25 to 75 times as much as that transmitted through the same plate from air to water. When the heat was passing from steam to water there was a sensible difference, due to the material and thickness of the plates used, but when passing from air to water this difference wholly disappeared. In passing from steam to water the rate of transmission increased very rapidly with increase in difference of temperature.

HEAT TRANSMITTED IN THERMAL UNITS FOR EACH SQUARE FOOT PER HOUR AND PER DEGREE DIFFERENCE OF TEMPERATURE.

Difference of Temperature of the Two Sides of the Plate, Deg. Fahr.	Steam to Water.				Lard Oil to Water.		Air to Water.	
	Clean Wrought Iron ⅛ inch thick.	Clean Cast Iron ⅞ inch thick.	Wrought-iron Plate and Scale 1½ inches thick.	Cast-iron Plate and Scale 1⅜ inches thick.	Clean Cast-iron Plate ⅞ inch thick.	Cast-iron Plate and Scale.	Clean Cast Iron ⅞ inch thick.	Cast-iron Plate and Scale.
25	28.8	21	2.7	1.8	6.5	4	1.2	0.15
50	60.0	48	5.5	3.6	13	8	2.5	0.3
75	96.0	84	8.2	5.4	19.5	12	3.7	0.45
100	150.0	127	11	7.3	26	16	5	0.6
125	228	185	13.7	9.1	31.5	20	6.2	0.75
150	348	255	16.5	10.9	39	24	7.5	0.9
175			19.2	12.7	45.5	28	8.7	1.05
200			22	14.6	52	32	10	1.2
300			33	21.9	78	48	15	2.8
400			44	36.2			20	2.4
500							25	3.0
600								3.6

48. Methods of Testing Radiators.—So far as the writer knows, no standard method has been adopted for use in the testing of radiators, and while numerous tests have been made by different engineers and experimenters, they are often not concordant either as to the method of testing or as to the results obtained. The results in the testing of radiators are greatly affected by small variations in temperature, by irregular air-currents, and by the amount of moisture contained originally in the steam. Obscure conditions of little apparent importance and often disregarded greatly influence the results. The heat emitted by the radiator is in all cases to be computed by taking the difference between that received and that discharged. This result is accurate, and easily obtained. This heat is utilized in warming the air and objects in the room, and to supply losses from various causes, which take place constantly, and is diffused so rapidly, and used in so many ways, that it is practically impossible to measure it, although it is, of course, equal to that which passes through the radiator. The radiating surface is almost invariably heated either by steam or by hot water. In the case of a steam radiator the heat received may be determined, by ascertaining the number of pounds of dry steam condensed in a given time, multiplying this by the heat contained in one pound of steam, and deducting from this product the weight of condensed water, multiplied by its temperature. To make a test of this kind with accuracy requires, first, a knowledge of the amount of moisture contained in the original steam; second, the pressure of the steam or its temperature; third, an arrangement for permitting

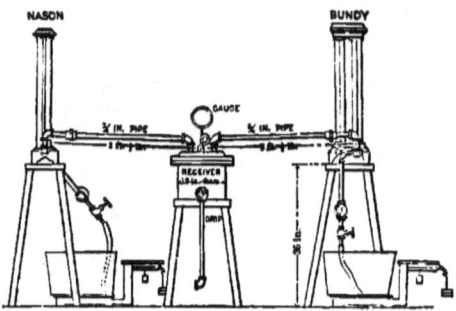

FIG. 29.—RADIATORS ARRANGED FOR TESTING.

water of condensation to escape from the radiator without the loss of steam, and means of accurately weighing this water, and also of determining its temperature. The radiator can be located in any desired position in the room; on the floor, or slightly elevated therefrom. The temperature of the room during the test should be maintained as nearly constant as possible, and no test should be less than from 3 to 5 hours in length. The method adopted by Mr. George H. Barrus in making a radiator test is shown in Fig. 29. The one adopted by the author, in many respects similar, is shown in Fig. 30.

In some recent tests of steam radiators made at Sibley College * the author adopted the following plan of operation

---

* See Transactions vol. i., American Society Heating and Ventilating Engineers.

for measurement of the heat discharged and for operating the radiators:

First, the steam supplied to the radiator to be passed through a separator and a reducing-valve to remove entrained water and maintains a constant pressure during any given run. Second, the amount of moisture in the steam to be measured by a calorimeter, and corrections made to the result for the entrained water. Third, the pressure and temperature of the steam in the radiator to be measured by accurate gauges and thermometers. Fourth, the amount of heat passing through the radiator to be obtained by weighing the condensed

FIG. 30.—RADIATOR ARRANGED FOR TESTING.

steam, measuring its temperature, and computing by this means the heat discharged.

Fifth, the air from the radiator to be effectually removed. Large errors are caused by leaving varying amounts of air in the radiator. The ordinary air-valve is often very unsatisfactory for this purpose; if used, it must be closely watched, or the results may be seriously affected.

The heat supplied was computed by knowing the weight, the percentage of moisture, and the heat contained in one pound of steam. Various methods were tried for drawing off the condensed water: in some tests a trap was used, but better results were obtained by employing a water-column with gauge-glass and drawing off the water of condensation by hand, at such a rate as to maintain a constant level in the glass. To prevent loss by evaporation, this water needs to be received either into a vessel containing some cold water, or else into one with a tight cover, the latter being generally preferred.

Methods of Testing Indirect Steam Radiators.—For this case the general methods of testing should be the same as those previously described, and in addition the volume of air which passes over the radiator should be measured; also, its temperature before and after passing the radiator. For measuring the velocity of air the most accurate instrument at present known is the anemometer, which has been described and illustrated in Article 30, page 37. In measuring the velocity the anemometer should be moved successively to all parts in the section of the flue, and the average of these results should be used. The velocity in feet per minute multiplied by the area of section in square feet should give the number of cubic feet. The number of cubic feet of air heated can also be computed, as explained in Article 30, page 40, by dividing the heat emitted by the radiator by the product of specific heat of air and increase in temperature.

The heat which is absorbed by the air can be computed by multiplying that required to raise one cubic foot one degree, as given in Table VIII, by the total number of cubic feet warmed multiplied by the increase in temperature. Fig. 31 shows an arrangement adopted by the author in testing indirect radiators, the air-supply being measured by an anemometer not shown.

FIG. 31.—ELEVATION OF APPARATUS.

Testing Hot-water Radiators.—The amount of heat transmitted through the surfaces of a hot-water radiator can be determined in either of two ways: first, by maintaining circulation at about the usual rate, measuring the temperature of the water before entering and after leaving the radiator; also, measuring or weighing the water transmitted. The heat transmitted would be equal in every case to the product of the weight of water, multiplied by the loss of temperature. In making these tests the same precautions as to removing the air from the radiator must be adopted as in testing steam radiators.

These radiators can also be tested by filling with water at any desired temperature and noting the time required for the water to cool one or more degrees. In this case the iron which composes the radiator would cool the same amount, and a correction must be added. The easier way to correct for the metal composing a radiator is to consider the weight as that of the water increased by that of the iron multiplied by its specific heat. The specific heat of wrought iron is, practically, 1 divided by 9; that of cast iron, 1 divided by 8; hence for a cast-iron radiator the effect would be the same as though we had an additional amount of water equal to $\frac{1}{8}$ of the weight of the radiator.

In the practical operation of this test the water in the radiator must be kept thoroughly agitated by some sort of stirring device.

49. Measurement of Radiating Surface.—The amount of radiating surface is usually expressed in square feet, and the total surface is that which is exposed to the air, and includes all irregularities, metallic ornaments, etc., of the surface.

Where the surface is smooth and rectangular or cylindrical it is easily measured, but where it is covered with irregular projections the measurement is a matter of some difficulty and uncertainty. The only practical method of measuring irregular surface seems to be that of dividing it up into small areas and measuring each one of these areas separately by using a thick sheet of paper or a bit of cord, and carefully pressing it into every portion of the surface. The sum of all the small areas is equivalent to the total area.

This method is at best only approximate, and even when exercising the utmost care different observers are likely to differ three or four per cent in their results. The writer has tried several other methods of measuring surface, but so far without marked success. One method, which promised good results, was to cover the whole surface with a thin paint and compare the weights with that required for covering one square foot of plain surface. This method proved even more approximate than the other, and had to be abandoned, as the paint was not of equal depths on all portions of the surface.

The total contents of the radiator in cubic feet can be easily determined by filling it with a weighed amount of water of a known temperature and dividing the result by the weight of one cubic foot. The volume displaced by the whole radiator can be determined by immersing it in a tank whose cubic contents can readily be measured. The difference between the cubic contents when the radiator is in the tank and when taken out is the volume of the radiator. For this test the openings in the radiator must be tightly stopped.

The same method applied with the radiator immersed in both cases; but in one case with the radiator filled with air and the other with water would give as a result the water displaced by the metal actually used in the construction, or, in other words, the cubic volume of the metal. This could no doubt be more accurately obtained by dividing the weight of the metal by the weight of one cubic inch or cubic foot. These methods give accurate means of measuring the total external and internal volume of the radiator, but not the surface.

50. Effect of Painting Radiating Surfaces.—In the experiments of Péclet which have been given in Article 46 the effect of different surfaces has been fully considered. From these experiments it would appear in a general way that the character of the surface affects the heat given off by radiation only, and not that given off by convection. In ordinary cases of direct radiation, because the surfaces are closely massed together, the radiant heat does not probably exceed on an average 40% of the total emitted, and is nothing in indirect heating. From the experiments quoted, on page 64, it would appear that if we consider the radiant heat given off as 100 from a new surface of cast iron, that from wrought iron would be 87, from a surface coated with soot or lampblack 125, from a surface with a lustre like new sheet lead $20\frac{1}{2}$, from a polished silver surface $13\frac{1}{2}$. These results make very much less difference, when applied to total heat emitted, since the total radiant heat is only a small portion of the whole heat given off. Calling the radiant heat as 40% of the total, we should have the following numbers as representing the heat emitted from various surfaces:

Cast iron, new........	100	Rusty surface........... ...	102
Wrought iron.	93	Bright iron surface........	72
Dull lampblack........	106	White lead, dull..........	106

The writer had some experiments made in Sibley College, the results of which showed that the effect of painting was to increase the amount of heat given off.

It was found that two coats of black asphaltum paint increased the amount 6%, two coats of white lead 9%. Rough bronzing gave about the same results as black paint.

On the other hand a coat of glossy white paint reduced the amount of heat emitted about 10%.

51. Results of Tests of Radiating Surface.—The results of the experiments of Péclet have been given quite fully, and they will be found to agree well with best modern tests when the conditions are similar. The radiating surface ordinarily employed for steam or hot-water heating consists of a number of pipes closely grouped together so as to occupy as little space as possible. In some instances long coils or series of parallel rows of pipe are employed arranged horizontally, but ordinarily the pipes are vertical, and grouped together in two to four rows. The usual height of radiator is 36 to 40 inches with the bottom placed about 3 inches from the floor, making the actual height of radiating surface about 3 feet. In some instances radiators are lower, in which case the results per unit of surface are considerably increased.

The value of a radiator in which the surface is grouped so as to prevent the free escape of radiant heat will depend largely upon the effectiveness with which the air-currents strike the heating surfaces. There is a tendency for heated air to move in a vertical current in contact with the radiator surface, and thus to keep the upper portion in a very hot atmosphere, which has the effect of materially lessening its efficiency. The practical effect of these restrictions is to reduce the heating power of radiators which are composed of a large amount of surface closely grouped. The following summary of a series of radiator tests made by J. H. Mills shows that with very small radiators the results are in practical accordance with those of Péclet's experiments, but as the radiators increase in size they fall off about in proportion to the loss of effective radiating surface.

Sq. Ft. of Radiating Surface.	Difference of Temperature.	B. T. U. per Sq. Ft. per Hour per Degree Difference of Temperature.	
		Péclet's Formula.	Actual
10	155	1.86	2.10
20	150	1.84	2.08
30	158	1.87	2.06
40	175	1.92	1.75
50	155	1.86	1.73
60	165	1.89	1.67

The following experiments were made by Tredgold* for the time of cooling of water in vessels of various kinds. The writer has reduced the results to heat-units given off per square foot of surface per hour.

SUMMARY OF TREDGOLD'S EXPERIMENTS.

Material Cooling.	Material of Radiator.	Temperature of Hot Body.	Temperature of Room.	Difference of Temperature.	Heat-units Emitted per Sq. Ft. per Hour.		
					Total Heat-units.	Per Deg. Diff. Temp.	By Péclet's Formula.
Hot water...	Tinned iron cylinder....	180	55.5	124.5	255	1.43	1.17
Hot water....	Glass.................	180	56.5	123.5	426	2.37	2.36
" "	Wrought-iron block....	180	57	123	434	2.41	2.36
" "	Rusty wrought iron....	180	57	123	486	2.70	2.5

Prof. C. L. Norton, Boston, Mass., reported in Transactions of American Society of Mechanical Engineers, 1898, that the heat transmitted from a body of hot oil was proportional to the following numbers:

New pipe.......... 100	Painted dull white 115	Coated with cylinder oil......... 116
Fair condition...... 115	Painted glossy " 100.4	Painted dull black 120.5
Rusty and black.... 118	Cleaned with potash............ 115	Painted glossy " 101
Cleaned with caustic potash........... 118		

* Tredgold's Warming and Ventilating of Buildings, second edition, pages 56 to 60.

The following table is abstracted from one published in "Warming and Ventilation of Buildings," by J. H. Mills:

EXPERIMENTS ON DIFFERENT STEAM-HEATED SURFACES AND DIFFERENT MATERIALS.
WROUGHT-IRON, CAST-IRON, STEEL, AND BRASS PIPES. CORRUGATED, RINGED, AND PLAIN SURFACES.

	Description of heating surface.	Square Feet of Surface. A	Steam. B	Air of Room. C	Difference. B−C. D	Lbs. of Water Condensed per Sq. Ft. per Hr. F	Heat-units per Degree Diff. Steam and Air. I
J. H. Nason, 1862.	Plain wrought-iron pipe, 1″, 100′ in a single horizontal line.	33.30	212	70	142	.41	2.89
	Plain cast-iron pipe 3″ diameter outside, 5′ long.	4.02	212	70	142	.344	2.42
	Plain cast-iron pipe 3″ " " " 5′ " but thinner than above.	4.02	212	70	142	.344	2.42
	Ribbed cast-iron pipe, S. Williams; core 3″ in diam., 5′ long; ribs ⅜″ deep, ¼″ between cylinders, 1″ cylinders	8.87	212	70	142	.24	1.69
	Ribbed cast-iron pipe, No. 1, J. Nason & Co., 3″ outside.	10.70	212	70	142	.275	1.95
	" " " " 2, " " " 3″ " heavier ribs than No. 1.	9.20	212	70	142	.280	1.98
	Plain cast-iron pipe, J. Nason & Co. } Placed side by side and tested together.	4.00	212	70	142	.36	2.54
	Ribbed " " " " }	10.70	212	70	142	.245	1.72
	Curved rib cast-iron radiator, Morris, Tasker & Co.	14.14	212	70	142	.269	1.89
	Box-radiator, cast iron, with straight vertical ribs	44.30	212	70	142	.21	1.66
	Vertical cast-iron ringed-pipe radiator, 7 Sec. "Clogston"	21.00	212	70	142	.237	1.67
J. H. Mills, 1888.	Cast-iron pipe 3″ diam., in single line.	25.00	222	70	152	.312	2.06
	Wrought-iron pipe 3″ diam., in single line.	26.00	228	70	158	.312	1.98
	Steel pipe 4″ diam., in single line.	25.00	230	68	162	.39	1.86
	Brass 1″ horizontal pipe; 4-branch circulation.	10.00	230	62	168	.351	2.00
	Wrought-iron, 1″ horizontal pipe, 4-branch circulation	10.00	228	65	163	.412	2.53
	Plain brass vertical tube-radiator ½″ diam., 2 × 4.	12.00	230	62	168	.269	1.60
	Corrugated brass vertical tube-radiator ½″ diam., 2 × 6.	48.00	228	60	168	.214	1.26
	Vertical wrought-iron tube-radiator, Walworth, 1 row of 20 pipes	20.00	228	77	151	.369	2.43
	" " " " " 3 " 16 "	32.00	228	72	156	.343	2.04
	" " " " " " " 16 "	48.00	228	72	156	.219	1.80
	"Union" radiator, cast iron, 6 sec., 29″ high	35.00	228	70	158	.32	2.03
	"Triumph" radiator, A. A. Griffing Iron Co., cast-iron, 8 loops	40.00	230	65	165	.275	1.67
	Peirce "Excelsior" cast-iron radiator, to sections.	40.00	228	63	165	.219	1.70
	"Art" radiator, cast iron, 6 panels	18.00	228	74	154	.264	1.71
	" " 12 " double	36.00	228	70	158	.199	1.25
	Detroit Radiator Exeter Machine Co., cast-iron, 8 loops	40.00	228	73	155	.256	1.65
	Single bar of Gold's Pin Indirect Radiator, 3″ × 6″ × 3½′	10.00	228	70	158	.375	2.37
	Howard Oxbow Radiator, 2 loops, cast iron. Date 1866.	25.00	228	70	158	.347	2.20

78 HEATING AND VENTILATING BUILDINGS.

The following table gives the abstract of a large number of radiator tests made under the supervision of the author: *

Name or Kind of Radiator.	Dimensions				Tests of Kelsey & Jackson.		Tests of Camp & Woodward.		Tests of Dunn & Mack.		Péclet's Coefficient.
	No. Sections.	Rows of Tubes.	Surface, Sq. Ft.	Height, Inches	Difference of Temperature. Deg. Fahr.	Coefficient.	Difference of Temperature, Deg. Fahr.	Coefficient.	Diff. of Temp., Deg. Fahr.	Coefficient.	
W. I. vertical pipes	12	4	53.6	30	94	1.62					
W. I. vertical pipes, Nason	16	3	47.94	36	90	1.669	145	1.70			1.81
					146.6	1.83	144	1.69			1.81
							133	1.62			1.78
W. I. hot water, Western No. 2..	12	4	41.19	32½					133.2	1.62	1.78
									130.1	1.56	1.77
									137.6	1.60	1.79
W. I. steam, Western No. 2......	12	4	43.33	32½					144.8	1.81	1.81
									148.2	1.68	1.82
									158.5	1.79	1.87
Steel, hot water, Western No. 1..	12	4	45.13	35					146.2	1.60	1.82
									147.6	1.79	1.82
									159.5	1.95	1.87
Steel, steam, Western No. 1......	12	4	47.24	35					144.6	1.59	1.81
									143.0	1.50	1.81
									155.0	1.55	1.86
Cast iron, Bundy................	16	1	45.11	37					153.2	1.76	1.85
									154.4	2.14	1.85
" " "	10	3	79	37	149	1.64			159.4	2.02	1.87
" " Bundy Elite....	9	3	41.8	36					153.1	1.88	1.85
									157.1	1.71	1.86
									171.1	1.96	1.91
" " "											
" " Reed.............	13	1	48.7				151	1.688			1.83
							147	1.627			1.82
							136	1.523			1.78
" " Royal Union..	11	3	49.12	37	151	2.08	151	1.688	150	1.73	1.83
							139	1.565	137.5	1.88	1.79
" " " "	26	3	52.81	17			130	1.582	157	1.88	1.87
									153	2.46	
									152	2.37	
									159	2.75	
" " Perfection Steam	13	1	49.9		91	1.63	147.8	1.456			
							147	1.374			
							144	1.433			
" " " "	12	1	48.17	37½					147.0	1.77	1.82
									156.3	1.59	1.86
									166.5	1.89	1.89
" " " "	10	2	40.2	37½					151.5	1.59	1.83
									145.4	1.51	1.81
									165.6	1.71	1.89
" " Perfection Hot Water.	14	1	48	37	89	1.664					
					150	1.55					
" " Ideal Steam.....	10	1	40	38					155.3	1.73	1.86
									158.7	1.70	1.87
									155.1	1.74	1.86
" " " Hot Water.. ...	10	1	40	38	140	1.61			154.5	1.91	1.85
									167.6	2.01	1.90
									158.4	1.99	1.87
" " National Steam.......	10	1	40	38					154	1.67	1.86
									153	1.69	1.85
									160	1.76	1.88
" " Whittier Ex. Surface..	3	1	38.65	30	142	1.13			152.6	1.51	1.83
									164.3	1.56	1.89
" " Michigan Indirect		1	58.2		91	1.434			151.0	1.45	1.83
					140	1.27					
2-inch pipe, single, horizontal...									155.6	3.3	
									167.1	3.7	
									194.5	4.3	
1-inch pipe, single, horizontal.,.									213.2	4.3	
									151.9	5.5	
									165	5.7	
									182.4	5.8	

* Vol. I., Transactions American Society Heating and Ventilating Engineers.

TESTS OF RADIATORS WITH EXTENDED SURFACE SO AS TO FORM AIR-FLUES, COMPARED WITH PLAIN CAST-IRON RADIATORS.*

	Description of Radiator.	Number of Loops.	Height above Floor, Inches.	Width of Loop, Inches.	Area Surface, Sq. Ft.	Steam-pressure.	Temperature. Steam.	Temperature. Room.	Temperature. Difference.	Wt. Steam Condensed per Hour per Sq. Ft., Lbs.	Ditto, per Deg. Diff. Temp.	B.T.U. per Sq. Ft. and per Deg. per Hr.	B.T.U. per Sq. Ft. per Hr.	B.T.U. by Peclet's Rules
		1	2	3	4	5	6	7	8	9	10	11	12	13
A	Extended surface, Joy flue....	9	37	$8\frac{1}{4}$	57.8	3.96	225	52.1	173	3.12	0.00170	1.65	302	312
a	Do. do. do.	1	37	$8\frac{1}{4}$	6.40	4.0	226	67.6	158	0.332	0.00212	2.05	323	312
A'	Same as *A* with flues planed off.	9	37	$8\frac{1}{4}$	40.4	3.9	224	57.8	172	0.329	0.00197	1.97	318	388
a'	Do. do. do.	1	37	$8\frac{1}{4}$	4.24	3.9	224	70.5	154	0.379	0.00247	2.39	369	288
B	Crescent Flue Radiator	9	$36\frac{1}{4}$	$8\frac{1}{4}$	60.8	3.81	223	73.6	149	0.245	0.00136	1.30	248	280
b'	Do. do. do.	1	$36\frac{1}{4}$	$8\frac{1}{4}$	6.23	4.0	225	68.8	156	0.360	0.00231	2.24	350	296
C	Plain Bundy, single row....	14	$39\frac{1}{4}$	$2\frac{1}{4}$	40.25	3.94	224	65.7	158	0.345	0.00243	2.33	335	312
c'	Do. do. do.	1	$39\frac{1}{4}$	$2\frac{1}{4}$	2.83	4.1	226	66.2	159	0.375	0.00237	2.26	365	312
D	Princess flue radiator	9	38	$8\frac{1}{4}$	63.1	3.96	225	71.5	153	2.21	0.00145	1.39	214	285
d	Do. do. do.	1	38	$8\frac{1}{4}$	7.18	4.1	226	70.5	155	0.301	0.00194	1.9	292	294
D'	Same as *D* with extended surface removed.	9	38	$8\frac{1}{4}$	41.2	3.97	225	71.7	153	0.292	0.00191	1.85	284	285
d'		1	38	$8\frac{1}{4}$	4.50	4.0	222	66.2	159	0.365	0.00231	2.24	355	312

52. Tests of Indirect Heating Surfaces.—The tests which have been made on indirect heating surfaces show very great difference in results, varying from those given by Peclet for the loss due to convection alone, to results which are 8 or 10 times as great. This difference in result is no doubt due in each case to the velocity of air which comes in contact with the surface. When the indirect radiators are not freely supplied with air, or the velocity is low, the amount of heat which is discharged is small; when the velocity of the air is high, the amount of heat taken up is proportionally greater. According to experiments made by the writer, the coefficient of heat transmission increases as the square root of the velocity of the air.

The amount of air passing over a given surface of the radiator can be estimated quite accurately by the amount of heat given off, which we can reasonably suppose in this case to be

* Test by Denton & Jacobus, July, 1894.

all utilized in warming the air. At a temperature of about 60 degrees, 1 heat-unit will warm 55 cubic feet of air 1 degree (see Table VIII), so that the number of cubic feet of air warmed is equal to 55 times the total number of heat-units given off from 1 square foot of heating surface per hour, divided by the difference of temperature of entering and discharge air.

NOTE.—Let T = temperature discharge air, t' that of entering air, H = total number of heat-units given off per square foot of surface, a the number of square feet of surface. Then,

$$\text{Cubic feet of air per square foot heating surface} = \frac{55H}{(T - t')a}.$$

The following tests, made under the direction of the writer, give actual results obtained in testing steam-pipes in a current of air moving at different velocities:

SUMMARY OF RESULTS.—TEST OF 2" STEAM-PIPE WITH BLOWER.

Steam-pressure by Gauge.	Average Difference of Temperature of Steam and Air of Room.	Velocity of Air Passing over Pipe, Feet per Second.	Heat Transmission in B. T. U. per Square Foot per Hour for each Degree of Temperature.	Increase in Temperature of Air, Deg. Fahr.	Cubic Feet of Air per Square Foot per Hour.
4.45	123.72	9.8	6.32	26.7	148
5.09	120.30	9.4	6.37	28.4	142
5.38	113.68	4.1	4.29	42.0	63
5.86	113.44	4.5	4.72	42.4	69
5.27	119.32	6.7	5.46	34.9	102
5.15	116.20	5.5	5.46	37.4	83
5.20	117.77				
12.48	134.29	7.1	5.53	35.9	112
13.70	132.73	6.7	5.19	37.3	101
12.10	127.84	6.0	5.24	40.9	91
12.25	125.75	5.5	5.19	43.1	83
13.73	125.93	4.3	4.53	48.3	65
13.55	122.87	4.4	4.99	51.4	66
12.97	128.24				
25.35	157.05	8.6	5.67	37.1	130
27.10	158.27	9.1	5.91	37.7	136
27.54	153.70	6.7	5.36	44.8	101
28.21	153.28	6.3	5.41	45.4	100
27.10	146.68	4.3	4.20	52.6	65
26.70	147.19	4.6	4.61	53.7	70
26.97	152.69				

HEAT GIVEN OFF FROM RADIATING SURFACES. 81

EXPERIMENTS ON INDIRECT RADIATORS.*

Number.	Names of Radiators, Engineers' and Dates of Experiments.		Square Feet Surface.	Gauge-pressure, Steam.	Temperatures.			Diff. Temp.		Oz. Water condensed per Foot per Hour.	Air. Cubic Ft per ft. per Hour.	Units of Heat.	
					Steam.	Entering Air.	Exit Air.	Enter and Exit Air.	Steam and Enter Air			Per Ft. per Hour.	Per Deg. Diff. Temp Stm. & Air.
1	C. B. Richards, 1873-4.	Gold's pin		1	215	0	160	160	215	5.44	111	340	1.58
2		Novelty		1	215	0	156	156	215	5.09	102	318	1.48
3		G. Whittier		1	215	0	135	135	215	4.40	106	275	1.28
4		Pipe coil		1	215	0	147	147	215	4.88	108	305	1.42
5	W. J. Baldwin, 1885.	Gold's pin	60	10	239	71	168	97	168	3.83	128	239	1.42
6		Compound coil	60	10	239	71	170	98	167	3.84	126	240	1.43
7	W. Warner, 1880.	Gold's pin	70	3	222	42	145	103	180	4.60	145	288	1.60
8	J. H. Mills, 1879.	Walworth	40	5	227	33	142	109	194	5.00	149	313	1.61
9		Mills	60	5	227	78	162	84	139	4.08	158	255	1.71
	100 Cubic Feet of Air per Foot per Hour, Average.										126	286	1.50
10	Dr. Gray, 1875.	Gold's pin	90	20	259	33	125	92	226	6.54	231	409	1.81
11	J. R. Reed, 1875.	Whittier	68	3	222	45	129	84	177	5.09	197	318	1.80
12	C. B. Richards, 1873-4.	Gold's pin		1	215	0	139	139	215	9.15	214	572	2.66
13		Novelty		1	215	0	132	132	215	8.70	214	544	2.53
14		G. Whittier		1	215	0	102	102	215	6.66	242	416	1.94
15		Pipe coil		1	215	0	106	106	215	6.98	214	436	2.03
	200 Cubic Feet of Air per Foot per Hour, Average.										214	449	2.13
16	J. R. Reed, 1875.	Whittier	68	3	222	52	110	58	170	5.50	308	344	2.02
17		G. Whittier	68	3	222	52	114	62	170	5.86	307	366	2.15
18		Gold's pin	58	3	222	52	127	75	170	7.92	343	495	2.91
19	C. B. Richards, 1873-4.	Gold's pin		1	215	0	129	129	215	12.65	319	791	3.68
20		Novelty		1	215	0	121	121	215	11.90	320	744	3.46
21		G. Whittier		1	215	0	87	87	215	8.53	319	533	2.48
22		Pipe coil		1	215	0	89	89	215	8.64	323	540	2.51
23	J. H. Mills, 1876.	Gold's pin	76½	10	239	81	159	78	158	8.49	354	531	3.36
24	W. J. Baldwin, Nov., 1885.	Gold's pin	60	5	227	82	150	68	145	8.16	390	510	3.52
25		Compound coil	60	5	227	82	152	70	145	8.16	379	510	3.52
	300 Cubic Feet of Air per Foot per Hour, Average.										336	536	2.96
26	J. H. Mills, 1876.	Gold's pin	76½	10	239	90	158	67	148	8.91	433	557	3.76
27	W. J. Baldwin, 1885.	Gold's pin	60	5	227	70	137	67	158	8.93	433	558	3.55
28		Compound coil	60	5	227	70	135	65	158	8.40	420	525	3.34
29	C. B. Richards, 1873-4.	Gold's pin		1	215	0	121	121	215	15.92	428	995	4.63
30		Novelty		1	215	0	113	113	215	14.86	428	929	4.32
31		G. Whittier		1	215	0	77	77	215	10.14	428	634	2.95
32		Pipe coil		1	215	0	76	76	215	10.02	428	626	2.91
	400 Cubic Feet of Air per Foot per Hour, Average.										428	689	3.64
33	J. H. Mills, 1876.	Gold's pin	77	6	230	88	158	70	142	10.04	467	628	4.42
34		Walworth	67½	6	230	88	142	54	142	8.88	534	555	3.91
	500 Cubic Feet of Air per Foot per Hour, Average.										501	592	4.17
35	J. H. Mills, 1876.	Walworth	85	20	259	90	160	70	169	13.69	636	856	5.06
36		Gold's pin	76½	20	259	90	166	76	169	15.16	649	948	5.61
	600 Cubic Feet of Air per Foot per Hour, Average.										643	902	5.34
37	J. H. Mills, 1876.	Walworth	85	3	222	90	142	52	132	11.61	726	726	5.50
38		Gold's pin	76½	3	222	90	145	55	132	12.54	741	784	5.94
	700 Cubic Feet of Air per Foot per Hour, Average.										734	755	5.72
39	J. H. Mills, 1876.	Gold's pin	77	5	227	94	145	51	133	13.43	855	839	6.31
40		Nason	85	7½	233	79	135	56	154	15.30	888	956	6.21
	800 Cubic Feet of Air per Foot per Hour, Average.										872	898	6.26

*From John H. Mills' work on Heat, by permission.

COMPARISONS OF WATER AND STEAM CIRCULATION, WITH INDIRECT RADIATORS, NATURAL AND FORCED AIR-SUPPLY. By J. H. MILLS.

No. for Reference.	Engineers, Radiators, and Dates.		Square Feet of Surface.	Square Inch of Outlet.	Temperatures.					Difference of Temperature.			Cubic Feet of Air.		Units of Heat per Sq. Ft. per Hour.	
					Water or Steam.		Air to be Warmed.			Steam or Water and Cold Air.	Warm and Cold Air.	$t - t'$	Velocity of Hot Air, Ft. per Sec.	Per sq. ft at Uniform Den.	Total.	Per Deg. Dif. Steam or Water & Air.
					Flow.	Return.	Cold.	Warm.								
	Radiators boxed in Stacks, open below, and with Outlet above for heated Air, the duty of Radiator being determined by the Volume and Temperatures of the heated Air.									Radiators with Water Circulation.						
1	W. J. Baldwin, Box coil, nat. draught, water		74	192	192	176	85	134		99	49	1.97	106	99	1.00
2	1886. Compound coil, nat. draught, water		48	192	195	1.5	85	143		105	58	1.5	117	131	1.25
3	J. H. Mills, Albany cast, nat. draught, water		60	78	202	172	38	128		149	90	4.5	123	213	1.43
4	1885. Box coil, nat. draught, water		55	82	205	162	45	135		138	90	4.7	145	251	1.82
5	W. J. Baldwin, Box coil		74	144	206	186	78	150		118	72	3.8	149	206	1.75
6	1886. Compound coil		48	144	206	194	78	166		122	88	2.5	148	250	2.05
7	J. H. Mills, Gold's pin		100	144	214	178	28	143		168	115	5.	146	323	1.92
8	1885. Mills' indirect		100	144	214	181	28	138		169	110	5.3	157	332	1.96
9	Staggered Tube Coil Radiator, Shakelton's, water		55	50	198	159	34	116		144	82	7.8	158	249	1.73
10	J. H. Mills, Mills' indirect, Shakelton's, water		100	144	196	155	40	111		135	71	6.6	204	279	2.07
11	1885. Gold's pin, Shakelton's, water		100	144	196	155	40	114		135	74	7.	211	300	2.22
	Averages				202	173	53	134		135	82	4.6	151	239	1.75
	Comparison of Steam and Water under similar Conditions and Temperatures.															
12	J. H. Mills, 1884, Box coil, nat. draught, water		55	50	210	163	50	146		137	96	5.8	107	198	1.45
13	C. B. Richards, 1874, steam				215		0	147		215	147		106	395	1.42
14	J. H. Mills, 1884, Mills' indirect, water		100	144	220	180	28	136		172	108	4.9	146	393	1.76
15	J. H. Mills, 1879, steam		60		227		78	162		149	84		151	255	1.71
16	W. J. Baldwin, Compound coil, water		48	144	206	194	78	166		122	88	2.5	148	350	2.05
17	W. J. Baldwin, steam		60	144	220		62	170		158	108		146	303	1.92
18	J. H. Mills, 1884, Gold's pin, water		100	144	240	200	26	157		104	128	5.7	163	401	2.07
19	W. J. Baldwin, 1885, steam		60	144	220		62	157		158	95		161	295	1.87
	Averages water				219	184	46	157		156	105	6.7	141	288	1.83
	steam				221		51	159		170	109		141	289	1.73
20	J. H. Mills, Walworth 10-foot flue { In-direct Radia- tor.	steam	66	217	231	192	84	138		147	54	6.7	461	479	3.26
21	forced draught	"	66	217	228	180	94	135		134	41	22.1	1360	1077	8.04
22	June 16, 1876, Gold's pin, 10-foot flue	"	76	217	231	184	84	155		147	71	6.3	360	492	3.35
23	" " forced dr't	"	76	217	228	150	94	145		134	51	19.7	1143	1121	8.36
24	Nason, 10-foot flue		85	217	228	178	93	131		145	50	10.1	534	513	3.54

NOTE.—For Nos. 5, 15, 17, and 19 the heat recorded is that due to the amount of steam condensed (see Table XII).

From the general results shown in the table page 80 it is seen that the heat-units given off per square foot per degree difference of temperature equals very nearly the square root of four times the velocity in feet per second. That is,

$$h = \sqrt{4v}.$$

The tables pages 81 and 82 contain an extensive summary of tests of indirect radiators, abstracted from Mills' work on Heating and Ventilation, and are of especial interest as showing the close agreement in results, whether water or steam is used. The higher results in this table agree fairly well with the rule stated; those for natural draught are much smaller, and approximately equal to the square root of the velocity in feet per second.

53. Conclusions from Radiator Tests.—The general results of radiator tests can be summed up as follows: First, that the values for heat transmission in recent tests of direct radiators vary greatly and differ more from an average result than from those given by Péclet, and consequently his results can be used with confidence as applying to modern radiators. Second, the results of the test show greater differences in favor of low radiators as compared with high ones than was shown in the experiments of Péclet. Third, the experiments do not show any sensible difference for different materials used in radiators or for hot water or steam, provided the difference in temperature between the air in the room and that of the fluid in the radiator is the same. Fourth, the internal volume of radiators is of value only in lessening the friction of the fluid. It has no special influence on the results. Fifth, the extended surface radiators, or radiators in which the cast iron projects from the surface into the air, show large results when estimated on the basis of projected or plain surface, but show very small results when estimated on the basis of measured surface. Sixth, thin radiators, or those with one row of tubes, always show higher efficiency than thick ones or those with numerous rows of tubes. Seventh, comparative tests of radiators should only be made between radiators of similar forms, or at least those which have about the same amount of surface.

54. Probable Efficiency of Indirect Radiators.—The velocity with which the air will move over radiators when heated a given amount can be readily computed as explained in Article 33. With a given velocity we can determine from the experiments cited the probable amount of heat that will be given off per degree difference of temperature per hour for natural and for forced circulation. The results deduced from experiments are given in the following tables:

TABLE FOR NATURAL CIRCULATION.

Height in Feet.	Temperature of Entering Air above Room.	Velocity in Feet per Second.	Units of Heat per Degree Difference of Temperature, Average per Square Foot per Hour.	Corresponding Story of Building.
5	50	2.97	1.72	1
10	50	4.17	2.02	1
17	47	5.3	2.3	2
20	45	5.6	2.36	2
25	45	6.3	2.52	2
30	42	6.6	2.58	3
35	42	7.2	2.68	3
40	40	7.5	2.72	4
50	40	8.4	2.81	5

TABLE SHOWING THE HEAT-UNITS PER DEGREE DIFFERENCE OF TEMPERATURE BETWEEN THE ENTERING AIR AND THAT OF THE HEATING SURFACE FOR DIFFERENT VELOCITIES OF AIR APPLICABLE IN FORCED CIRCULATION.

Velocity in Feet per Second.	Velocity in Feet per Minute.	Gauge-reading, Inches of Water-pressure.	Heat-units per Degree Difference of Temperature per Square Foot per Hour.
1	60	0.002	2
2.5	150	0.014	3.13
5	300	0.064	4.5
7.5	450	0.124	5.5
10	600	0.22	6.33
12.5	750	0.37	7.1
15	900	0.50	7.75
17.5	1050	0.65	8.35
20	1200	0.82	9
22.5	1350	1.08	9.5
25	1500	1.28	10

55. Temperature produced in a Room by a given Amount of Surface when Outside Temperature is High.—

Guarantees are often made respecting heating apparatus that it shall be sufficient to maintain a temperature of 70 degrees when the external air is at some fixed point, as zero, or 10 below. As under the exact conditions of the guarantee the trial can only be made when the external temperature corresponds with that specified, it becomes of some importance to establish an equivalent temperature which would indicate the efficiency of the heating apparatus for any specified condition. The following method applicable for such computations and is expressed in the shape of a formula:

Let T equal temperature of radiator, t' that of room, and t that of outside air for the conditions corresponding to the guarantee. Let B equal loss from room for 1 degree difference of temperature; let c equal the heat-units from 1 square foot of radiator per 1 degree difference of temperature for conditions corresponding to the guarantee; let c' denote the same values for other conditions; let x equal resulting temperature of room, t'' outside air for the actual conditions, R equal square feet of radiation.

For guaranteed conditions,

$$(t' - t)B = c(T - t')R. \quad \cdots \cdots \quad (1)$$

For actual conditions,

$$(x - t'')B = c'(T - x)R. \quad \cdots \cdots \quad (2)$$

Dividing (1) by (2),

$$\frac{t'-t}{x-t''} = \frac{c(T-t')}{c'(T-x)}. \quad \cdots \cdots \quad (3)$$

When $t' = 70$, $T = 220$, $t = 0$, and $c = 1.8$, we have

$$c'\left(\frac{T-x}{x-t''}\right) = 3.86.$$

The coefficient of heat transmission c' grows less as the temperature in the room becomes higher, as already shown in Art. 46; so the equations can only be solved in an approximate manner. The following table gives the temperatures in column 4, which a room would have for various temperatures outside, provided there was sufficient radiating surface to heat the room to 70 degrees in zero weather. The temperature of the radiator in all cases is assumed to be that due to 3 pounds pressure of steam by gauge, or 220 degrees.

TABLE.*

Temperature Outside Air.	Coefficient.† Heat per Square Foot per Hour per Degree	Total Heat per Square Foot per Hour.	Resulting Temperature of Room.	Difference Temperature Radiator and Room.
−10	1.85	288	64.7	155.3
0	1.8	270	70	150
10	1.75	253	75.1	144.9
20	1.7	236	81	139
30	1.65	218	86.5	133.5
40	1.6	203	93.1	126.9
50	1.55	188	98.7	129.3
60	1.5	172	104.7	115.3
70	1.45	158	110.5	109.5
80	1.4	142	117.1	102.9
90	1.35	130.5	123.5	96.5
100	1.3	117	130.3	89.7

Example showing Application of Table.—To determine by a test of the apparatus, when weather is 60°, whether a guarantee to heat to 70° in zero weather is maintained, operate the apparatus as though in regular use and note the average temperature of the room. If the room has a temperature equal to or in excess of 104.7° F., it would have a temperature of 70° in zero weather, all other conditions, such as wind, position of windows, etc., being the same as on the day of the test.

* This table, although calculated for steam with radiator at temperature of 220° F., is practically correct for hot-water radiation or for steam at any pressure and temperature.

† Value of c' in formulæ.

‡ Vol. I, Transactions American Society Heating and Ventilating Engineers.

CHAPTER V.

PIPE AND FITTINGS USED IN STEAM AND HOT-WATER HEATING.

56. General Remarks.—In this chapter will be found a concise description of pipe and fittings to be had regularly of most dealers. Such a description is entirely unnecessary to those familiar with current practice in the industry of steam and hot-water heating; but as the writer has found by experience detailed knowledge on this subject is often required, the following descriptions are deemed necessary.

It may be remarked in a general way, that for conveying heated air, galvanized or tin pipe or brick flues are usually provided, but for the purposes of conveying steam or hot water wrought-iron pipe is used almost exclusively.

57. Cast-iron Pipes and Fittings.—Cast-iron pipe was used very largely at one time for both supply-pipe and radiating surface in hot-water heating, but at present it is used only to a limited extent in greenhouse heating. For this purpose one size of pipe only is used, and this is $4\frac{1}{2}''$ outside diameter. The pipe weighs about 12 lbs. to the foot, and has a capacity of $\frac{1}{2}$ gal. per foot. The pipes are usually joined by socket-joints, for which purpose a socket is cast on one end of each pipe. The joints are formed by inserting one end of one pipe into the

FIG. 32.—CAST-IRON PIPE WITH SOCKET.

socket of another and filling the interspace either with melted lead, iron-filings and sal-ammoniac, sulphur, or cement, and calking thoroughly. The lead joint, which is ordinarily used, is formed by making a mould, by wrapping a hemp rope covered with clay around the joint, with a pouring-place on top, into which the melted lead is run. After the joint cools the lead is driven into place with a calking-iron. The rust-joint is a very excellent joint, and often used. It is made with a cement formed by saturating for ten or twelve hours iron turnings or filings with sal-ammoniac. This cement is pressed into the socket, and then pounded tightly into place with a calking-iron. Joints made with Portland cement are sometimes used, but they are likely to crack from the heat, and cannot be recommended.

The regular form of pipe and some of the principal fittings are shown in Figs. 32 to 36.

FIG. 33.—ELBOW FOR CAST-IRON PIPE.

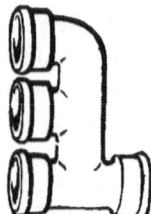

FIG. 34.—ROUND TEE FOR CAST-IRON PIPES.

FIG. 35.—RADIATING SURFACE AND PAN FOR HOLDING WATER TO MOISTEN AIR

Two or more lengths of pipe, supported on special brackets are usually run in parallel lines with a slight descent in the direction of the flow, and thus serve both for radiating surface and circulating pipes. For greenhouse heating, where the air is to be kept moist, a special pan to be filled with water, as shown in Fig. 35, supported by the pipes, is used at intervals.

FIG. 36.—VALVE OR STOP FOR CAST-IRON PIPE.

For the purpose of checking or stopping the flow a stop consisting of a flat plate, which can be set at any angle with the pipe, and of a form as in Fig. 36, is used. Each length of cast-iron pipe is sometimes provided with flanges, and joints are made by bolting the pipes together, packing being inserted to prevent leaks. These are inferior to the calked joints.

58. Wrought-iron and Steel Pipe.—Pipe made of wrought iron is now almost exclusively used for the purposes of conveying steam or hot water in heating systems. This pipe is made in a number of factories and of standard sizes, so that the pipe obtained from one is reasonably certain to fit that from another. Wrought-iron pipe is manufactured from iron of the proper thickness, which is rolled into pipe shape, and raised to a welding heat, after which the edges are welded by drawing through a die. The smaller sizes, $1\frac{1}{4}$ inch and under, are butt-welded; the larger sizes are in all cases lap-welded.*

This pipe is put on the market in three different grades of thickness: first the *standard grade*, which is used principally for heating purposes; this is tested to a pressure of 250 lbs. per sq. in. and has the dimensions given in Table XV; it is manufactured in sizes from $\frac{1}{8}$ in. to 15 in. in diameter. Thicker pipe, called *extra strong*, and still heavier pipe called *double-extra strong*, is manufactured, and can be obtained if required. The thick piping has the same distinguishing name as pipe of standard weight, having the same external diameter, which is in all cases that of the internal diameter of the standard pipe. The extra-strong and double-extra strong have smaller diameters than would be implied by the name; thus, for instance, inch pipe, standard size, has an inside diameter of

---

* The process of lap-welding is as follows: The sheet of iron is rolled to the desired thickness, width, and length. The edges are then scarfed. It is then drawn while red-hot by means of an endless chain through a bell-shaped die, which rounds it up and laps one edge over the other. The whole length is put into the furnace and heated to a welding heat, and afterward pushed out of the furnace at the opposite end into grooved rolls of a size corresponding to the size of the pipe. The inside lap is supported by a ball attached to a large bar of iron. The ball, the iron, and the groove in the roll all correspond so that the roll shall produce a sufficient pressure upon the iron and the ball to force the laps of the iron firmly together, thus producing the weld.—From paper by R. T. Crane, *Early History Wrought-iron Pipe*, Fifth Annual Convention Master Steam-Fitters' Association.

about one inch, an outside diameter of 1.315 inches, while the extra-strong pipe of the same nominal size has the same outside diameter and an inside diameter approximately 0.951 inch, while the double-extra strong has the same outside diameter and an inside diameter of 0.587 inch.

FIG. 37.—SECTION OF STANDARD PIPES ⅛ TO 3 INCHES INTERNAL DIAMETER.

The following table gives the diameters, external and internal, and weights per foot, of the various kinds of pipe. In the table * the normal inside diameter is the actual diameter, or nearly so, for the standard pipe; sizes to 1¼ inch are butt-welded, larger sizes lap-welded:

Nominal Diameter (Name), Inches. †	Actual Outside Diameter, Inches.	Actual Inside Diam., In.†		Thickness of Iron, Inches.†			Weight per Foot, Pounds.†			Threads per Inch.
		Extra Strong.	Double Extra Strong.	Standard.	Extra Strong.	Double Extra Strong.	Standard.	Extra Strong.	Double Extra Strong	
⅛	0.405	0.205	....	0.068	0.100	....	0.24	0.29	... ...	27
⅜	0.54	0.294	....	0.088	0.123	....	0.42	0.54	....	18
¼	0.675	0.421	....	0.091	0.127	....	0.56	0.74	... ...	18
½	0.84	0.542	0.244	0.109	0.149	0.298	0.84	1.09	1.70	14
¾	0.105	0.736	0.422	0.113	0.157	0.314	1.12	1.39	2.44	14
1	1.315	0.951	0.587	0.134	0.182	0.364	1.67	2.17	3.65	11½
1¼	1.66	1.272	0.885	0.140	0.194	0.388	2.24	3.00	5.20	11½
1½	1.9	1.494	1.088	0.145	0.203	0.406	2.68	3.63	6.40	11½
2	2.375	1.933	1.491	0.154	0.221	0.442	3.61	5.02	9.02	11½
2½	2.875	2.315	1.755	0.204	0.280	0.560	5.74	7.67	13.68	8
3	3.5	2.892	2.284	0.217	0.304	0.608	7.54	10.25	18.56	8
3½	4	3.358	2.716	0.226	0.321	0.642	9.00	12.47	22.75	8
4	4.5	3.818	3.136	0.237	0.341	0.682	10.66	14.97	27.48	8
4½	5	....	....	0.246	....	...	12.49	....	... ..	8
5	5.563	4.813	4.063	0.259	0.375	0.75	14.50	20.54	38.12	8
6	6.625	5.75	4.875	0.280	0.437	0.875	18.76	28.58	53.11	8

Steel Pipe.—For nearly every purpose of manufacture, soft steel has replaced wrought iron, and this will doubtless be the case some time so far as piping is concerned. Up to the present time, however, the pipe made of steel has not been as soft as that of wrought iron, and is more likely to dull and injure the dies and cutters used by workmen. It is often not so well welded, and is more likely to split.‡ Solid-drawn pipe has been made to a limited extent, and is very likely at no distant date to supersede welded pipe of all descriptions.

Each length of pipe as sold is provided with a collar or coupling screwed on to one end and has a thread cut on the other end. Connections are made by screwing the threaded end of one pipe into the coupling on the other. There is no standard length of pipes, the range usually being from 16 to 24 feet, with occasional short pieces. It can be ordered in lengths, cut as desired for slightly extra prices; but it can be readily cut any length, and right- or left-handed threads may be cut as desired. It is quite malleable, and when heated may

* See more extended table in Appendix.

† Approximate, outside diameter only is exact.

‡ 1898. Steel pipe can be purchased equal in every respect to wrought iron.

be bent into almost any shape by a skilful v
materially changing the form of its cross-secti

59. Pipe Fittings.—Fittings for connecti
giving them any required direction with resp(
are regularly on the market. These fittings
of cast and malleable iron, the prominent
straight couplings with right-handed thread
which are usually of wrought iron.

Cast-iron fittings are generally preferred t
able iron in any system of piping for heatin
that, being harder than the pipe and less ela:
likely to stretch and yield sufficiently to perr
the pipes are connected; if broken, a fractui
detected and a new fitting supplied. Malle
frequently stretch if pipes are screwed somev
that future expansion and contraction is quite
a leak. If it is necessary to take down a l
in which no removable joints occur, a cast-ir
easily broken, thus often saving more time tha
fitting, while the malleable fitting cannot be
It is quite true that malleable fittings are st
iron when of equal weight, but those on the ɪ
lighter than the cast-iron ones; and, moreov
fittings are abundantly strong for any press
sustained in ordinary systems of heating.

The standard pipes are considerably st
standard fittings, and if extra heavy pressures
100 to 150 pounds per square inch, it is advisal
fittings, which differ from the ordinary on(
weight.

The fittings which are on the market can
various classes, depending upon their use.

Pipe Connections.—For joining pipes in th(
is provided, first, the *wrought-iron coupling* sh
to 40.

The coupling, usually with plain exterioɪ
threads cut in both ends, and is used principɛ
pipe line where the construction is continuous
the other. A reducing coupling, Fig. 40, is

for uniting pipes of different sizes. In cases where it is necessary to "make up" or unite lines of piping which come together from different directions, a left-hand thread can be cut on the end of one of the pipes and the junction formed by

Fig. 38.
Coupling.

Fig. 39.
Right-and-left Coupling.

Fig. 40.
Reducing Coupling.

using a coupling similar to the above, but with a right-hand thread cut in one end and a left-hand thread cut in the other, such a coupling being known as a *right-and-left coupling.* To use this coupling room is required for end motion of one of the pipes sufficient to insert it.

In making up right-and-left couplings care must be taken that both threads on the pipe engage with those in the coupling at about the same instant. This can be done by screwing the coupling by hand on the end of each pipe, and counting the number of turns that can be made, noting the number of threads in sight after the joint is made up. This coupling, while sometimes difficult to use, forms the most certain method of uniting two pipe lines so that they will not leak. For joining pipes a coupling which separates into three pieces, termed a *union,* is often employed. The parts of the union are

Fig. 41.—The Union.

Fig. 42.—Section of Union.

screwed onto the ends of the pipe, and are drawn together by a revolving collar which engages with the thread on one of the pieces. The joint is formed either by drawing flat faces in the union against some elastic and soft material, as packing, or else by producing contact of ground and fitted metallic surfaces. Pipes are also held together by screwing flanges to the pipes, and drawing these flanges either in contact or against a ring of packing by bolts (Fig. 43). Such a joint is called a *flange union*.

FIG. 43.
FLANGE UNION.

FIG. 44.
LONG-THREADED NIPPLE AND LOCK-NUT.

Lengths of pipe are frequently made up by a short piece of pipe with a long screw-thread cut on one end, onto which is screwed a very short collar or lock-nut, Fig. 44. The junction is made between two ordinary pipe couplings by first screwing the long thread into one pipe coupling until the piece is short enough to be slipped into position, then it is screwed into the other coupling by unscrewing from the first. When screwed home, the collar or lock-nut is turned tightly against the first coupling, forming a steam-tight joint either by metallic contact or by use of packing.

Pipe Bends and Elbows.—For changing the direction of pipe lines there can be purchased elbows with bends of 45 or 90 degrees, also reducing elbows in which one opening is for smaller size of pipe than that of the other. The 90-degree elbow can be had either with right threads in both ends or with right and left threads, as required. The right-and-left threaded elbow can be used for making up two pipe lines in a manner similar to that described for a right-and-left coupling.

FIG. 45.
90° CAST-IRON ELBOW.

FIG. 46.
45° CAST-IRON ELBOW.

FIG. 47.
90° REDUCING ELBOW.

The internal diameter of elbows is somewhat in excess of that of the external diameter of the pipe, and the radius of the bend is, according to Briggs' table (Van Nostrand Science

Series, No. 68), equal in nearly every case to the diameter of the pipe plus a constant which varies from ⅜ inch for the smallest size of pipes to ½ inch for the largest size. For the sizes of pipes used in heating the radius of curvature is practically equal to that of the diameter of the pipe plus ½ inch.

Where the friction caused by a standard elbow is detrimental, special fittings (Figs. 48 and 49) can be obtained

 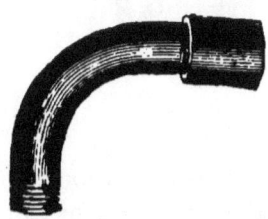

FIG. 48.—LONG-RADIUS ELBOW.   FIG. 49.—QUARTER BEND OF PIPE.

in which the radius of curvature is from two to three times that given. Such fittings are especially desirable in heating by hot-water circulation, and often permit the use of smaller pipes than would be possible with standard fittings.

Pipe Junctions, Tees, Y's, etc.—For the purpose of taking off one pipe line from another special fittings can be had,

FIG. 50.
PLAIN TEE—OPENINGS ALL SAME SIZE, THREADS RIGHT-HANDED.

FIG. 51.
REDUCING TEE—OPENINGS VARIOUS SIZES. (In describing state diameter of branch last.)

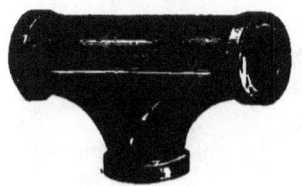 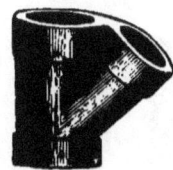 

FIG. 52.
LONG-RADIUS TEE.

FIG. 53.
Y FITTING.

FIG. 54.
LONG-RADIUS Y.

which are designated, according to their shape, as *tee, cross, side-outlet elbow*, and *Y-branch*, all of which can be bought with the openings for the same or different sized pipes in any combination required.

These various fittings are shown in the annexed engravings.

FIG. 55. SIDE-OPENING ELBOW.   FIG. 56. CROSS.   FIG. 57. REDUCING CROSS.

Miscellaneous Fittings.—For reducing the size of opening in a fitting, bushings of cast (Fig. 58) or malleable iron can be used; for closing up the end of fittings a screwed plug (Fig. 59) can be employed; and for closing the end of a pipe a screwed cap (Fig. 60) can be used. Where a coil of pipe is desirable, it can be formed by screwing pipes into U-shaped fittings, called return bends. These can be had with either right threads or right-and-left threads, and in close (Fig. 61) or open pattern (Fig. 62), and with the threads tapped so as to give nearly any pitch or rake of the pipe. For slightly changing the position of a pipe an offset (Fig. 63) can be used. To prevent leaking where a long-threaded nipple has been used, a lock-nut can be screwed on against a grummet, or ring of packing.

FIG. 58.—BUSHING.   FIG. 59.—PLUG.   FIG. 60.—CAP.

FIG. 61.   FIG. 62.   FIG. 63.   FIG. 64.
RETURN BENDS.   OFFSET.   LOCK-NUT.

Fittings can also be had for erecting parallel lines of pipe, as shown in Figs. 65 and 66; they are termed *branch tees*, and

FIG. 65.—BRANCH TEE, PLAIN.

FIG. 66.—BRANCH TEE, WITH BACK OUTLET.

can be had for almost any number of pipes, and for sizes varying from three-quarter to three inches. The distance between centres of branches is varied somewhat, but is usually 2 inches for three-quarter-inch pipe, $2\frac{1}{2}$ inches for one-inch pipe, 3 inches for one-and-a-quarter-inch pipe, and $3\frac{1}{2}$ inches for one-and-a-half-inch pipe. The branch tees are fitted with opening for supply-pipe and discharge-pipe either in end or side as specified. In those made for circulation the holes are tapped with right-hand threads; those made for box-coils are tapped for left-hand thread on branches.

Short pieces of pipe called *nipples* can be had of any length required, provided with right-hand threads cut on both ends, or with right thread on one end and left thread on the other. Short pieces of pipe called quarter or eighth-bends (Fig. 49) may be used in place of elbows when a long-radius turn is required.

FIG. 67. SHOULDER NIPPLE. 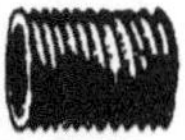   FIG. 68. CLOSE NIPPLE.

In addition to the fittings mentioned there can be had, for supporting the pipes to side walls, hooks and hook-plates with curved or straight arms, ringed plate, and coil-stand, as desired.

FIG. 69.—HOOK-PLATE.

There can also be had hangers of various patterns for supporting and holding pipes from ceilings. These are of great variety of pattern, and are made so that, if desired, they can be put on after the piping is in place.

The principal standard fittings as above described are also made of brass.

FIG. 70.—EXPANSION-PLATE.

FIG. 71.—RING-PLATE.

FIG. 72.—COIL-STANDS.

Ceiling and Floor Plates are collars used to hold the pipes in place, and to prevent overheating of woodwork by the steam or hot water. These are often made in halves, which may be slipped on over the pipes, and are fastened to the woodwork by screws, thus holding the pipe in position and keeping it from contact with wood.

60. Valves and Cocks.—The fittings used for the purpose of stopping the passages in pipes are operated by moving a disk across the pipe with or without rotation, or by simply turning through an angle. The first class have been generally called *valves*, the second *cocks*.

Valves are of two classes: the globe valve (Fig. 73), which closes an opening in a diaphragm parallel to the direction of flow, and the gate valve (Fig. 74), which closes an opening at right angles to the pipe.

The globe valve forms a serious obstruction, since any fluid in passing through it must make two turns, each nearly a right angle; while the gate valve when open presents little or no resistance.

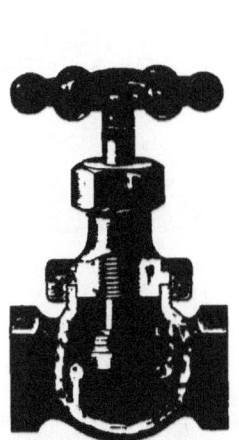 

Fig. 73.—Globe Valve.   Fig. 74.—Gate Valve.

The globe valve is much more simple in construction than the gate valve, is cheaper, and often will answer all requirements for steam-heating, but will seldom do for hot-water heating. It should be set so that the valve closes against the flow; when set in the opposite way accidents might happen—for instance, if the valve should be detached from the stem it could not be opened, although the stem would move apparently all right. It will be noted that the diaphragm of the globe valve forms an obstruction in the pipe, which extends to the centre, and if the stem of this valve be set vertical when used for a horizontal pipe it is likely to cause the pipe to stand half full of water. Whenever used in steam-heating, on a horizontal pipe, the stem should be placed in a horizontal position, so that it will not interfere with the drainage of water of condensation from the pipe.

The construction of the gate valve varies in detail as made by different manufacturers, but it in general consists of a gate which is moved across the opening in the pipe by turning the stem. When the gate reaches the bottom of the pipe it moves laterally sufficient to bring a strong pressure on the seat.

These valves are made with a stem which rises with the gate as shown in Fig. 74, or with one which remains in one position, the gate travelling up the stem. This latter form is objectionable, as one cannot tell by looking, whether the valve is open or closed.

Globe valves are made with a solid metallic seat, as in Fig. 73; or with a seat made of soft metal or packing, as in Fig. 75, of such a form that it can be replaced whenever the valve begins to leak.

FIG. 75.—GLOBE VALVE WITH DISK SEAT.   FIG. 76.—ANGLE VALVE.

Angle Valves (Fig. 76) are made in the same general way as globe valves, except that the openings are at right angles to each other. They cause a slightly greater resistance to motion than the ordinary elbow, but not sufficient to prevent their use for any system of heating. The seats are either metallic or of soft material, which can be removed.

Stuffing-boxes.—In all classes of valves a cavity is left around the stem, which must be filled with some packing material by turning back a cap-screw. Hemp, lamp-wicking, asbestos fibre, well oiled and, if possible, covered with plumbago, will make satisfactory packing for this purpose. Patent ring packing can be purchased, usually made of asbestos fibre soaked in oil, and serves an excellent purpose.

Radiator Valves.—These are forms of angle valves with fittings making them especially convenient for radiator connections, being plain as shown in Fig. 77 or with an attached union as in Fig. 78. These are often nickel-plated.

Radiator valves can be had with pedal attachment, so that they can be opened or closed with the foot.

The various kinds of valves which have been described are made with sockets for screwed connections to the pipes, or with flanges which are to be bolted to similar flanges screwed on the pipes as desired. They can also be had, especially for the larger sizes, with either brass or iron bodies.

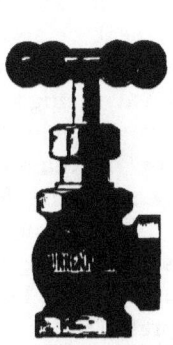

FIG. 77.—RADIATOR VALVE.   FIG. 78.—HOT-WATER VALVE.

Cross Valves.—A form of angle valve with one supply and and two opposite discharge openings is sometimes convenient, and is termed a *cross* valve. (See Fig. 83.)

Corner Valves, in which the openings are at the same level but at right angles, can be purchased if desired.

Cocks.—A plug, slightly conical, provided with one or more ports or holes through it, and arranged so that it can be turned in any direction, is termed a *cock*. When there is but a single hole it is called a plain cock. When two or more holes at angles to each other, it is called a two-way or three-way cock, since water can be directed in two or more directions by varying the angle through which the plug is turned. Cocks are very little used in steam-heating; as ordinarily made they are apt to leak, and, besides, do not provide a full opening for the fluid.

Improved cocks with larger openings and with packed ends are now much used on the blow-off pipes from boilers, and are for this purpose superior to valves.

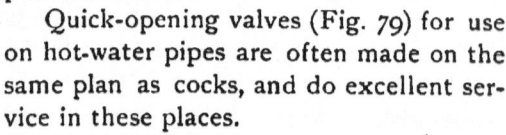

Quick-opening valves (Fig. 79) for use on hot-water pipes are often made on the same plan as cocks, and do excellent service in these places.

Check Valves.—Where it is necessary that the flow should always take place in the same direction and there is danger of a reverse flow, check valves are employed. These are usually of a similar pattern to the globe valve, the seat being at right angles to the direction of flow, with either a flat or ball valve (Figs. 80, 81).

FIG. 79.—QUICK-OPENING RADIATOR VALVE FOR HOT-WATER.

In this class the valve is held in place by its own weight or by the weight of the fluid in case of reverse flow. They are made for horizontal pipes, vertical pipes, or angles. One known as the swinging-check valve, in which the seat is at an angle of about 45 degrees to the direction of flow (Fig. 82), offers less resistance to the fluid, and is generally to be preferred.

61. Air-valves.—It is necessary to provide means for allowing the air to escape in systems of steam and hot-water heating. Air is heavier than steam, and although it will mix with it to a great extent, it will finally settle at or near the bottom of a radiator or pipe filled with steam. Air is, however, much lighter than water, and it will gather in any bends that are convex upward and in the upper part of radiators filled with water, and unless removed it will prevent the circulation.

For removal of the air several forms of valves and cocks have been especially manufactured. These are usually made of $\frac{1}{4}$- or $\frac{1}{8}$-inch pipe size, and vary in quality and design from the simplest valve to be opened by hand to a complicated automatic pattern, which permits the escape of air, but not of water or steam.

One of the simplest patterns of air-valves is shown in Fig.

FIG. 80.—HORIZONTAL CHECK WITH BALL CLACK.
FIG. 81.—HORIZONTAL CHECK VALVE.
FIG. 82.—SWINGING CHECK.

Globe Valve.
Angle Valve.
Cross Valve.

Horizontal Check Valve.
Angle Check Valve.
Vertical Check.
Steam Cock, Flanged Ends.

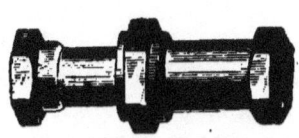

Expansion or Slip Joint.
Steam Cock, Screwed Ends.

FIG. 83.—PRINCIPAL VALVES AND STOPS USED IN HEATING.

84. This can be had with a bibb if desired, also with various forms of handles or keys, and with nickel or brass finish.

Automatic air-valves are made of a great variety of patterns. Those for steam-radiators are all closed by the expansion of some material. Fig. 85 shows an expansion air-valve, in which the valve is closed by the expansion of a curved metallic strip. The valve will remain open until this curved

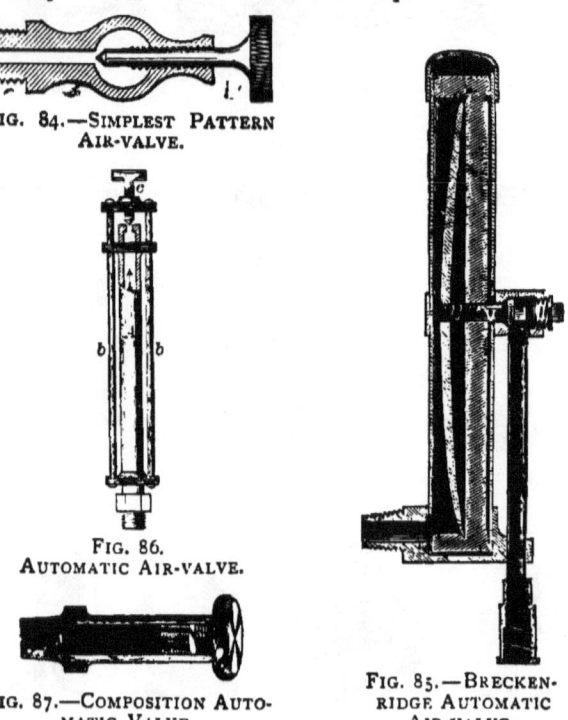

FIG. 84.—SIMPLEST PATTERN AIR-VALVE.

FIG. 86. AUTOMATIC AIR-VALVE.

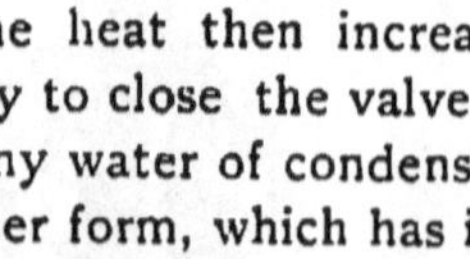

FIG. 87.—COMPOSITION AUTOMATIC VALVE.

FIG. 85.—BRECKENRIDGE AUTOMATIC AIR-VALVE.

strip becomes nearly equal in its temperature to that of the steam; the heat then increases its length and it bends out sufficiently to close the valve. A drip-pipe is provided for removing any water of condensation escaping from the air-valve.

Another form, which has in the past been extensively used, is shown in Fig. 86. In this case the interior tube A is heated more than the frame bb; this serves to press the valve c against the end of the tube when it is heated, thus closing the orifice. This is best adapted for use in a vertical position.

A form of air-valve now in extensive use is shown in Fig. 87. In this a composite material which expands rapidly when heated is used instead of metal. It is claimed for some of these valves that with suitable adjustment of the top screw the temperature of the radiator will be automatically maintained at any desired point—a mixture in any required proportion of air and steam being maintained in the radiator by this action.

To prevent escape of water and injury to furniture a radiator-valve with a float attachment is often used, as shown in Fig. 88. The valve is closed when heated, as in Fig. 87, by the expansion of a composite substance; it is connected to a float, so that if water passes into the air-valve the float will rise and close the orifice regardless of the temperature.

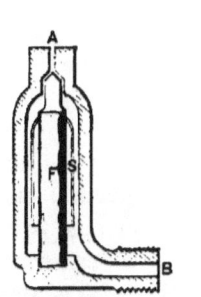
Fig. 88.—Radiator Air-valve with Float.

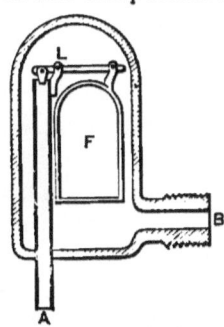
Fig. 89.—Hot-water Air-valve.

An automatic air-valve for hot-water radiators is shown in the sketch, Fig. 89. The air escapes at A, the orifice being closed by the float F acting on the lever L. So long as only air surrounds the float it sinks and keeps the orifice open, but

Fig. 90.—Flanged Expansion-joint.

as soon as water surrounds it it rises and closes the orifice.

62. Expansion-joints.—In the erection of any system of piping means must be provided so that the elongation of the pipe due to expansion will not cause a leak.* For all ordinary purposes of heating the expansion can be provided for by the use of elbows and right-angled offsets, of such length that the expansion will simply cause one pipe to slightly unscrew in one or more joints. This requires the use of two or three elbows, and so causes a slight increase of resistance to flow due to friction; but it is a very satisfactory arrangement, and will stand for years without developing leaks, even with high-pressure steam, if properly erected.

It is sometimes necessary to provide for expansion in a long line of straight pipe, in which case expansion-joints of some kind must be used. The ordinary expansion-joint, Fig. 90, consists of a sleeve sliding into an exterior pipe, provided with a stuffing-box. This joint, when heavy and provided with a catch to prevent it pulling apart, is a very durable and satisfactory construction. The packing will have to be renewed occasionally, and one part needs to be solidly anchored to prevent motion.

Expansion-joints are often used constructed of copper pipe in form of a U-shaped bend; also of one or more diaphragms connected to each other at the edges and to the pipes near the centre (Fig. 91). The copper bend is always satisfactory. The last-named device works very well if means can be adopted to thoroughly drain off any water lodging against the diaphragm. If used in a horizontal position, and on large pipes it is likely to gather sufficient moisture to form a water-hammer that may produce rupture when steam is turned on.

FIG. 91.—BUNDY ELASTIC COUPLING.

---

* The expansion of iron is one part in 148,000 of length per degree. This is equivalent to about 1.45 inches per 100 feet in changing from temperature of freezing to boiling.

CHAPTER VI.

RADIATORS AND HEATING SURFACES.

63. Introduction.—The amount of heat which will pass through various kinds of radiating surface is determined largely by experiment, and has been fully discussed in Chapter IV. In this chapter we will consider briefly the methods of construction.

When steam and hot water-heating were first employed the radiating surface consisted almost entirely of cast-iron pipe arranged in horizontal lines, as shown in Fig. 35, page 88. With the invention and use of wrought-iron pipe, cast-iron pipe was superseded by coils of this pipe, and at a somewhat later day largely by the radiator with vertical surfaces made either of cast or wrought iron. The change from pipe surfaces to radiators was, no doubt, largely due to the attempt to economize space in the room, as well as to improve the appearance.

64. Radiating Surface of Pipe.—Very efficient radiating surfaces can be made of coils of piping arranged as shown in Figs. 92 and 93. The return-bend coil shown in Fig. 92 is made by connecting return-bends, Fig. 61, page 96, with lines of straight pipe. The pipe mostly used is one inch in diameter, although, when the bends are numerous, $1\frac{1}{2}$ or 2-inch pipe should be used to reduce the friction. In use the flow is continuous, the fluid entering at the top and thence with a gradual descent flowing to the right and left alternately, finally discharging at the bottom. There is a great deal of friction in coils of this class, and air is likely to gather in the bends and stop circulation. The writer would, therefore, recommend that they be employed only when other forms will not answer.

The branch-tee or manifold coil is constructed by connecting branch-tees with parallel lines of pipe. In each pipe-line one or more elbows must be placed to counteract the effect of unequal expansion.

The coil may be arranged on a flat wall-surface so as to form a *mitre* branch-tee coil as in Fig. 93, lower part, or with both branch-tees at one end and elbows and nipples at the opposite end; the fittings at ends being connected by pipes having the proper pitch. Such a construction is called a *return* branch-tee coil, see upper part Fig. 93. The coil may be arranged on two sides of a room with the elbows placed in the intervening corner, in which case it is called a *corner coil*.

The various types of branch-tee or manifold coils as described present small frictional resistance to the flow of steam or water and give satisfactory service for either steam or hot-water heating.

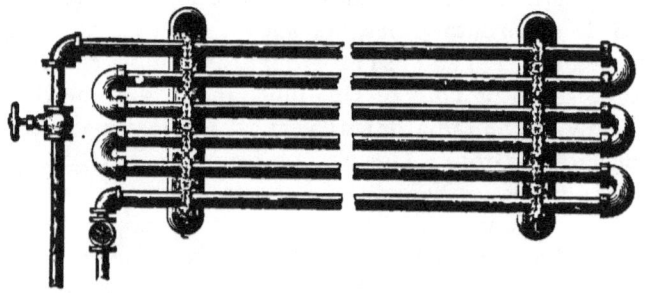

FIG. 92.—RETURN-BEND COIL.

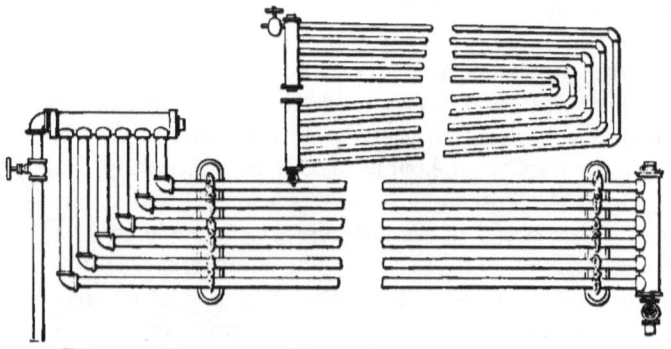

FIG. 93.—BRANCH-TEE MITRE COIL AND RETURN-COIL.

If two connections are used the steam should be supplied at the highest point of the coil, and the return taken off at the lowest; if one connection, steam is to be supplied at the lowest point. The horizontal portion should be given a pitch of one inch in ten or twelve feet, and an air valve or cock should be connected to each coil. When several return-bend coils are grouped together, as in Fig. 94, the construction is termed a *box coil*. This has all the faults in an aggravated manner that were ascribed to the return-bend coil, and in addition causes a loss of efficiency due to close grouping of surface.

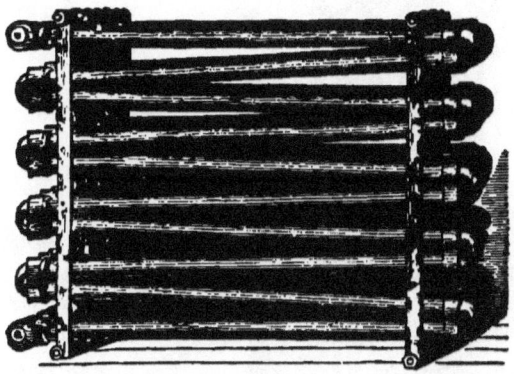

Fig. 94.—The Box Coil.

The pipe coils, Figs. 92 to 94, will do equally well for steam or hot-water circulation.

65. Vertical Pipe Steam-radiators.—These were at one time used extensively, and were made by screwing short pieces of vertical pipe into a cast-iron base and connecting the pipes in pairs at the top with return-bends, which were usually screwed but sometimes pressed on. One form still in extensive use was made by screwing pipes, having the upper end closed and provided with an internal diaphragm, into a cast-iron base. The diaphragm being so placed as to produce the same circulation in one pipe that was obtained in two pipes with the other form.

The pipes are arranged in two or more rows as necessary to secure the desired radiating surface. In early radiators of this class the base was provided with a diaphragm, and each return-pipe was trapped by a cavity filled with water so as to insure a continuous circulation of the steam through each pipe. In some of the recent radiators the return-pipes are trapped as explained above; but in nearly every case the base is entirely open and arranged so that it will drain freely, no attempt being made to force circulation in any direction. In some of the recent radiators of this type the base, instead of being in one piece, is made up of sections connected by nipples, so that it can be lengthened or shortened at pleasure. An air-valve must always be provided with these radiators; the best location for which is at about one third the height of the radiator, and on the end opposite the admission.

FIG. 95.—PIPE RADIATOR.

The wrought-iron radiator is constructed in nearly every case of one-inch pipe, taken of such length that there is one square foot of exposed radiating surface for each pipe in the radiator. The form being quite regular its surface can be accurately measured.

66. Cast-iron Steam-radiators.—Cast-iron radiators are now mostly used in direct heating.

Those principally used have vertical radiating surfaces, and are made either by screwing loops or sections into a hollow base provided with the requisite openings, or by connecting at the bottom a series of parallel vertical sections by nipples screwed from the outside or inside of the base. The first form of radiators, having a base of fixed dimensions, is often called the *standard* form; the latter, which can be increased or diminished in length by adding or taking off sections, is called a *sectional radiator*.

The radiator is in some instances provided with a flat top which is held in place by screws, but the greater portion of those of recent design have a highly ornamented surface and are used without top or screen of any description. The illustrations, Figs. 93 to 106, give a very fair idea of the appearance of those in use. They are painted in various colors, enamelled or bronzed, as may be required by the house owners or architects.

The efficiency of direct radiation is somewhat increased by painting or bronzing, but is lessened by varnishing or enamelling: but that of indirect is not so affected.

These radiators are made in great variety of forms, and can be had of such shape as to surround columns, or fit in corners; and of almost any height desired. Some of the radiators are fitted with warming closets. (See Fig. 98,* frontispiece, for illustration of styles in use.)

FIG. 96.
STANDARD CORNER RADIATOR.

FIG. 97.—SECTIONAL RADIATOR.

FIG. 99.—CRESCENT FLUE RADIATOR.

* With permission from *Heating and Ventilation*.

The sectional radiators are in many cases built in such a manner as to form flues for the passage of air from the bottom to the top of the radiator for the purpose of increasing the air-heating capacity. Such radiators are termed *flue radiators* (Fig. 99).

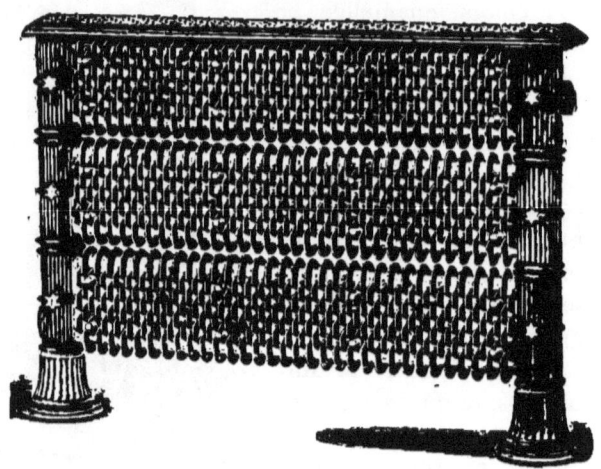

FIG. 100.—WHITTIER EXTENDED SURFACE RADIATOR.

Radiators are sometimes built with projecting fins or ornaments of cast iron for the purpose of greatly extending the surface in contact with the air. Such a radiator is termed an *extended surface radiator*, and is now little used for direct heating (Fig. 100).

The radiators in principal use are constructed as described, but radiators have been built by many other methods and in many other shapes. They have been constructed of one solid casting, and by uniting sections of various forms by bolts and packed joints.

67. Hot-water Radiators.—Hot-water radiators differ essentially from the steam-radiators in having a horizontal passage at the top as well as at the bottom. This construction is necessary in order to draw off the air which gathers at the top of each loop or section. Aside from this the construction may be the same in every particular as that for steam-radiators; in general the hot-water radiator will be found well adapted for

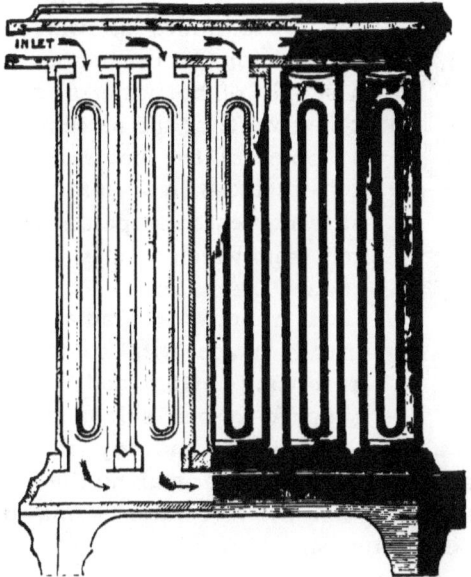

Fig. 101.—Section of Hot-water Radiator.

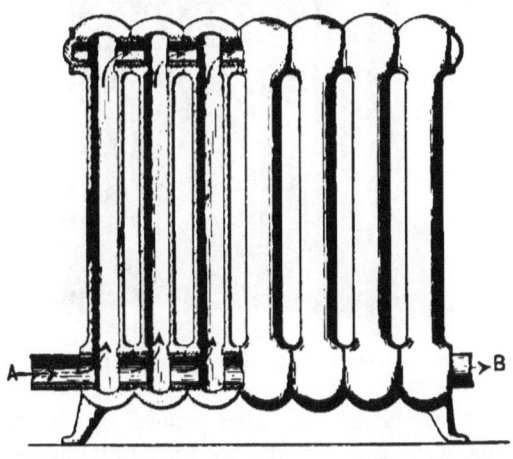

Fig. 102.—Sectional Hot-water Radiator.

steam circulation, being in some respects superior to the ordinary form.

Many of the hot-water radiators, as shown in Fig. 101, are made with an opening at the top for the entrance of water and at the bottom for its discharge, thus insuring a supply of hot water at the top and of colder water at the bottom.

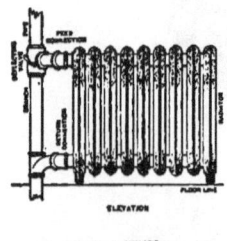

Fig. 103.*—Radiators with Top and Bottom Connections.

Some of the hot-water radiators are constructed with a cross-partition so that all water entering passes at once to the top, from which it may take any passage toward the outlet.

The hot-water radiator, is however, usually made with continuous passages at top and bottom, and the warm water is supplied at one side and drawn off on the other, as shown in Figs. 102 and 105 (right hand). The action of gravity is depended on for making the hot and lighter water pass to the top and the cold water to sink to the bottom and flow off in the return.

Fig. 104.—Sectional Hot-water Radiator.

Hot-water radiators are also made by joining vertical pipe sections with nipples at top and bottom, as shown in Fig. 106.

---

* Heating and Ventilating of Residences, by Willet.

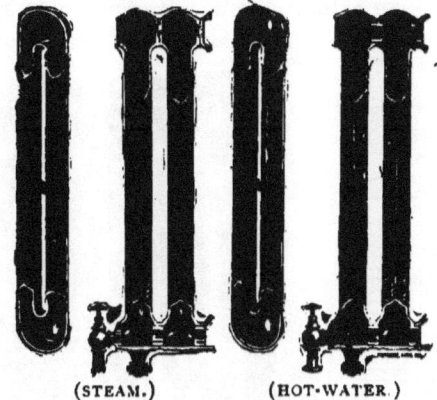

(STEAM.)   (HOT-WATER.)

Fig. 105.—Section of Cast-iron Radiator.

Fig. 106.—Sectional-pipe Hot-water Radiator.

68. Direct-indirect Radiators.—Radiators arranged with a damper under the base and located so that air from the out-

Fig. 107.—Direct-indirect Radiator in Position.

Fig. 108.—Direct-indirect Radiator.

side will pass over the heating surface before entering the room are often used to improve the ventilation. The surface of these radiators should be about 25 per cent greater than that of a direct radiator for heating the same space. The styles and kinds either for steam or hot water are the same as the direct.

69. Indirect Heaters.—Radiators which are employed to heat the air of a room in a passage or flue which supplies air are termed indirect. These heaters are made in various forms, either of pipe arranged in return bend or in manifold coils, as in Fig. 93, or of cast-iron sections of various forms united in different ways. When cast-iron surfaces are used, they are generally covered with projections like the extended surface radiator. The sections, or, as they are sometimes called, the *stacks* for indirect heating, are usually held together by bolts. The joints being formed by inserting packing between faced surfaces. The sections are sometimes united by nipples screwed into branch-tees above and below, as shown in Fig. 109, which is an excellent form for hot-water circulation.

Indirect radiators should be placed in a chamber or box as

FIG. 109.—INDIRECT HEATING SURFACE.

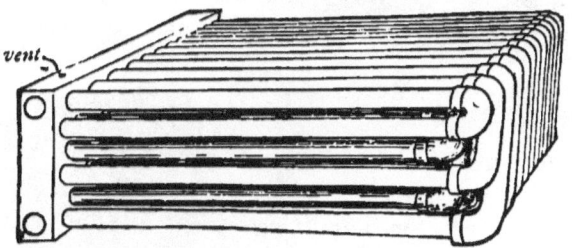

FIG. 110.—INDIRECT PIPE COIL.

nearly as possible at the foot of a vertical flue leading to the room to be heated.

Air is admitted through a passage from the outside provided with suitable dampers to a point beneath the indirect stacks. It is taken off generally on the opposite side, and directly into the flue leading into the room to be heated.

The chamber surrounding the indirect radiator is usually built of

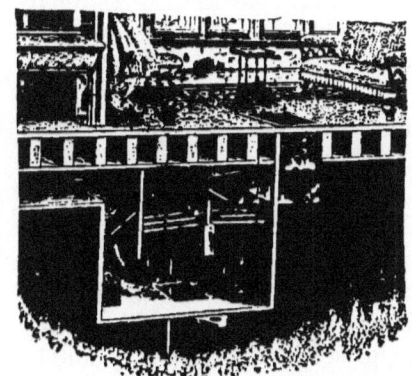

FIG. 111.—ARRANGEMENT OF INDIRECT HEATER.

a casing of matched wood, as in Fig. 111 and Fig. 112, suspended from the ceiling of the basement, and lined inside with bright tin; but a small chamber of masonry at the bottom of a flue is a better and more durable construction. The flue leading from the chamber is of masonry or galvanized iron; that supplying the cold air, of matched wood and sheet iron. There should be a door in the chamber so that the indirect heater can be examined and cleaned when required. It is often of advantage to have a passage and deflecting damper so arranged that air can be drawn into the

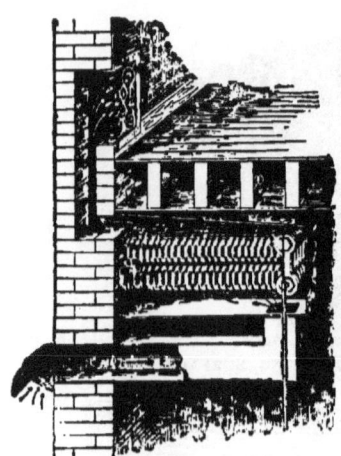

FIG. 112.—ARRANGEMENT OF INDIRECT HEATING SURFACE.

room for ventilation without passing over the heater.

The registers for admitting the heated air into the rooms can be located as desired, either in the walls or the floor; for ventilation purposes it is preferable to admit the air near the ceiling, and as shown in Fig. 113.

The size of registers and air-flue will be given in Chapter XIII.

Setting of Indirect Heaters.—The indirect heating-surface is supported usually by bars of iron or pieces of pipe held in place by hangers fastened at the ceiling (Fig. 111). This heater should be set so as to give room for the freest possible circulation of air, and so that all parts will be at least ten inches from top or bottom of casing, and arranged so

FIG. 113.—INDIRECT HEATER ARRANGED FOR VENTILATION.

that no air can pass into rooms without being warmed. An automatic air-valve should be used to remove the air from the sections of the heater.

If the sections are of proper form, one connection will be sufficient for steam; but in nearly every case two connections, one for the supply and one for the discharge, will be required for water circulation.

70. Proportions of Parts of Radiators.—There is great difference regarding the relative volume of radiators of different make as compared with the surface; but the practice is quite uniform as regards the sizes of supply-pipes for either steam or hot water. Because of the high efficiency of a radiating surface formed of one-inch horizontal pipe, it has been argued that this should form a standard for relation of contents to surface. It is seen, however, by consulting the tests given in Chapter IV, that inch-pipe vertical radiators are not more efficient than cast-iron radiators with larger volume; so that it is doubtful if the relative ratio of volume to surface is of importance.

It is of importance that the steam or water should circulate through the radiators with the least possible friction, and that in the case of steam-radiators the base should be of such a form as to perfectly drain; otherwise the water which remains in will be certain to cause the disagreeable noise and pounding known as water-hammer.

The following table gives the standards which are almost universally adopted by the different makers for the size of inlet and outlet to the direct radiators; those for indirects are to be taken one size larger:

Size of Radiator, Sq. Ft.	Diameter of Openings.	
	Two Openings.	One Opening.
0 to 50	1 inch.	1¼ inches.
50 to 125	1¼ inches.	1½ "
125 to 200	1½ "	2 "
200 to 300	2 "	2½ "

CHAPTER VII.

STEAM-HEATING BOILERS AND HOT-WATER HEATERS.

71. General Properties of Steam—Explanation of Steam-tables.—Steam has certain definite properties which always pertain to it and distinguish it from the vapor of other liquids than water.

Steam, at any given pressure above a vacuum, possesses a definite temperature. The atmospheric pressure is different at different localities and for different conditions of the weather, thus causing slight changes in temperature of the boiling-point. The pressure which is read by any steam-gauge is that in excess of the atmosphere; *the pressure which is given in the steam-tables is that which is reckoned from a perfect vacuum,* and is usually called *absolute;* hence, in order to use the steam-table which is given in the back of the book, the pressure as determined by a steam-gauge reading must be increased by the atmospheric pressure. The atmospheric pressure is given accurately by a barometer, but it will be sufficiently accurate, for most cases, to consider it as 14.7 pounds. To use the table add this quantity to the gauge-reading and the result will be the absolute pressure. For approximate purposes the atmospheric pressure may be considered as 15 pounds. The steam-tables referred to give, in the first column, the pressure above a vacuum; in the second column, the temperature Fahrenheit; in the third, the heat, expressed in heat-units, required to raise one pound of water from zero Fahrenheit to the required temperature. If the specific heat of water were unity at all temperatures, the heat contained in one pound of water would be numerically the same as the temperature. The difference is not great in any case.

The fourth column gives the value in heat-units of the latent heat of evaporation for each pound of steam. This quantity expresses the amount of heat which is stored, without change of temperature or pressure, during the physical change of condition from water to steam; and it has been termed *latent* because it cannot be measured by a thermometer (see Art. 13, page 15). It will be noted that this quantity is relatively large as compared with the sensible heat. It is of importance, since it expresses the amount of heat which is contained in one pound of steam in excess of that in one pound of water at the same temperature.

The fifth column gives the total heat contained in one pound of steam; this is the sum of the sensible and latent heat.

The sixth column gives the weight in pounds of one cubic foot of steam for various pressures. In many instances steam-tables are arranged so as to give the heat in one pound of steam above 32° Fahr., the freezing-point of water, instead of above zero.

It should be noted that the temperature of steam corresponding to different pressures, as given in column (2), is also the boiling-point of water corresponding to the same pressure.

As the temperature and absolute pressure of steam always bear definite relation to each other, it is quite evident that a steam-table could be arranged giving the properties of steam from measurements of temperature. This is generally not so convenient as the present arrangement. If temperatures are known, the corresponding pressure can be determined by inspection and interpolation in the present table.

72. General Requisites of Steam-boilers.—The steam-boiler is a closed vessel, which must possess sufficient strength to withstand the pressure to which it may be subjected in use; but it may have almost any form, and may be constructed of various materials.

It is used in connection with a furnace, from which the heat required for evaporation is obtained by combustion of fuel. The heat is received on the surface of the boiler, and passes by conduction through the metallic walls to the water or steam. The surface which receives this heat is called *heating surface*, and is partly situated so as to receive the direct or radiant heat, and partly located so as to receive the convected or indirect heat from the gases only. The heating surface in most modern boilers is made relatively great, as compared with the cubic contents, by the use of tubes containing water or heated gases, or by subdividing the boiler so as to make the surface large with respect to the cubic contents and weight. The steam generated rises in the shape of bubbles through the water in the lower part of the boiler, and is liberated from the surface of the water at the water-line.

The power of the boiler depends upon the amount and form of heating surface, upon its capacity for holding water and steam, and upon the extent of fire-grate surface. Its economy depends upon the relative proportions of these, and the character and amount of fuel burned. Its ability to produce dry steam depends upon the circulation of its liquid contents, and also upon the extent of surface at the water-line.

For safety, the boiler must be provided with safety-valve, pressure and water gauges. For convenience automatic damper-regulators, water-feeding apparatus, etc., are desirable.

73. Boiler Horse-power.—As a boiler performs no actual work, but simply provides steam for such purposes, a boiler horse-power is entirely an arbitrary quantity, and may be transformed into a lesser or greater amount of work, as the character of the engine which uses the steam varies.

The standard established by the Committee of Judges at the Centennial Exhibition in 1876 as a boiler horse-power has been universally adopted, and would, no doubt, in absence of other stipulations, constitute a legal standard of capacity. This committee defined a boiler horse-power as the evaporation of 30 pounds of water from feed-water at 100° Fahr. into steam at 70 pounds pressure; this is equivalent to the evaporation of 34.5 pounds of water from a temperature of 212° Fahr. into steam at atmospheric pressure.* Engines require from 12 to 40 pounds of steam per horse-power per hour, depending upon the grade or class to which they belong; hence the steam required to perform one horse-power of work in an engine bears no definite relation to a boiler horse-power.

---

* The condition of evaporating from water at 212° into steam at the same temperature will be referred to hereafter as *evaporation*, without other qualification.

Since the evaporation of one pound of water from and at 212° Fahr. requires 966 heat-units, one boiler horse-power is equivalent to 33,327 heat units.

For heating purposes a more convenient standard of power is the square foot of radiating surface. Each square foot of direct steam-radiating surface gives off 270 to 330 heat-units per hour when the difference of temperature is 150 degrees (see Art. 51), which is that usually existing in low-pressure steam-heating. About two thirds as much is given off by one square foot of hot-water radiating surface. As the evaporation of one pound of water requires 966 heat-units, there is needed about one third of a pound of steam for each square foot of steam-radiating surface per hour, hence one boiler horse-power will be sufficient to supply somewhat more than 100 square feet of direct radiating surface; that is, we can consider the boiler horse-power as equivalent to 100 square feet of direct steam radiation, with sufficient allowance to meet ordinary losses.

74. Relative Proportions of Heating to Grate Surface.—The relative amount of grate surface and heating surface required in a steam-boiler depends, to a large extent, upon the nature and amount of coal burned per unit of time. That part of the heating surface which is close to the fire and receives directly the radiant heat is much more effective than that which is heated by contact with hot gases only; but it will be found that considerable indirect heating surface will in every case be required, in order to prevent excessive waste of heat in the chimney. Power-boilers have been rated for a long time not on their actual capacity, but on the amount of heating surface; and this would seem to be a fair standard of rating for heating-boilers. It is the general practice to consider 11.5 square feet of heating surface in water-tube boilers or 15 square feet in plain tubular boilers as equivalent to one horse-power.

The actual power of the boiler depends more upon the method and management of the fires than upon the size; and either of the above classes of boilers can be made to develop under favorable circumstances from two to three times the capacity for which they are rated.

A rating of 15 sq. ft. of heating surface to one horse-power requires an evaporation of 2.3 lbs. of water per square foot of heating surface per hour, and a rating of 11.5 sq. ft. per horse-power requires an evaporation of 3 lbs. Experience for a number of years with power-boilers—20 horse-power and larger—indicates these proportions to be safe ones and to result in durable construction. With the small boilers often used in house-heating the waste due to loss of heat from the heating surfaces, imperfect combustion, and bad management generally are much greater, so that it is necessary to use boilers somewhat larger than would be required by the data given. Knowing the amount of coal per hour and the evaporation per pound of coal, we could readily calculate the steam produced in pounds. This result multiplied by three would give very closely the extent of direct radiating surface which could be supplied.

With perfect combustion and no waste, one pound of pure carbon would evaporate about 15 lbs. of water; all coal contains considerable ash and refuse, on account of which the best results are lower, so that one pound of best anthracite coal might evaporate 13 lbs. of water, and of bituminous from 10 to 14 lbs. Our average evaporation in power-boilers is probably about 9 lbs. when served by good firemen, and in heating-boilers it is usually much less, not because of faulty construction of the boiler, but for lack of proper and careful management. The amount of coal burned per square foot of grate per hour is rarely less than 15 lbs. with power-boilers, and in some cases is very much greater, but is usually less than 10 lbs. and is sometimes as small as 3 or 4 with heating-boilers.

For these reasons no hard and fast rule can be given for the proportions of different boilers and heaters, and a considerable variation may be expected in the relative proportions of heating, grate, and radiating surface existing in successful plants. By making allowance for the probable loss of efficiency in small heaters, we can, by starting with the proportions which have been found to be satisfactory in large plants where power-boilers are used, compute a table which will be based on the results of actual trial and experiment. This table will give dimensions which are well within the limits of those in actual use, but it should not be inferred that satisfactory plants cannot be constructed with proportions varying ten or twenty per cent from those given. The table is computed from the following data, which were assumed for reasons already stated: 1st, one pound of steam will supply 3 sq. ft. of direct steam-radiating surface; 2d, 15 sq. ft. of heating surface (one horse-power) in the boiler will supply 100 sq. ft. of steam or 150 sq. ft. of hot-water direct radiating surface, when the boilers contain 450 sq. ft. and above of heating surface; 3d, loss in efficiency assumed to be 10 per cent for reduction in capacity of 50 per cent; 4th, rate of evaporation for steam-boilers is taken so as to agree with the experience of the writer. Two cases are considered in the table: (A) when the rate of coal consumption is 10 lbs. and (B) when the rate of coal consumption is 8 lbs. per sq. ft. of grate per hour. The latter in every case gives a somewhat larger grate, and for hot-water heating is no doubt to be preferred.

PROPORTION OF PARTS OF STEAM-HEATING BOILERS.

Radiating Surface, Square Feet.	250	500	750	1000	1500	2000	3000	4000	5000	7500	10000
Nominal horse-power	2	5	7.5	10	15	20	30	40	50	75	100
Ratio radiating to heating surface	4.5	5.1	5.4	5.6	6	6.2	6.7	6.9	7 / 9*	7 / 9*	7 / 9*
Probable evaporation per lb. coal	5.5	5.7	6	6.5	*7	7.5	8	8.5	9	9.5	10
Pounds of steam per sq. ft. grate (A)	55	57	60	65	70	75	80	85	90	95	100
" " " " (B)	44	46	48	52	56	60	64	68	72	76	80
Ratio radiating to grate surface (A)	165	171	180	195	210	225	240	255	270	285	300
" " " " (B)	132	138	144	156	168	180	192	204	216	228	240
" heating to grate surface (A)	36.5	33.2	33.2	34.8	35	36.2	36.5	37	38.5	40.5 / 31.5*	42.5 / 33.3*
" " " " (B)	28.5	27	26.7	27.7	28	29	28.5	29.6	30.8	32.7 / 25.6*	34.5 / 26.5*
Heating surface, sq. ft.	55	98	138	178	250	322	447	580	710	1071 / 833*	1430 / 1111*
Grate surface, sq. ft. (A)	1.52	2.92	4.15	5.68	7.15	8.9	12.4	15.7	18.5	26.5	33.3
" " " (B)	1.88	3.88	5.4	6.37	8.92	11.2	15.5	19.5	23.2	32.5	41.5
Diameter safety-valve, inches†	1.5	2.25	2.50	2.75	3	3.25	3.5	4	4	2 of 3	2 of 4
" smoke-flue, inches	7	10	11	12	15	17	19	23	25	28	34

HOT-WATER HEATERS.

Ratio radiating to heating surface	6.8	7.6	8.1	8.4	9	9.3	10	10.4	10.5	10.5 / 13.5*	10.5 / 13.5*
" " " grate (A)	247	256	270	292	315	337	360	382	405	427	450
" " " (B)	193	207	216	232	252	270	288	306	324	342	360
" heating to grate (A)	36.5	33.2	33.2	34.8	35	36.2	36.5	37	38.5	40.5 / 31.5*	42.5 / 33.3*
" " " " (B)	28.5	27	26.7	27.7	28	29	28.5	29.6	30.8	32.7 / 25.6*	34.5 / 26.5*
Heating surface, sq. ft.	36.5	65	91.5	118	166	215	296	385	470	703	905
Grate surface, sq. ft. (A)	1	1.96	2.75	3.75	4.75	5.9	8.2	10.4	13.3	17.6	22.2
" " " (B)	1.25	2.58	3.6	4.25	5.9	7.1	10.3	13	15.3	21.5	27.5
Diameter smoke-flue, inches	7	10	11	12	15	17	19	23	25	28	34

* Water-tube boiler.

† Safety-valves by Board of Trade rule. Smaller boilers figured to blow at 5 and 10 lbs.

126   HEATING AND VENTILATING BUILDINGS.

A very interesting comparison of relative proportions of various boilers used for steam-heating was made by S. Q. Hayes from published statements of manufacturers in *Heating and Ventilation*, April 15, 1895, and from which the following table is abstracted. It will be seen that the proportions of radiating surface to grate surface agree well, when the fact is considered that many published statements are far from accurate, with the values recommended for a coal consumption of 8 lbs. of coal per hour, per square foot of grate.

TABLE SHOWING PROPORTIONS CLAIMED BY MAKERS FOR STEAM-HEATING BOILERS.

Steam-heating Boilers.	Ratio Radiating to Grate Surface.				Ratio Heating to Grate Surface.				Ratio Radiating to Heating Surface.			
Square Foot of Radiation.	250	500	750	1500	2000	250	500	750	1500	2000	250	500	750	1500	2000
KIND OF BOILER.															
Tubular, vertical, magazine....	134	167	167	180	228	23	30	23	24	30	5.8	5.6	7.3	7.4	7.7
" " surface......	180	190	200	190	220	30	32	33	34	36	6	6	6	6	6
" " steel	170	147	138	..	..	30	24	25	....	...	3.5	4.7	....	..
Vertical shell, drop- and fire-tubes,........................	140	170	170	180	....	26	32	30	30	....	5	5.3	5.5	5.7	.
Pipe boiler..	180	180	180	180	180	20	23	25	27	30	6.2	6.5	6	6	6
Pipe-coil boiler.	75	75	75	75	75	25	25	25	....	...	3	3	3	....	..
Drop-tube, wrought iron......	185	193	205	....	....	42	36	33	32.7	...	4.4	5.4	6	....	6.5
" " cast-iron magazine.	198	172	204	....	....	36	28.7	35.7	..	...	5.5	6	5.7	....	..
" " surface..	240	186	224	....	....	43	32	40	....	...	5.4	5.8	5.7	....	..
" " magazine., ...	...	180	170	170	180	...	36	27	23	22	..	5	5	6	7
" " wrought iron..	165	165	....	...	...	35	37	...	...	...	4.5	4.5	...	...	..
Coil and drop-tube............	150	150	..	....	....	22	25	....	...	...	6.3	6	...	...	..
Horizontal sectional...........	...	240	200	210	230	...	44	40	35	27	...	5	6.2	7.3	8.3
" "	160	200	216	216	...	32	35	36	36	...	4	5.7	6	6	..
" "	...	180	200	300	...	...	36	26	36	...	4	5	6	6.5	..
" "	170	147	138	...	...	30	24	25	....	...	5	4.7	...	...	..
Vertical sectional......	130	112	130	130	....	16	24	24	24	...	5.7	5.7	4.7	6	..
" "	...	135	150	155	155	28	30	25	25	24	...	5.3	5.7	6	..
" sectional, tubular.....	130	145	138	163	167	...	23	20	25	30	6	6.5	7	5.7	6.3
" " "	...	192	200	230	250	...	32	33	37	40	..	6	6	6	6

75. Water Surface—Steam and Water Space.—The surface on the water-line from which ebullition takes place should be so large that the velocity of steam will not be great enough to project particles of water into the main steam-pipes. Practice is variable in this respect; in successful plants it will be found that from one third to one square foot of surface is provided per horse-power or per 100 square feet of radiating surface. The greater this surface the less water will be carried out of the boiler with the steam, other things being equal.

There is much variation in the amount of water and steam space provided in various kinds of boilers: in the fire-tube and shell boilers there is much more space than in water-tube and sectional boilers. A large amount of water and steam absorb the heat slowly, but on the other hand they require less frequent attention and are more regular in operation. The following rules have been given:

Tredgold * states that the volume of steam space should be sufficient to prevent variations in pressure exceeding 1 in 30, by irregular use.

The Artisan Club allowed 5 cubic feet of water space and 3.2 cubic feet of steam space per horse-power for Cornish boilers.

In the ordinary tubular boilers to-day there will be found about 2.0 cubic feet of water and 1.0 cubic foot of steam per horse-power, and about one third the above amounts for the water-tube boilers.

76. Requisites of a Perfect Steam-boiler.—The late Mr. George H. Babcock of Plainfield, N. J., gives as the results of his experience the following requisites for a perfect steam-boiler for power purposes:

1st. The best materials sanctioned by use, simple in construction, perfect in workmanship, durable in use, and not liable to require early repairs.

2d. A mud-drum to receive all impurities deposited from the water in a place removed from the action of the fire.

3d. A steam and water capacity sufficient to prevent any fluctuation in pressure or water-level.

4th. A large water surface for the disengagement of the steam from the water in order to prevent foaming.

5th. A constant and thorough circulation of water throughout the boiler, so as to maintain all parts at one temperature.

6th. The water space divided into sections, so arranged that should any section give out, no general explosion can occur, and the destructive effects will be confined to the simple escape of the contents; with large and free passages between the different sections to equalize the water line and pressure in all.

7th. A great excess of strength over any legitimate strain; so constructed as not to be liable to be strained by unequal

---

* Thurston's Steam-boilers.

expansion, and, if possible, no joints exposed to the direct action of the fire.

8th. A combustion-chamber, so arranged that the combustion of gases commenced in the furnace may be completed before they escape to the chimney.

9th. The heating surface as nearly as possible at right angles to the currents of heated gases, and so as to break up the currents and extract the entire available heat therefrom.

10th. All parts readily accessible for cleaning and repairs. This is a point of the greatest importance as regards safety and economy.

11th. Proportioned for the work to be done, and capable of working to its full rated capacity with the highest economy.

12th. The very best gauges, safety-valves, and other fixtures.

The same requirements apply equally well to a boiler for heating, but the relative importance of the various requirements might be different, and some might be omitted as unimportant; thus, for instance, the mud-drum, which is of importance in a boiler for power, because it is receiving constant accessions of water with more or less impurities, is seldom on heating boilers when they are supplied with water of condensation. The importance of provisions for cleaning is less in heating than in power boilers, but should not be neglected.

77. General Types of Boilers.—*Power-boilers.*—It seems necessary to consider boilers built for high-pressure steam and of large sizes as a separate class from those used principally in heating small buildings, although boilers of similar structure may be constructed for heating. These boilers will be spoken of as *power-boilers*, and are required to fulfil conditions as to strength and capacity not needed in heating-boilers.

The principal boilers of this type now in use can be grouped into two classes, viz., *fire-tube* and *water-tube* boilers, and one or the other of this type must be used for heating purposes, with the present condition of the market, whenever high-pressure steam is required.

The fire-tube or common tubular boiler consists of a cylindrical boiler with plain heads, connected by a large number of tubes which serve as passages for the smoke or heated gases. The fire is built underneath, and the smoke passes horizontally either twice or thrice the length of the boiler. The general form of this boiler is shown in Fig. 114. This boiler is also

FIG. 114.—HORIZONTAL TUBULAR BOILER.

used sometimes in a vertical position with the fire beneath one head, in which case it is called a vertical tubular. The water-tube boilers have the water in small tubes, and the heated gases pass out between the tubes. In this class of boilers the steam is contained in drums or horizontal cylinders, which are located above the heating surface. The tubular boilers are made in small sizes, 10 horse-power and larger, while the water-tube boiler for power is seldom less than 60 horse-power capacity.

Heating-boilers.—The boilers which are used for steam-heating are designed in a multiplicity of forms, and present examples of nearly every possible method of producing extended surfaces, both of the water-tube and fire-tube types. They are generally built for low-pressure steam, and are expected to be used mainly in buildings where the condensed water is returned by gravity to the boiler without pumps or traps. They are usually built in small sizes having a capacity of 250 to 2000 ft. of radiating surface ($2\frac{1}{2}$ to 20 H.P.), and are fitted with safety-valves, water and steam gauges and damper regulators.

The limits of this book prevent a detailed description of any make of heating-boiler, but the leading general types are described. Several types of the power-boiler are described quite in detail, and much that is said with respect to them will apply in a general way to heating-boilers.

The following classification of steam-heating boilers was suggested by one presented by Mr. A. C. Walworth in a paper before the New York Convention of Master Steam and Hot water Fitters, June, 1894:

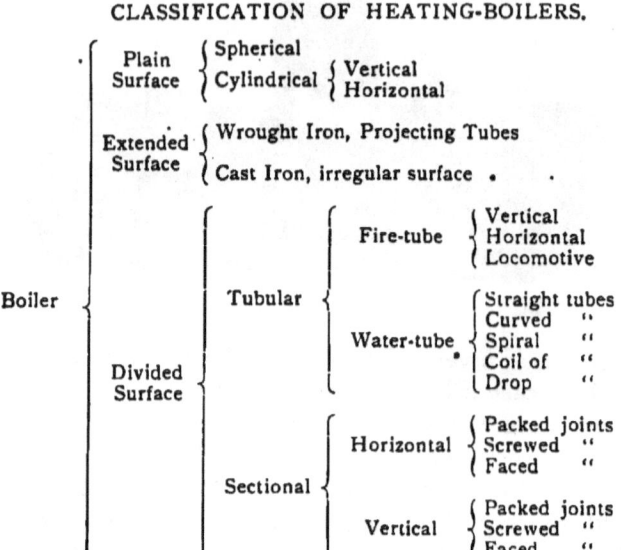

CLASSIFICATION OF HEATING-BOILERS.

78. The Horizontal Tubular Boiler. — This boiler manufactured in many places, so that in many respects it is standard article of commerce, and it can be purchased in nearly every market for a slight advance over the cost of materials and labor used in its construction. In the construction of this boiler the shell is now almost invariably made of soft steel of a thickness depending upon the pressure which the boiler is expected to sustain. The heads of the boiler are made of flange steel and are generally $\frac{1}{16}$ inch thicker than the material in the shell. Lap-welded iron tubes are almost invariably used, the standard sizes being as given in Table XVII. The tubes are expanded into the heads of the boiler and may or may not be beaded, and are generally arranged in parallel vertical rows in the lower two thirds part of the boiler. In some instances the middle row of tubes is omitted with good results. It is not a good plan to stagger the tubes, since in that case they are difficult to clean and also act to impede the circulation of the water. The boiler should be provided with manholes, with strongly reinforced edges, so that a person can enter for cleaning. The heads of the boiler above the tubes should be thoroughly braced in order to sustain safely any pressure from the inside of the boiler.

Domes are often placed above the horizontal part of the boiler, and serve to increase the capacity for the storage of steam and also provide ready means of drawing off dry steam. The dome is always an element of weakness, and if used it should be stayed and reinforced in the strongest possible manner. The dome is frequently omitted, and steam taken directly from the top of the shell or drawn through a long pipe with numerous perforations, termed a *petticoat pipe*.

In construction this boiler must be strongly braced wherever any flat surfaces are exposed to pressure, and the girth and longitudinal seams must be riveted in such a manner as to secure the maximum strength.

The following table gives principal dimensions for a series of horizontal tubular boilers designed for a working pressure of 80 to 100 pounds per square inch:

Radiating surface	1000	1200	1600	2000	2500	3000	4000	5000	6000	8000	10000
Horse-power	10	12	16	20	25	30	40	50	60	80	100
Diameter of boiler, inches	32	32	36	36	42	42	48	54	54	60	66
Length of boiler, feet	6	7½	8	10	10	12	12	12	14	16	16
Thickness of shell, inches	1/4	1/4	1/4	1/4	9/32	9/32	9/32	5/16	5/16	11/32	3/8
Thickness of heads, inches	5/16	5/16	5/16	3/8	3/8	3/8	3/8	3/8	3/8	1/2	1/2
Length of flues, feet	6	7½	8	10	10	12	12	12	14	16	16
Number of flues	32	32	30	32	40	40	52	70	70	83	104
Diameter of flues, inches	2½	2½	3	3	3	3	3	3	3	3	3
Square feet of heating surface	155	192	239	310	385	462	600	765	901	1206	1504
Proper diam. of smoke-pipe (20' chimney), inches	13	14	15	17	18	20	24	26	28	32	37
Approximate weight, lbs	1800	2000	2700	3100	4000	4600	5600	7000	8000	10500	12500
Wt. of grate and fixtures, lbs	1200	1400	1600	1800	2100	2200	2800	5200	5400	7200	7500

Fifteen square feet of surface to each horse-power.

79. Locomotive and Marine Boilers. — Boilers of the horizontal tubular type with a fire-box entirely enclosed and surrounded by heating surface are usually termed locomotive boilers from the fact that such construction is common on locomotives. Boilers of this style are sometimes used for stationary power purposes, and possess the advantage over the plain tubular boiler of requiring no brick setting. They are not, however, as strong in form as the plain tubular, since large flat surfaces have to be used over the fire-box.

Marine Boilers.—A cylindrical boiler with an internal cylindrical fire-box is principally used on large boats. The fire-box

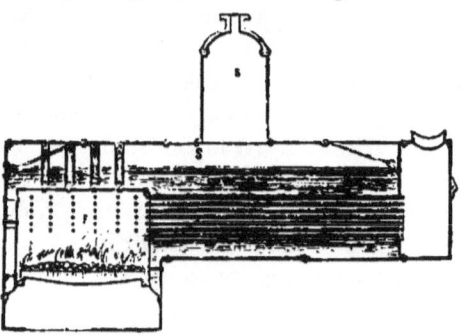

FIG. 115.—LOCOMOTIVE BOILER.

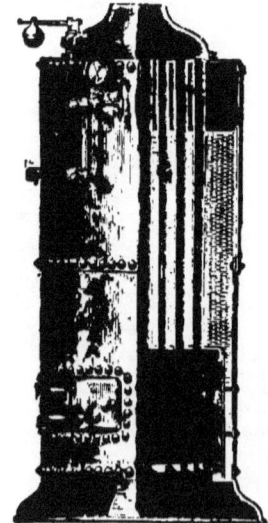

FIG. 116.—UPRIGHT TUBULAR BOILER.

is often corrugated. This form of boiler is very strong and efficient, but because of cost of construction has been little used for stationary purposes.

79. Vertical Boilers.—Vertical boilers of large size are made in every respect like the horizontal tubular boiler, but are set so that the flame plays directly on one head and the heated gases pass up through tubes. These boilers are generally provided with a water-leg which extends below the lower crown sheet and is intended to receive deposits of mud, etc., from the boiler. They are usually made so that the heat passes directly out of the top of the flue, but in some cases the heat is made to pass down a portion of the length of the external shell before being discharged.

They are economical in the use of fuel and occupy very small amount of floor-space; they require, however, a great deal of head-room, are very easily choked up with deposits and sediment, very difficult to clean, and very likely to leak around the tubes in the lower crown-sheet, and consequently have a short life.

Vertical boilers with horizontal radial tubes projecting outward with ends closed, known as porcupine boilers, are also on the market, and quite recently a vertical boiler of the water-tube type has been constructed.

80. Water-tube Boilers.—The water-tube boilers, which are used for power purposes, are designed to withstand great pressures, and can be purchased in sizes ranging from 60 to 500 horse-power per boiler. The general construction of these boilers is such as to have the water on the inside of the tubes and the fire without. There are two general forms: first, those with straight tubes, and second, those with curved tubes.

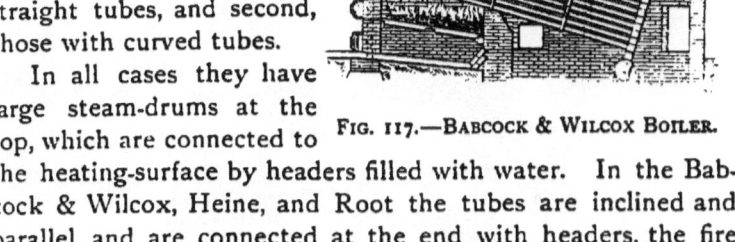

FIG. 117.—BABCOCK & WILCOX BOILER.

In all cases they have large steam-drums at the top, which are connected to the heating-surface by headers filled with water. In the Babcock & Wilcox, Heine, and Root the tubes are inclined and parallel, and are connected at the end with headers, the fire being applied in each case under the elevated portion of the inclined tube, so as to insure circulation uniformly in one direction.

In the Babcock & Wilcox boiler, cast-iron zigzag headers are used; in the Root boiler, the tubes are connected together by external U-shaped bends; in the Heine boiler (Fig. 120), the tubes are connected to large, flat-stayed surfaces. In the Babcock & Wilcox and Heine boilers, feed-water is supplied at the lower part of the top drums; while, in the Root boiler, it is supplied to a special drum in the down-circulation tubes at the back end of the boiler. The Stirling boiler has three horizontal drums at the top connected by curved tubes to a single lower drum at the back end of the boiler; the Hogan has one drum at top and two at bottom, which are parallel and connected by curved tubes, and also a series of down-circulating tubes connecting the same drums, but not exposed to the heat of the fire. In the Stirling boiler, the feed-water is intro-

FIG. 118.—ROOT BOILER.

duced in the top drums; in the Hogan boiler, into a special heater and purifier arranged as a part of the downward circulation.

FIG. 119.—STIRLING BOILER.

The Harrison boiler consists of an aggregation of spheres of cast iron or steel connected by necks, forming what is to be considered rather as a sectional, than a water-tube boiler. These spheres are held in place by bolts, which will stretch and act as safety-valves in case of excessive pressure.

In addition to the water-tube boilers for power purposes which have been mentioned here, there are many others which cannot be described in the space at our command, but of which we may name the National, Campbell & Zell, and the Caldwell as worthy of notice.

All the water-tube boilers are provided with mud-drums, which are usually cast-iron cylinders removed from the circu-

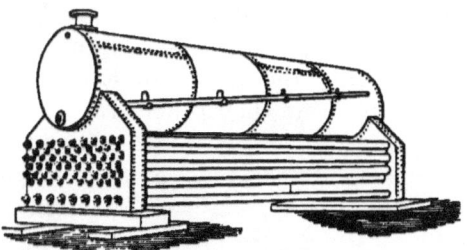

FIG. 120.—HEINE BOILER.

lation and intended to receive any deposits of scale or material which is loosened in the process of circulation.

81. Hot-water Heaters.—Hot-water heaters differ essentially from steam-boilers, principally in the omission of a reservoir or space for steam above the heating surface. The steam-boiler might answer as a heater for hot water, but the large capacity left for the steam would tend to make its operation slow and quite unsatisfactory.

The passages in a hot-water heater need not extend so directly from bottom to top as in a steam-heater, since the problem of providing for the early liberation of the steam-bubbles does not have to be considered. In general, the heat from the furnace should strike the surfaces in such a manner as to increase the

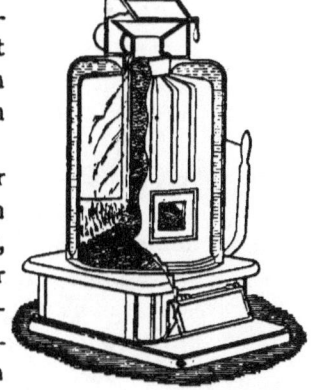

FIG. 121.—VERTICAL MAGAZINE HOT-WATER HEATER.

natural circulation, and not act to produce a backward circulation. This may be accomplished in a certain measure by ar-

ranging the heating-surface so that a large p
direct heat will be absorbed near the top of th

There is a great difference of opinion a*
merits of horizontal and vertical heating-surfa
pose, but the writer cannot find that any experi
made which satisfactorily decide this question
face is very much divided, and the fire is main
temperature, considerable steam is likely to be
always acts in a certain measure to increase c
heating-pipes and diminish it in the heater; it
produce a disagreeable crackling noise.

Practically, the boilers for low-pressure ste
water differ from each other very little as to
the heating-surface, and in describing the gene
are in use no attempt will be made to make a:
to whether the apparatus will be used for hot
heating. If designed for steam-heating, a rese:
connected with the circulating system is in
vided, containing water in its lower part a
steam capacity above the water-line, also s
water-surface to permit the separation of the
water without noise and violent ebullition.

82. Classes of Heating-boilers and Hot

—*Plain-surface Boilers.*—There are probabl)
heaters built at the present time with a pla
spherical or cylindrical, since the expense of a
surface in that form would practically preclud

Extended-surface Heaters (Figs. 122 and 1:
this class with extended and irregular surface
extensively in hot-water heating, and with
domes are used to some extent in steam-he
heaters the water is received at the lowest po
is heated as it gradually rises, receiving the e
at various projections, and is finally dischar
grate is at G, the smoke being discharged at
and heated gases move in nearly a direct line
in a sinuous course in Fig. 123.

A form which is in extensive use, and
and smoke are each grouped in one bod

STEAM-HEATING BOILERS.—HOT-WATER HEATERS. 137

Fig. 124. In this case the extended surface is produced by the wedge-shaped hollow prisms extending over the fire-space. The heated gases have a return circulation around the lower portion of the heater, and also come in contact with a top dome from which the heated water is drawn off.

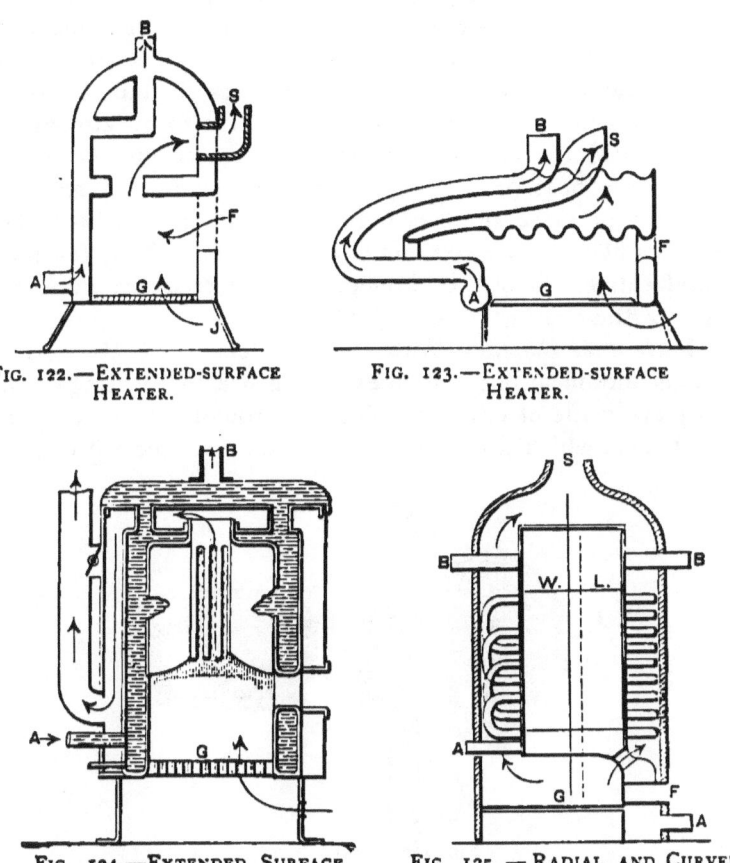

FIG. 122.—EXTENDED-SURFACE HEATER.

FIG. 123.—EXTENDED-SURFACE HEATER.

FIG. 124.—EXTENDED SURFACE, VERTICAL PRISMS.

FIG. 125. — RADIAL AND CURVED WITH EXTENDED SURFACE.

Heaters belonging to the extended-surface class made with vertical cylinders, into which are connected either straight horizontal tubes with closed end, as shown on the right-hand side of Fig. 125, or U-shaped projections of pipe either horizontal or slightly inclined, are in use for both water- and steam-heating. In case they are used for steam-heating the water-line is carried at sufficient distance from the top of the cylinder to give the required steam-space, and the heater is supplied with both pressure- and water-gauges. The heated gases pass around the cylindrical part of the boiler and may be made to circulate among the projections by means of baffle-plates.

Tubular Boilers.—Heating-boilers with fire-tubes and with a steel shell similar in construction to the horizontal and vertical tubular boiler described in Articles 76 and 78, are in use for heating to considerable extent in the forms already described. Modifications of these, with return flues arranged so that the heat passes both upward and downward, and also with two or more short cylindrical shells connected together by tubes filled with water, are in extensive use. Very few horizontal tubular boilers, or boilers of the locomotive type, are used for the heating of small buildings.

Water-tube Boilers.—Water-tube boilers of all classes and various modifications are in extensive use for heating. The tubes are made of either cast-iron or wrought-iron pipe. The pipe-boilers which are in the market are arranged with nearly

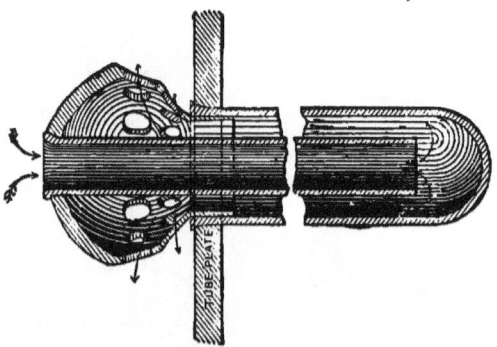

FIG. 126.—FIELD TUBE.

every form of heating-surface; some are built with heating-surface in the form of the pipe-coil, as shown in Fig. 92, page 108, and others in the form of a manifold coil, as shown in Fig. 93, page 108. Still other boilers have the pipe arranged in the form of a spiral connecting with a receiving-drum below and a steam-drum above. The heated gases are arranged to move in some cases parallel with the surfaces, and in other cases at right angles.

The Field tube is used extensively for the purpose of increasing the heating-surface; in its original form it consisted of a tube with a closed end projecting downward and expanded into the boiler-shell; into this extended another tube which did not reach quite to the bottom, and was held in position by an internal perforated support, as shown in Fig. 126. This is used in heating-boilers with various modifications both projecting downward and horizontally. When used projecting downward, it is termed a drop-tube, and is supplied either with an internal tube, as shown, or a partition; when used horizontally the internal tube is frequently supplied from a compartment separated from that to which the external tube is attached. Fig. 127 illustrates a type of heating-boiler which is quite extensively used for both hot water and steam, and is built by different manufacturers, either of steel or cast iron. The heater consists of a cylindrical drum, the lower surface of

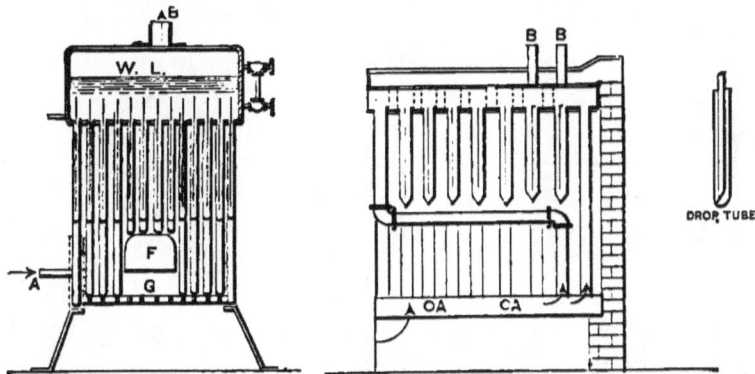

FIG. 127.—DROP-TUBE SURFACE.   FIG. 128.—DROP-TUBE AND COIL-HEATER.

which is covered with tubes of the type described which project downward. The tubes directly over the fire and over the fire door are short, while those around the fire are sufficiently long to form the external walls of the heater. The return water is received in one of the long pipes near the bottom of the heater, and the steam or heated water is taken off at the top. The drum in one of these heaters is provided with a baffle-plate connected to the diaphragm in the drop-tube, so that the circulation must take place in a vertical direction in the tube.

Fig. 128 shows a heater in which the surface is made up partly of pipe-coils and partly of drop-tubes. The return water is received in the lower concentric drum, and as it is warmed passes to the top drum of the heater, from which it flows to the building; a type of heater in many respects similar is made without drop-tubes, the whole surface being obtained by use of pipe-coils, made either with return bends or with branch tees.

Sectional Boilers.—The greater number of cast-iron boilers are made by joining either horizontal or vertical sections. These sections are joined in some instances by a screwed nipple, in other cases by a packed or faced joint, and are held in place with bolts. The sections generally contain water and

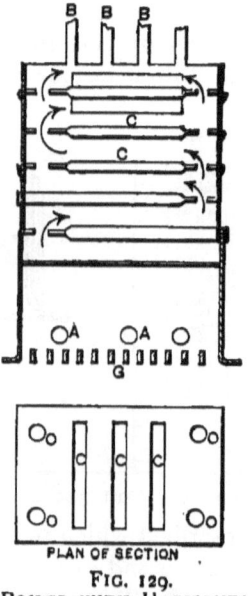

FIG. 129.
BOILER WITH HORIZONTAL SECTIONS.

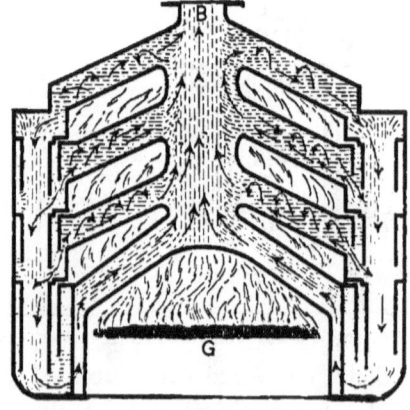

FIG. 130.
BOILER WITH HORIZONTAL SECTIONS.

steam, and the heated gases circulate around the sections in flues provided for that purpose. The joints in the flues are usually made tight enough to prevent the escape of smoke by the use of an asbestos or similar cement.

Horizontal Sections.—Fig. 129 represents a type of heater in which the various sections are horizontal, the surface being increased to any amount by adding sections. This form is used extensively in a number of hot-water heaters. Fig. 130 shows another form of boiler made in a similar manner, but with the sections of such form as to produce both an up and down circulation within the heater. The up circulation takes place over the hottest portion of the fire, the down circulation in special external passages which are not heated.

Vertical Sections.—Boilers with vertical sections are made in the same manner in many respects, the sections being united by internal or external connections. When united by external connections, screwed nipples connecting the sections to outside drums, of the general form as shown in Fig. 131, are usually employed. In this case the return-water is received into horizontal drums *AA*, which extend the full length of the heater, and flows into the lower part of each section. The steam or hot water is drawn off from a similar drum, *B*, which extends over the top of the heater and is connected with each section by a screwed nipple. Fig. 130 shows methods of attaching steam- and water-gauges.

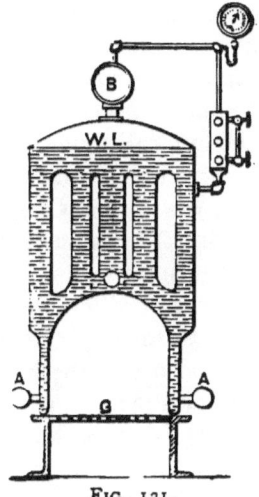

FIG. 131.

This form is used quite extensively in steam-heating and to some extent for hot-water heating.

83. Heating-boilers with Magazines.—Nearly all of the heating-boilers are manufactured as required with or without a magazine to hold a supply of coal. The magazine in most cases consists of a cylindrical tube opening at or near the top of the heater and ending eight to twelve inches above the grate. The magazine is filled with coal, which descends as combustion takes place at the lower end, and provides fuel for further combustion (see Fig. 121). The magazine works successfully with anthracite coal, which is that ordinarily employed in domestic heating, but it takes up useful space in the heater, decreases the effective heating surface for a given size, and in that respect is objectionable. The writer's own experience would lead him to believe that the magazine heater, except in very small sizes, requires as much attention as the surface burner, and consequently has no special advantage.*

84. Heating-boilers for Soft Coal.—It is quite probable that no furnace, either for power or heating boilers, has yet been produced which will consume soft coal without more or less black smoke. This smoke is due principally to the imperfect combustion of the hydrocarbons contained in the coal. The hydrogen burning out after the gases have left the fire leaves solid carbon in the form of small particles, which float with and discolor the products of combustion. The amount of loss as found by experiment in Sibley College,† even when dense black smoke is produced, seldom reaches one per cent, and is of no economical importance. The sooty matter produced in the combustion of this coal is likely to adhere to the water-heating surfaces, and if these are minutely divided it will be certain to choke the passages for the gases of combustion. For the combustion of soft coal those heaters have been the most successful which have a grate with small openings, and with an area 50 to 70 per cent as large as that needed for anthracite coal, also with the heating-surface of comparatively simple form and arranged so as to be easily cleaned. It is considered important that the air-flues be so arranged as to keep the products of combustion as hot as possible. This coal is likely to swell when first heated, and cannot be fed successfully by a magazine.

---

* Magazine heaters have been constructed with a magazine set obliquely above and to the side of the grate, and in that position are not open to all the objections stated.

† See Table XII, page 390.

CHAPTER VIII.

SETTINGS AND APPLIANCES—METHODS OF OPERATING BOILERS AND HEATERS.

85. Brick Settings for Boilers.—Horizontal tubular boilers and a few heating-boilers require to be set in brickwork, of which the general arrangement is shown in Fig. 132. The horizontal tubular boiler is usually supported from cast-iron flanges which are riveted to the sides of the shell, and which rest

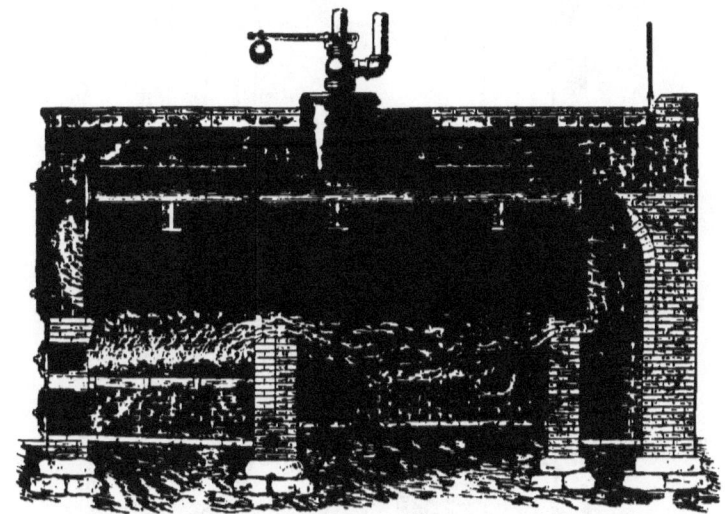

FIG. 132.—PERSPECTIVE VIEW OF TUBULAR BOILER SET IN BRICKWORK.

directly on the walls of brickwork, or are supported by suspension-rods from above. In some instances the boiler-lugs rest on cast-iron columns embedded within the brickwork, and of such a length that all the brickwork above the grates can be removed without affecting the setting. In setting the boiler

143

144 HEATING AND VENTILATING BUILDINGS.

the back end should be slightly lower than the front, in order that the entire bottom of the boiler may be drained at the blow-off pipe. One of the lugs of the boiler on each side should be anchored in the brickwork; the others should rest on rollers, which in turn rest on an iron plate embedded in the brick

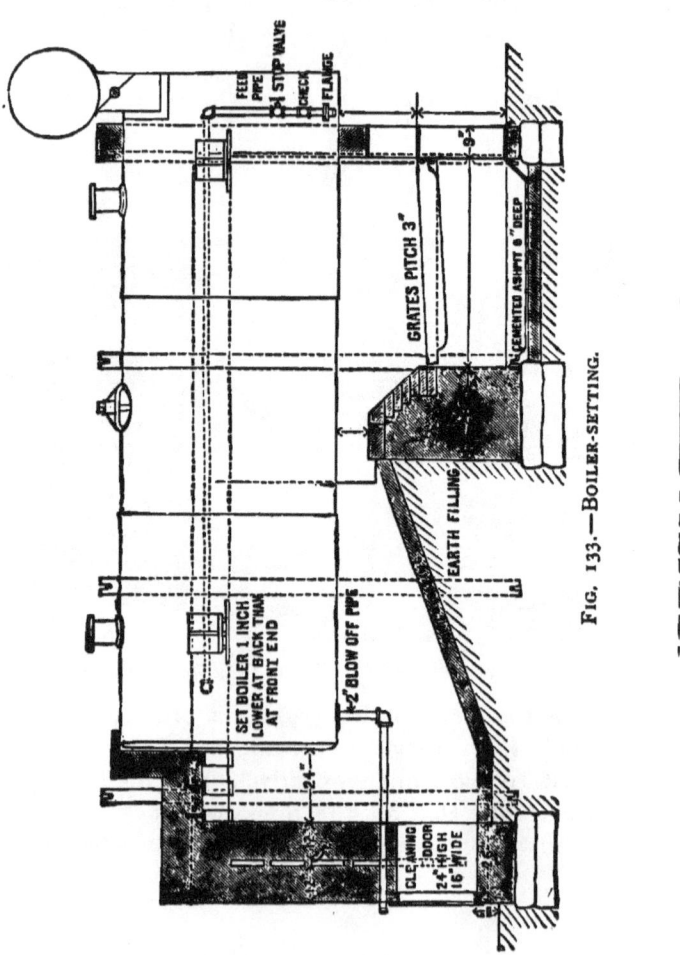

FIG. 133.—BOILER-SETTING.

walls. This permits expansion due to heating and cooling to take place without straining the boiler. If the boiler is not over 14 feet in length, two lugs on a side will be sufficient to sustain it, but if it is of greater length, more lugs will need to be supplied. The brickwork surrounding the boiler is more durable if built with an air-space, as shown in Fig. 134. It must be thoroughly stayed, by means of cast-iron braces, connected with tie-rods at top and bottom of wrought iron to

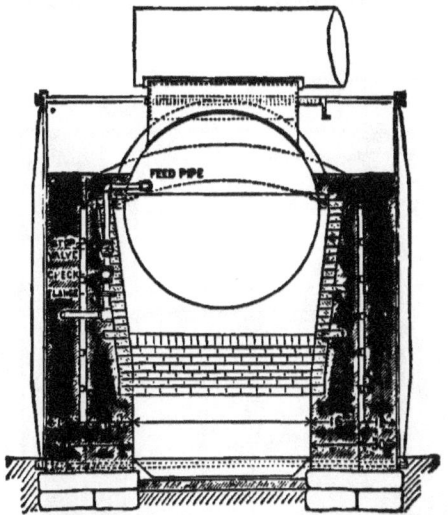

FIG. 134.—SECTIONAL VIEW OF BOILER-SETTING.

prevent transverse or longitudinal motion. The top may be arched over so as to leave a passage for the hot gases directly over the shell, as in Fig. 132, or made to rest directly on the boiler, and the hot gases taken away at the front end by means of a flue, usually termed a *breeching*, which extends to the chimney. The practice of taking the heated gases from the front end of the boiler is rather more common than that of returning them to the back end over the top, and there are many engineers who believe that the hot gases injure the boiler when coming in contact with the shell above the water-line. Figs. 133, 134, and 135 show longitudinal and transverse sections of a boiler-setting, with smoke-pipe or breeching in front, which can be highly commended as representing the best practice.

The depth of foundation to be used in boiler-setting will depend upon the character of the soil and the weight of the boiler. For large tubular and water-tube boilers it should generally be not less than 3 feet. Fire-brick of the best quality should be used to line the brick walls for a height equal to that from the grate to the water-line of the boiler, and these should be arranged so that if necessary they can be relaid without disturbing the outer brickwork. In the setting shown in Figs. 133–134 the top of the boiler is covered with a coating of some

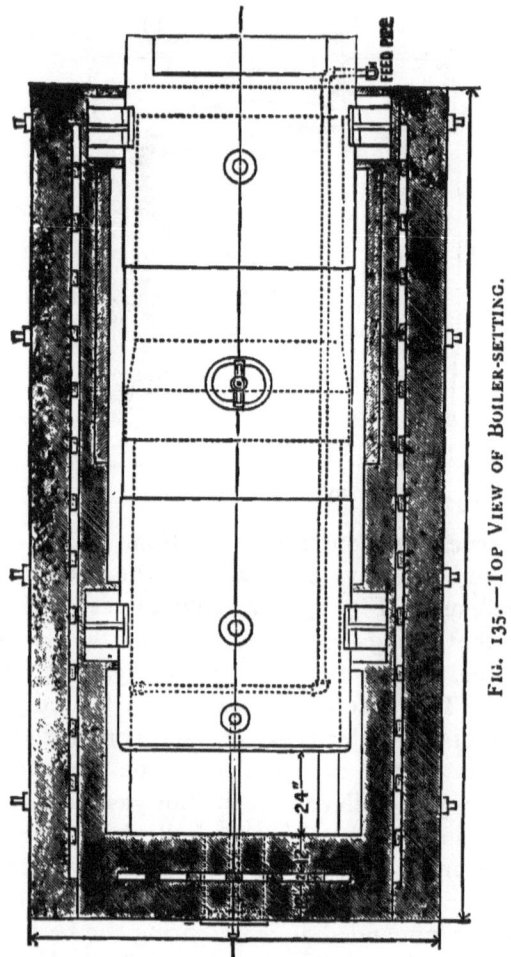

FIG. 135.—TOP VIEW OF BOILER-SETTING.

good, non-conducting material, for which magnesia, asbestos, or mineral wool may be recommended, put on while in a plastic condition to the depth of 2 inches with a mason's trowel. Brickwork is often used; but it is heavier, and quite liable to crack from the effects of heat.

86. Setting of Heating-boilers.—If heating-boilers are to be set in brickwork, the special directions which have already been given can be applied, with such modifications as may be needed for the boiler in question. Nearly all heating-boilers are now set in what is called a *portable setting*, in which no brick whatever is used. Some of the heaters are made by the system of manufacture adopted so that no outside casing is required, as in Fig. 138; others require a thin casing of galvanized or black iron which is lined with some non-conducting material, as magnesia, asbestos fibre, or rock wool, which is placed outside the heater and arranged so as to enclose a dead-air space, as in Fig. 137. These coverings are nearly as efficient in preventing the loss of heat as brickwork, and they form a more cleanly and neater appearing job.

FIG. 136.
BRICK-SET MAGAZINE BOILER.

The slight amount of heat which escapes from such a setting is seldom more than that required to warm up the basement or room in which the heater is located.

The boiler must in all cases be provided with a steam-gauge, safety-gauge, and damper regulator, all of which are specially described later. The steam-gauge should be either connected below the water-level or else provided with a siphon to prevent dry steam entering the interior tube. A safety-valve of the single-weighted type is preferable and should be connected at the top of the heater. The damper regulator usually consists of a rubber diaphragm which is acted on by pressure so as to open and close the dampers as required. It will prove more durable, generally, if connected below the water-line and located about on a level with the top of the heater, as this will insure the contact of water against the rubber diaphragm. Fig. 137 represents a boiler with portable setting with external iron casing and equipped with all appliances, and Fig. 138 represents a portable setting without enclosing case.

Hot-water heaters are set in the same general manner as steam-boilers. Each should be provided with thermometers showing both the temperature of the flow and the return water,

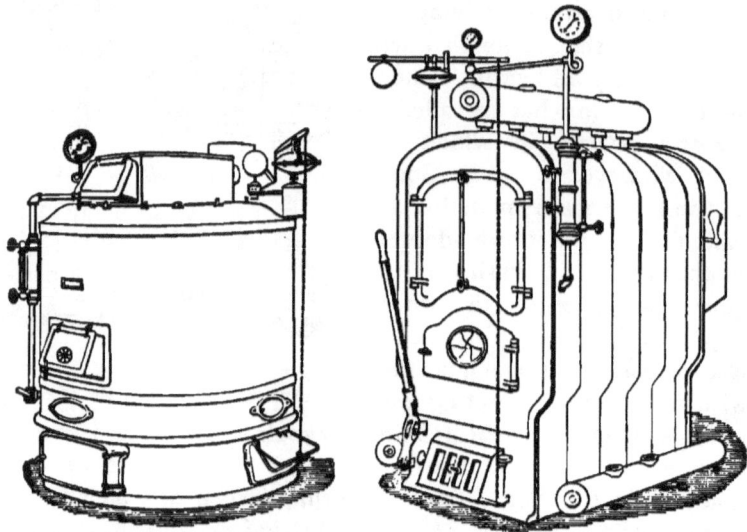

FIG. 137.—HEATING-BOILER WITH PORTABLE SETTING.

FIG. 138.—HEATING-BOILER WITH PORTABLE SETTING.

and with a pressure-gauge graduated to show pressure of water in feet and sufficiently large to show any variation in height in the open expansion tank. The dampers to a hot-water heater

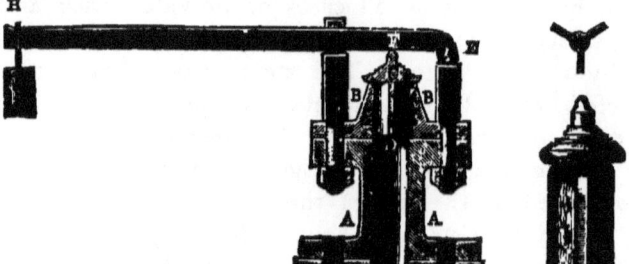

FIG. 139.—SECTION OF LEVER VALVE, OLD FORM.

cannot be opened and closed by variation in pressure, but reliable thermostats are now on the market which will operate the dampers by change of temperature in the various rooms of the building.

87. The Safety-valve.—The safety-valve has been used since the earliest days of boiler construction for reducing the pressure when it reached or exceeded a certain limit. It has been built in various forms, but in every case has consisted essentially of a valve opening outward and held in place by a weight or a spring. One form in common use consists of a valve held in place by a weight on the end of a lever, shown in Fig. 139 in section and in Fig. 140 in elevation. In this form of safety-valve the force required to lift the valve

FIG. 140.—LEVER SAFETY-VALVE, MODERN FORM.

can be regulated by sliding the weight to different positions on the lever. The form shown in Fig. 141 consists of a single weight suspended from the valve and hanging in the upper

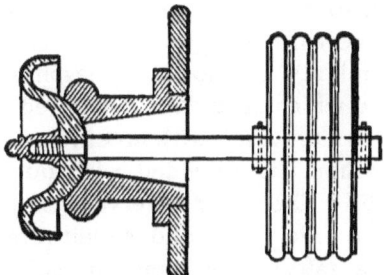

FIG. 141.—DEAD-WEIGHT SAFETY-VALVE—WEIGHT INSIDE OF BOILER.

part of the boiler. This form is to be commended, since it cannot be adjusted without opening the boiler.

A form used very extensively for low-pressure heating-boilers consists of a single weight resting on a valve, as shown in Fig. 142; its principle of operation is the same as that of the other valves. A form much used on power-boilers, and frequently called, from the suddenness with which it opens, a *pop-valve* consists of a very quick-opening valve held in place with a spring, one form of which is shown in **Fig. 143.**

 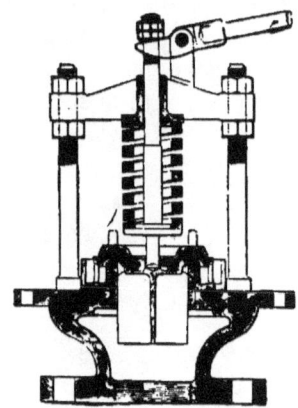

FIG. 142.—EXTERNALLY WEIGHTED SAFETY-VALVE.

FIG. 143.—SECTION OF SPRING OR POP SAFETY-VALVE.

It is desirable that the safety-valve be made in such a manner that the engineer or attendant to the boiler cannot manipulate it at pleasure so as to maintain a higher pressure on the boiler than prescribed.

Serious accidents have been caused by excessive weighting of the safety-valve through ignorance or carelessness on the part of the attendants, and for this reason a class of valves should be selected which cannot be tampered with. Some of the safety-valves are provided with an external case which can be locked, and others are provided with internal weights, as already described. The lever safety-valve offers the most temptation for extra weighting and should rarely be used.

The area of a safety-valve must be sufficiently large to effectually reduce the boiler pressure when the valve is open and when a brisk fire is burning on the grate. It may be computed from the following considerations:

The steam which will flow through one square inch of opening in one hour of time was found by Napier[*] to equal in

---

[*] Rankine's "Steam Engine."

pounds nearly 50 times the absolute pressure of the steam; further, it has been found by experiment that the safety-valves in ordinary use open only to such an extent as to make $\frac{1}{5}$ of the total area of the valve effective in reducing the pressure. From these considerations it will be seen that the area of the safety-valve in inches should be $\frac{9}{60}$ the weight of steam generated per hour, divided by the absolute pressure. Considering that 100 lbs. of steam can be generated from each square foot of grate per hour, this would be equivalent to the following rule: The area in square inches is equal to 18 times the grate surface in square feet, divided by the absolute pressure.

The following table gives the area of safety-valve in square inches per square foot of grate required on marine boilers by the English Board of Trade:

Boiler Pressure.		Area in square inches for each sq. ft. grate.	Boiler Pressure.		Area in square inches for each sq. ft. grate.
Absolute, Pounds per sq. inch.	Above atmosphere.		Absolute.	Above atmosphere.	
15	0	1.25	60	45	0.50
20	5	1.07	70	55	0.44
25	10	0.94	80	65	0.40
30	15	0.83	90	75	0.36
35	20	0.75	100	85	0.33
40	25	0.68	110	95	0.30
45	30	0.625	120	105	0.277
50	35	0.576	130	115	0.258

The following formula gives results very closely in accord with the English Board of Trade table. Let A = area of safety-valve in square inches, P = absolute pressure = gauge pressure plus 15, G = number of square feet of grate surface.

$$A = \left(\frac{18}{P} + 0.002P\right)G.$$

Various rules quite different from the above are given in treatises on boiler construction, but it is believed that the above table represents the best practice of to-day and forms a safe guide for estimating the size of safety-valves.

Safety-valves are liable to stick fast to the seat, through corrosion, in which case they fail to raise with excess of pressure; for that reason they should be periodically lifted from their seats and otherwise inspected.

In case the area of the valve required is greater than 4 inches in diameter, two safety-valves should be provided for each boiler.

88. Appliances for showing Level of Water in the Boiler. —In the first boilers constructed floats were used, and such appliances are still common on European boilers. In this country water-gauge glasses and try-cocks are now used, to the exclusion of all other devices. The water-gauge (see Fig. 144), consists of two angle-valves, one of which is screwed into the boiler above the water line; the other is screwed about an equal distance below, and these are connected by means of a glass tube usually $\frac{3}{8}$ to $\frac{5}{8}$ inch external diameter and strong enough to withstand the steam-pressure. When both angle-valves are open the water will stand in the gauge-glass the same height as in the boiler, but if either valve is closed the water-level shown in the glass will not accord with that in the boiler. Three try-cocks are usually put on a boiler in addition to the water-gauge. The try-cocks are made in various forms, one kind being shown in Fig. 145, these are located so that one is above, the other below, and the third at about the mean position of the water-line. When the top one is opened, it should show steam; when the bottom one is opened it, should

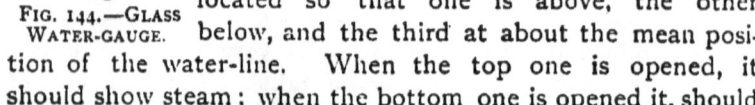

Fig. 144.—Glass Water-gauge.

Fig. 145.—Register Gauge-cock.

show water. Both try-cocks and gauge-glasses should usually be put on boilers, so that the reading as shown in the water-gauge glass can be checked from time to time. This is necessary, because if dirt should get in the angle-valves or passages leading to the gauge-glass the determination would be inaccurate.

Water-columns attached to the boiler by large pipes, both above and below the water-line, and fitted with try-cocks and water-gauge as shown in Fig. 146, are often provided. These columns frequently contain floats (Fig. 147), so arranged that steam is admitted into a small whistle if the water falls below or rises above the required limits, and thus gives an alarm.

FIG. 146.—WATER-COLUMN.

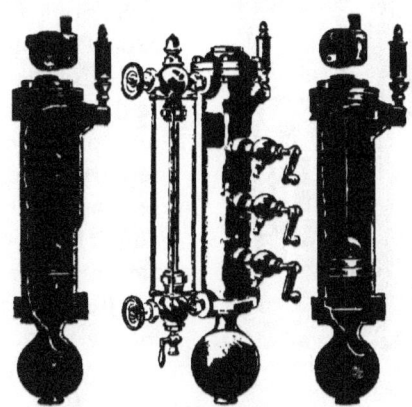

FIGS. 147.—RELIANCE ALARM WATER-COLUMN.

89. Methods of Measuring Pressure.—The excess of pressure above that of the atmosphere is measured by some form of manometer or pressure-gauge. Where the pressure is small in amount, a siphon, or U-shaped tube filled with some liquid is a very convenient means of measuring pressure. The method of using a simple manometer of this character is shown in Fig. 148, in which a U-shaped tube, $G\ F\ E\ D$, has one branch attached to the vessel containing the fluid whose pressure is to be measured; the other, as at D, is open to the air.

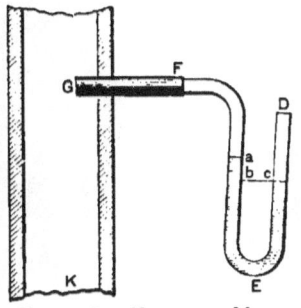

FIG. 148.—U-SHAPED MANOMETER.

If water, mercury, or other liquid be placed in the U-shaped tube it will be forced down on the side of the greater pressure and upward on the side of the less, a distance proportional to the pressure. The height of the fluid in one side in excess of that on the other will be a measure of the difference of pressure between that of the atmosphere and that in the vessel.

Various forms of manometers are used, of which several are shown in Fig. 149. For very low pressures water is the liquid generally employed; for moderate pressures up to 15 or 25 pounds mercury is very convenient, and often used; while for high pressures a pressure-gauge (Fig. 150), as described later, is commonly employed.

The Bourdon pressure-gauge is ordinarily used. This consists of a tube of elliptical cross-section bent into a circular form. The free end of the tube is attached by gearing to a hand which moves over a dial. Pressure on the interior of the tube tends to straighten it, and

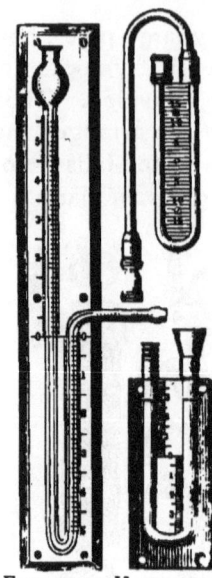

FIG. 149.—U SHAPED MANOMETER TUBES.

FIG. 150.—BOURDON GAUGE

moves the hand an amount proportional to the pressure.

Fig. 150 shows the interior of a pressure-gauge of this character with the dial removed. In place of the tube a corrugated diaphragm is sometimes employed. A section of such a gauge is shown in Fig. 151. In the use of gauges of the character just described it is necessary to protect them from extreme heat. For this purpose when they are connected to a steam-boiler a

FIG. 151.—DIAPHRAGM GAUGE.

siphon or U-shaped form of pipe is to be used in the connection, so that water and not steam will be forced into the interior of the gauge.

The manometers and gauges described in every case measure the pressure above or below that of the atmosphere. If they measure a pressure lower than that of the atmosphere they are commonly called vacuum-gauges, but the principle of construction is the same as described.

The relations of various units used in measuring pressure can be readily determined from the following table of equivalents: 1 inch of mercury = 13.619 inches of water = 1.134 feet of water = 0.49101 pound = 399.51 feet of air at 60 degrees Fahrenheit and barometer pressure 30 inches. The pressures are usually taken as acting on one square inch of a body.

90. Thermometers.—The methods of constructing various kinds of thermometers have been described in Articles 8 to 12. In any hot-water heating system it is quite important to know the temperature of the water leaving the heater, and in many cases also that of the return. This information, while not so vital to the safety of the heater as that given by a pressure-gauge on a steam-heating system, is of the same character, and will prove to be equally valuable in indicating the work done by the heater, and the heat absorbed by the system.

Any of the suitable forms described in Chapter I can be used, but special forms in which the thermometer-bulb sets in a cup of mercury (Fig. 152) are often used, the cup being screwed into the pipe whose temperature is required. These thermometers should be set so as to extend deep into the current of flowing water, and there should be no opportunity for air to gather around the bulb; otherwise the readings will not be the true temperature.

FIG. 152.—THERMOMETER FOR HOT-WATER HEATING.

91. Damper-regulators.—Nearly all steam-boilers are provided with an apparatus for opening or closing the dampers and draft-doors to the boiler as may be required to maintain a constant steam-pressure. For low-pressure steam-heating plants the regulator consists in nearly every case of a rubber diaphragm (Fig. 153), which receives the steam-pressure on one side, and acts against a counter-weight resting on a plate on the opposite side. The plate is connected by a rod to a lever pivoted to the external case, which in turn is connected to the various drafts by means of chains, and so arranged that if the pressure rises the lever is lifted and the drafts closed, while if the pressure falls the lever also falls, and the drafts are opened. By means of weights on the lever the regulator can be set to operate at any pressure. The regulator should be connected to the boiler below the water-line, or by means of an U-shaped pipe, arranged so that the part of the vessel below the diaphragm will remain full of water; otherwise the heat in the steam will cause the rubber to deteriorate rapidly. The form shown in Fig. 153 is so arranged that the diaphragm must in every case be in contact with water.

While rubber diaphragms are usually durable for low-pressure steam-regulators, still they occasionally are ruptured. In order to prevent accident from such a cause, the Nason Manufacturing Co. have devised a form of such a character that the draft-doors will close, instead of open, in case of rupture. This is done by using a link in the connecting-chain to the draft-doors of some metal that will be fused at a temperature below that of boiling water, and arranged so that in case of rupture the escaping steam and hot water will impinge upon and melt it; the damper will be closed by its own weight when the link breaks.

Damper-regulators for high-pressure steam are constructed so as to operate on the same principle as those described, but instead of a rubber diaphragm either a metallic diaphragm or a piston working in a cylinder, and operated by water-pressure, is employed.

The following cut shows the external appearance of one of the many forms in use.

FIG. 153.—DIAPHRAGM DAMPER-REGULATOR.

FIG. 154.—PISTON DAMPER-REGULATOR.

92. Blow-off Cocks or Valves.—Every steam-boiler should be provided with an appliance for emptying all of the water at any time. This may be done by leading a pipe from the lowest part of the boiler and providing a cock or valve so that it can be discharged at pleasure. The pipe leading from the boiler should have a visible outlet, so in case there is any leak it can be seen and stopped. The writer prefers a cock (Fig. 155) to a valve for use on the blow-off pipe, since it is less likely to be stopped by scale or sediment from the boiler.

In case the water of condensation from the heating coils is not returned to the boiler it is necessary to blow off some of the water very frequently in order to lessen the deposition of scale or dirt on the bottom of the boiler.

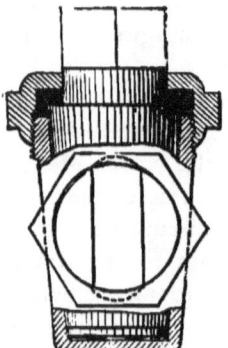

FIG. 155.—PACKED PLUG COCK.

93. Expansion-Tank.—An expansion-tank will be needed in hot-water heating systems. With increase of temperature from 40° F. to the boiling-point, water expands 4.66 parts in 100, or nearly 5 per cent. The force of expansion is nearly irresistible, and the increase in volume due to it must be provided for, so as not to produce a dangerous pressure.

The method ordinarily adopted consists in the use of a vessel called an *expansion-tank*, whose cubical contents must be somewhat greater than one twentieth of the total cubical contents of heater, pipes, and radiators. It must be connected to the heating system in such a way as to receive the increase in volume, and should be placed on a level somewhat above that of the highest radiating surface.

If there is to be no sensible increase in pressure due to expansion the tank is connected with the outside air by a vent-pipe, and in this case the pressure inside will be atmospheric; the pressure on the heating system will depend on the distance from the water-level in the tank, each foot corresponding to 0.435 pounds per square inch (2.4 feet being equivalent to one pound of pressure at 212° F.).

FIG. 156. EXPANSION-TANK.

In case a pressure in excess of the atmosphere is required, the vent pipe is closed and a safety-valve attached which will open when the pressure reaches the desired point. By increasing the pressure on the system the boiling temperature of the water will be much increased, and hence it will be possible to maintain a higher temperature throughout the system. As showing the increase in temperature of the boiling point with excess of pressure, the following table is inserted :

Pressure.		Temperature of Boiling Point (degrees F.).
Pounds per sq. in. above Atmosphere.	Equivalent Head, in Feet.	
0	0	212
5	12	228
10	24	240
15	36	250
20	48	259
25	60	267
30	72	274
35	84	280
40	96	287
45	108	292
50	120	297
55	132	302
60	144	307
70	168	316
80	192	324
90	216	332
100	240	338
125	300	352
150	360	365
175	420	378
200	480	388

Pressure systems of hot-water heating were used at one time to a considerable extent in England, under what was known as the Perkins* system, in which small pipes and exceedingly high pressures and temperatures were used. It has also been used to some extent in this country in the Baker system of car-heating.

The advantages of the pressure system are those which are due simply to the use of higher temperatures and smaller radiating surfaces; the disadvantages are the danger of an explosion

---
* Hood's " Heating and Ventilating of Buildings."

which would be likely to happen were the safety-valve inoperative, or did any part of the apparatus give way. The sudden liberation of a considerable body of water having a temperature above the boiling point would result in the instantaneous production of a large amount of steam, which might produce disastrous results.

With the open expansion-tank it seems hardly possible that any serious accidents could result even from the most careless management, since the escape of steam from the top of the expansion-tank would prevent the accumulation of pressure. To prevent accident the expansion-tank should be connected to the heater by a pipe protected from frost and without stop or valve, so as to render it impossible to increase the pressure on the system by stoppage of the connection.

It is desirable to provide the expansion-tank with a glass water-gauge showing the depth of water, and a connection to the supply-pipe for adding water to the system. In case the expansion-tank occupies a cold location where it might freeze in extreme weather, a small pipe connected with the circulating system, in addition to those described, should be run to the tank and connected at a higher level than the expansion-pipe, so as to insure circulation of warm water.

94. Form of Chimneys.—The form and size of the chimney is of great importance in connection with the satisfactory operation of a heating plant, and it should in every case receive the closest inspection before guarantees of capacity are made.

It will be found that for a specified area a round chimney will have the most capacity, but in ordinary building construction such a chimney is difficult to construct and is not ordinarily built. A square chimney of the same area has somewhat more friction, and one with a rectangular narrow flue very much more, so that an increase in area proportional to excess of perimeter should be made for such cases. The chimney should be as smooth as possible on the inside in order to prevent loss of velocity by friction, and, if of brick, the flue should in every case be plastered. In the construction of chimneys it is better that the inside be made with a thin wall not connected in any way with the outside, both in order to permit

SETTINGS AND APPLIANCES.   161

free expansion of the inner layer of the chimney with the heat and also to secure the advantage of the non-conducting power of an air space between the inside and outside walls. Such a construction is common for chimneys for power purposes, but is not ordinarily applied to those used in buildings.

95. Sizes of Chimneys.—The area of cross-section required for a given chimney will depend upon its height and also upon the amount of coal to be burned. The conditions which affect chimney draft are so numerous, and so difficult to consider in any theoretical discussion, that empirical or practical formulæ derived from the study of actually existing plants are probably more satisfactory than those obtained from purely theoretical computations. Of the various formulæ which have been given for the capacity of chimneys the writer prefers that of William Kent, from which the accompanying table is computed.

Kent's formula is computed on the assumption that the chimney shall have a diameter two inches greater than that required for passage of the air, in order to compensate for friction. The following is his formula:

$$E = \frac{0.3H}{\sqrt{h}} = A - 0.6\sqrt{A};$$

$$H = 3.33 E \sqrt{h};$$

$$S = 12 \sqrt{E} + 4;$$

$$h = \left(\frac{0.3H}{E}\right)^2;$$

in which A = actual area in square feet of the chimney, E = effective area, h = height in feet, S = side of the square in inches, H = horse-power of plant.

If we let R = number of square feet of radiating surface to be supplied, then, Article 73, page 173,

$$R = 0.1 H;$$

from which $E = \dfrac{.003R}{\sqrt{h}}$.

The table gives the diameter of round or side of square chimneys in inches for various heights computed from the above formulæ, with the diameter increased by 2, to allow for friction. A square chimney is considered the equivalent of the inscribed round one.

DIAMETER OR SIDE OF CHIMNEY IN INCHES REQUIRED FOR VARYING AMOUNTS OF DIRECT STEAM-RADIATING SURFACE.

Height of Chimney in Feet		20	30	40	50	60	80	100	120
Square Feet of Steam Radiation.	Horse-power.								
250	2.5	7.4	7.0	6.7	6.4	6.2	6.0	6.0	6.0
500	5.0	9.6	9.2	8.8	8.2	8.0	6.6	7.3	7.0
750	7.5	11.3	10.8	10.2	9.6	9.3	8.8	8.5	8.2
1,000	10.0	12.8	12.0	11.4	10.8	10.5	10.0	9.5	9.2
1,500	15.0	15.2	14.4	13.4	12.8	12.4	11.5	11.2	10.8
2,000	20.0	17.2	16.3	15.2	14.5	14.0	13.2	12.6	12.1
3,000	30.0	20.6	18.5	18.2	17.2	16.6	15.8	15.0	14.4
4,000	40.0	23.6	22.2	20.8	19.6	19.0	17.8	17.0	16.3
5,000	50.0	26.0	24.6	23.0	21.6	21.0	19.4	18.6	18.0
6,000	60.0	28.4	26.8	25.0	23.4	22.8	21.2	20.2	19.5
7,000	70.0	30.4	28.8	27.0	25.5	24.4	23.0	21.6	20.8
8,000	80.0	32.4	30.6	28.6	26.8	26.0	24.2	23.4	22.2
9,000	90.0	34.0	32.4	30.4	28.4	27.4	25.6	24.4	23.4
10,000	100.0	37.0	34.0	32.0	30.0	28.6	27.0	25.4	24.6
15,000	150.0	....	....	38.4	36.2	35.0	33.0	31.0	29.2
20,000	200.0	....	....	43.0	42.0	41.0	37.0	35.0	34.0
30,000	300.0	....	....	....	50.0	48.0	46.0	43.0	41.0

For other kinds of heating multiply the radiating surface by the following factors: Hot-water heating 1.5, indirect steam 0.7, hot-blast heating 0.2.

96. Chimney-tops.—The draft of a chimney is influenced to a great extent by the conditions of the surrounding space. If other buildings exist in the vicinity of such a form as to deflect the currents of air down the chimney, the draft will be impaired and may be entirely destroyed. The objects which tend to produce downward air-currents may sometimes be situated a considerable distance from the chimney and thus render the specific cause of poor draft very difficult to determine. The remedy for a smoky chimney is sometimes difficult to apply, but usually the draft will be improved, first, by increasing the height of the chimney; second, by adopting some form of chimney-top which utilizes the force of horizontal currents to aid by induction in increasing the draft.

The writer found that curved trumpet-shaped tubes located with the small ends projecting into the chimney in an upward direction increased the draft materially when the wind was blowing into the openings, and there is little reason to doubt but that a chimney-top may be constructed in such a manner as to materially increase the draft.

97. Grates.—For supporting the fuel during its combustion in such a manner as to allow a free passage of air, a perforated metallic construction of some sort is required. For burning very fine coal the perforation must be small and close together; for burning larger sized coal the perforations may be larger and further apart. The area of the air-spaces compared with the total area of the grate should be about 50 per cent in order to secure best results, but they will more generally be found to be 30 to 40 per cent. The grates are usually constructed of cast iron and in a very great variety of forms, as shown in Figs. 157 and 158. In some instances a series of parallel bars is used; in others the grates are made in one solid

FIG. 157.   DIFFERENT FORMS OF GRATES.   FIG. 158.

casting. This latter practice is never one to be recommended. The solid grate is likely to break from expansion strains due to heating unless made in such form that the various parts are free to expand independently.

Nearly all heating-boilers, hot-water heaters, and furnaces are supplied with some form of shaking- and dumping-grate. Many of these grates are known from experience of the writer to give most excellent satisfaction, and doubtless all present points of merit. The various shaking-grates operate in nearly every way, and it is hard to conceive either a form of grate-bar or a method of shaking which is not exemplified in some of these grates. Some of the bars are flat or rectangular in shape, and are operated by shaking backward and forward; others are triangular and are continually rotated so as to present successively new surfaces to the fire each time they are shaken. The shaking-grate will, in general, be found much superior to the fixed one, and a furnace fitted with such grates is more easily managed and more cleanly than one with a fixed grate of any description.

98. Traps.—In all systems of gravity steam-heating, the water of condensation returns directly to the boiler, and no appliance either for maintaining a water-line in the building or returning the condensed steam to the boiler is required. But there are cases in which it is necessary to maintain the water-line at a certain definite height, and also to prevent the escape of steam without interfering with the discharge of condensing water. For this purpose a steam-trap is required. One form of a steam-trap which has always been used to a greater or less extent for this purpose is a siphon made in the shape of a U bend, or its equivalent of pipe and fittings, as shown in Fig. 159. It consists of two legs, AB and BC, which may be close together or any distance apart, but the length of which must be sufficiently great to prevent pressure acting through the pipe FA forcing the water out of BC. CE is a vent-pipe extending to the air; D is the discharge for the condensed water. In ordinary operation the leg CB is filled with water which is constantly overflowing, and AB with steam and water; the total pressure in both legs being in each case equal.

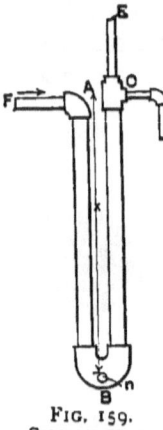

FIG. 159.
SIPHON-TRAP.

The siphon-trap may be open to the objection that it will require a great deal of vertical room if the pressure is great;

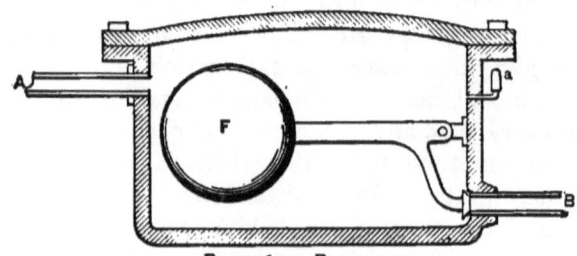

FIG. 160.—FLOAT-TRAP.

for this reason traps with mechanical movements of some kind are usually preferred. The simplest of these traps contains a float (Fig. 160) which rises and falls with change of level of the water in the vessel. Rising above a certain point, it opens a discharge-valve; falling below, it closes it. Traps of this class are made of a great many designs. In some instances traps are made as in Fig. 161, in which a weight W is used instead of

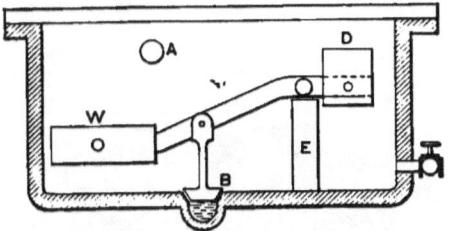

FIG. 161.—COUNTER-WEIGHTED TRAP.

a float and is nearly counter-balanced by the weight D. As the water rises in the trap it tends to lift the weight W an amount proportional to its volume, thus opening a discharge-valve at B. When the water falls, the valve is closed. It is noted that the counter-weight D is always above the water-line P.

A large number of traps are made with a hollow metallic float or bucket, so arranged as to open a valve when the bucket is full of water. One form is shown in Fig. 162, in which the water enters the trap at A, filling the

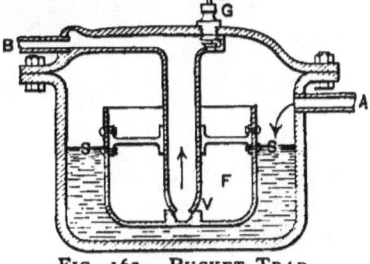

FIG. 162.—BUCKET TRAP.

space S between the bucket and the walls of the trap. This causes the bucket to float, and thus to close an orifice in the discharge-pipe V. When the water rises above the edges of the bucket it flows into it and causes it to sink, which opens the discharge-valve at V. The water is forced out through the pipe B by the steam pressure acting on the surface SS.

The bucket traps are made in great variety, both as to form of valve, guides for bucket, etc. Fig. 163 shows one of the traps which is in common use, with all details of construction.

Another extensive class of traps are made so as to be closed by the expansion due to increase in temperature. These traps differ from each other very much in form; the principle, however, is in all cases the same. Thus in the diagram, Fig. 164, steam is supplied at *A* and discharged at *B*. The bent springs *S* are prevented by guides from moving laterally, so that the expansion due to heat causes a motion which closes the orifice in the discharge-pipe *B*. When the water in the traps cools

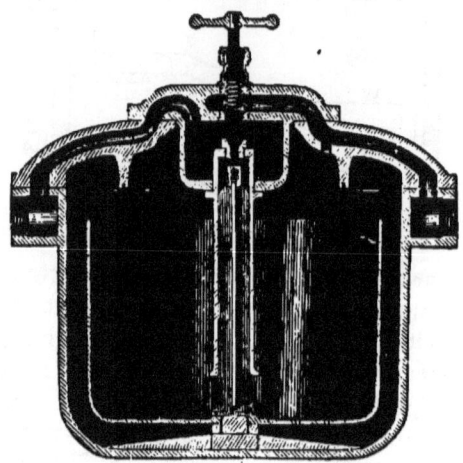

FIG. 163.—BUCKET TRAP.

the valve opens. The materials used for traps of this class can be metallic or some composition of material like that employed for air-valves. The discharge can be arranged to take place from the bottom or, as shown in the diagram, from the side.

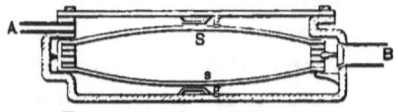

FIG. 164.—EXPANSION-TRAP.

Traps which combine one or more of the principles of operation as described are on the market. Thus Fig. 165 represents a trap with two valves in which one valve is opened by expansion, the other by a float.

The bucket traps have generally proved the most reliable and less likely to be injured by use. The float-traps have been liable to failure because of leakage of the float, but recent improvements in manufacture render this accident quite improbable. All traps need periodical inspection, as the valves are likely to become more or less choked up, in which case the trap may fail to operate. All of the traps described

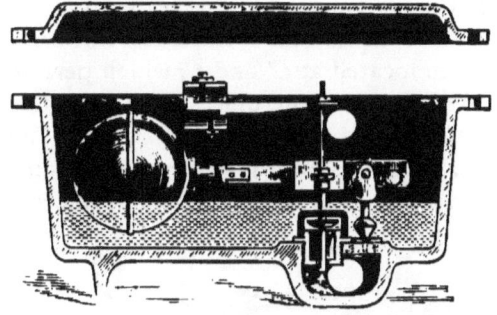

FIG. 165.—COMBINED FLOAT- AND EXPANSION-TRAP.

will discharge the water to a height which corresponds to the steam-pressure in use, and hence when used with high-pressure steam will lift water to a considerable distance; but in no case will they return the water into the boiler from which the steam was received. For this purpose a trap of considerable more complexity, known as a return-steam trap, must be used.

99. Return-traps.—Traps which receive the water of condensation and return it to a boiler having considerably higher-pressure steam than that acting on the returns, are known as

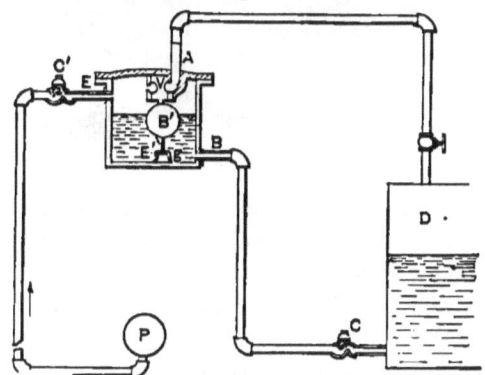

FIG. 166.—DIAGRAM SHOWING ACTION OF RETURN-TRAP.

return-traps. They are made in quite a variety of forms, but the general principle of operation is shown by the diagram Fig. 166. In this figure D represents the boiler and AB the trap, which is located above the boiler and is supplied with steam from the boiler at *A*. It is connected with the return system by a pipe leading from the tank or drum *P*, and pipe discharging into the trap at *E*. A pipe leads from the bottom of the trap *B* and connects below the water-line with the boiler. Check-valves are located at *C'* and *C*, which permit the flow to take place toward the boiler only. The essential method of operation of the trap is as follows: First, water flows into the trap from the return *P*, until it reaches a certain level, when it acts on the float *B* so as to open a balanced steam-valve,

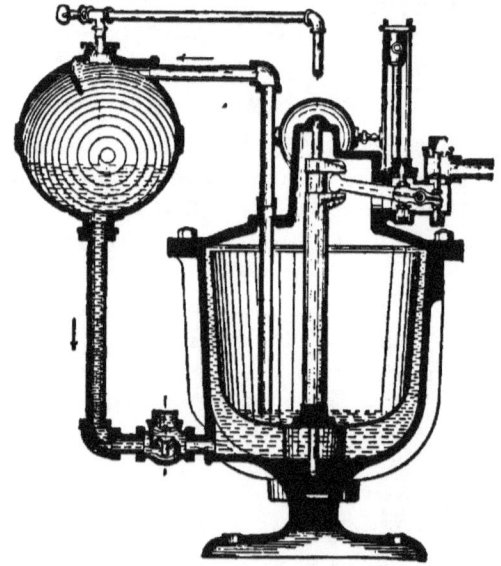

FIG. 167.—BUCKET RETURN-TRAP.

called an *equalizing-vavle*, connected to the main pipe *A*. This permits steam from the boiler to enter the trap, which equalizes the pressure of steam in the trap and boiler. The water in the trap, because of its greater density, then commences to flow out through the pipe *B*, and need only cease when the level becomes nearly the same as in the boiler. The discharge of the water causes the float *B* to fall, which closes the equalizing valve, and the operation as described is again repeated.

Instead of a float a bucket may be used to operate the

equalizing-valve, acting in a manner similar to that described for the ordinary bucket trap. A section of such a trap is shown in Fig. 167.

The bucket is probably superior to the float for this purpose, since it is less likely to be affected in its operation by change in density or pressure of the steam.

Various other systems for opening and closing the equalizing-valve have been adopted, of which one, shown in Fig. 168,

FIG. 168.—GRAVITATING RETURN-TRAP.

consists in mounting the trap so that it will move into one position when empty and into another when full, the motion so obtained being used to open and close the equalizing-valve.

A different construction for accomplishing the same purpose is shown in Fig. 169.

FIG. 169 —RETURN-TRAP.

100. General Directions for the Care of Steam-heating Boilers.—Special directions will be no doubt supplied by the

maker for each kind of boiler, or for those which are to be managed in a peculiar way. The following directions are general and should always be observed, regardless of the kind of boiler employed:

1. Before starting the fire see that the boiler contains water. Its surface should stand a distance of from one third to one half the height of the gauge-glass.

2. See that the smoke-pipe and chimney-flue are clean and that the draft is good.

3. Build the fire in the usual way, using a quality of coal which is adapted to the heater.

4. In operating the fire keep the fire-pot full of coal and shake down and remove all ashes and cinders as often as the state of the fire requires it. If a magazine heater is used it must be kept full of coal.

5. Hot ashes or cinders must not be allowed to remain in the ash-pit under the grate-bars, but must be removed at stated intervals to prevent burning out of the grate.

6. To control the fire, see that the damper regulator is properly attached to draft-doors and damper; then regulate the draft by weighting automatic draft-lever as required, lightly or not at all in mild weather, but increasing as the weather becoming colder.

7. Should the water in the boiler escape, by means of a broken gauge-glass or other mishap, it will be safer to dump the fire and let the boiler cool before letting in cold water.

In no case should an empty boiler be filled when hot. If the water gets low, but not out of sight, in the gauge-glass, extra water may be added at any time by the means provided for this purpose.

8. Occasionally lift the safety-valve from its seat to see that it is in good condition.

9. Clean the boiler, if used in a gravity system of circulation, once each year by filling with pure water and emptying through the blow-off pipe. If the steam is used largely for power, the boiler must be cleaned at frequent intervals. In case the boiler should become foul or dirty it can be thoroughly cleaned by adding a few pounds of caustic soda and allowing it to stand one day, then emptying and thoroughly rinsing. Kerosene oil will loosen boiler scale and not injure the boiler, but

its odor will be quite likely to penetrate the whole building in which the heating system is located.

10. During the summer months the writer would recommend that all the water be drawn off from the system and that air-valves and safety-valves be opened, to permit the heater to dry out and remain so. Good results are, however, obtained by filling the heater full of water, driving off the air by boiling slowly, and allowing it to remain in this condition until needed in the fall. The water should then all be drawn off and fresh water added.

11. Keep the fire surfaces of the boiler clean and free from soot. For this purpose a brush is provided with most heaters.

12. In case any of the rooms are not heated, look out for the steam-valves at the radiators. If a two-pipe system, both valves at each radiator must be opened or closed at the same time, as required. See that the air-valves are in proper condition. If a one-pipe system, one valve only has to be opened or closed.

13. If the building is left unoccupied in cold weather, draw all the water out of the system, which can only be done by opening blow-off pipe, all radiators, and air-valves.

101. Care of Hot-water Heaters.—The general directions for the care of steam-heating boilers, Article 100, apply in a general way to hot-water heaters as to the methods of caring for the fires and for cleaning and filling the heater. The special points of difference only need to be considered. All the pipes and radiators must be full of water and the expansion-tank should contain some water, as shown by the gauge-glass or by the pressure-gauge; and this condition should be determined before building a fire and whenever visiting the heater for the purpose of replenishing the fuel. Should any of the radiators not circulate, see that the radiator valve is open then open air-valve until the water runs out, after which it must be closed tight. Water must always be added at the expansion-tank when for any reason it is drawn from the system.

102. Boiler Explosions. — Boiler explosions sometimes occur with disastrous results. They are not limited to boilers in which high-pressure steam is employed, but also occur in some instances with low-pressure boilers employed in heating.

The cause of a steam-boiler explosion is in every case an excess of pressure above that of the strength of the boiler. The effect of this is primarily to rupture a part or portion of the boiler, relieving the pressure on the side of the rupture. This leaves unbalanced all the pressure acting on the opposite side of the boiler, which usually is sufficient to project the boiler into the air with considerable velocity. As showing the amount of force which exists even with small pressures we would have for each square foot of the boiler with 10 pounds pressure above the atmosphere a force of 1440 pounds per square foot of surface, applied to move it as a projectile. If the pressure were ten times as great the force would be ten times greater, and the effect many times worse. The disaster caused by the explosion would depend largely upon the suddenness with which this force was applied; if it were applied gradually no bad results might follow; if applied instantly the results might equal the explosion of a large amount of dynamite. Boilers sometimes explode because of defective material, poor construction, or overheating of parts; they also sometimes explode because of defects in the safety-valve or in the appliances for showing the true level of the water; but in all cases the immediate cause of the explosion is over-pressure. The causes which lead to the formation of steam with a pressure in excess of that of the strength of the boiler are various; one of them is the practice of permitting the water in the boiler to get low and then supplying feed-water, which because of the highly heated condition of the surfaces is rapidly converted into steam, causing the pressure to become excessively high.

It is not necessary to suppose that boiler explosions are caused by any mysterious force which is suddenly developed in the boiler. On the other hand, the amount of force which is stored in the hot water and steam is sufficient to produce at any time a terrific explosion, provided the necessary opportunity is presented. Dr. R. H. Thurston has computed the energy stored in various classes of boilers under the ordinary conditions of working, and the following table shows some of the principal results of that calculation and will give some idea of the enormous force stored in heated water and steam:

STORED ENERGY OF STEAM-BOILERS.*

Type.	Pressure, lbs. per sq. in.	Rated Power, H. P.	Total Stored Energy Available.	Energy per lb. of Boiler. Foot-lbs.	Maximum Ht. of Proj't'n of Boiler. Feet	Initial Velocity. Total
1. Plain cylinder...	100	10	47,281,898	18,913	18,913	606
2. Cornish cylinder.	30	60	58,260,060	3,431	3,431	290
3. Two-flue cylind'r	150	35	82,949,407	12,243	12,243	625
4. Plain tubular....	75	60	51,031,521	5,372	5,372	430
5. Locomotive.....	125	525	54,044,971	2,786	2,786	375
6. "	125	650	71,284,592	2,851	2,851	379
7. "	125	600	66,218,717	3,219	3,219	397
8. "	125	425	65,555,591	4,677	4,677	455
9. Scotch marine...	75	300	72,734,800	2,687	2,687	348
10. "	75	350	109,724,732	2,889	2,889	356
11. Flue and return..	30	200	92,101,987	1,644	1,644	245
12. " " "	30	180	104,272,264	1,862	1,862	253
13. Water tube	100	250	174,568,380	5,067	5,067	445
14. "	100	250	230,879,830	5,130	5,130	450
15. "	100	250	109,624,283	2,030	2,030	323

* "Steam-boiler Explosions, in Theory and Practice," by R. H. Thurston.

Considering the total number of heating-boilers in use in the United States the number of explosions is very small, so that if we suppose no improvement in construction over the ordinary methods, the risk which any person would run is very slight; and it seems quite probable that if one were to use a heating-boiler as safe as the average boiler, the chances would be that if he did not die until killed from this cause he would live to be 10,000 years old. That is, estimating from the total number of boilers in use for heating, as compared with the number of explosions of such boilers, the chances are that one per year in ten thousand would explode.

Some disastrous explosions of heating-boilers have, however, occurred in the United States, of which may be mentioned that at the Central Park Hotel, Hartford, Feb. 17, 1889, in which fifteen people were killed and the hotel entirely destroyed; also the boiler explosion at St. Mary's Church, Fort Wayne, Ind., in which the church and priest's house were nearly torn down, which occurred Jan. 13, 1886; another at Dell Brown's Hotel, Eagle Bridge, N. Y., Dec. 20, 1888, in which several people were injured and the building badly wrecked. Also various other explosions doing less damage.

It would seem, from a study of the boilers which are injured by explosions, that no boiler is entirely free from the disastrous effects of an explosion when it is badly managed; but on the other hand it also appears that the sectional boilers, or boilers in which the water occurs in small quantities, are subject to injuries which are comparatively slight and generally easily repaired. So far as the writer can find from a study of all the explosions recorded in the United States, the water-tube boilers, or those with small masses of water, are singularly exempt from disastrous explosion. They are, however, quite likely to have some part broken away, in which case the pressure on the boiler is relieved quickly enough to avert a serious explosion. The worst accidents which usually happen to the sectional boilers are those due to the burning out of a tube or some easily replaceable part. This results ordinarily in a very severe leak, which can, however, be repaired.

The total number of boiler explosions for the United States for all classes of boilers average about 255 per year, and, as reported by the *Locomotive*, they have been as follows for the last ten years:

BOILER EXPLOSIONS IN THE UNITED STATES.

Year.	Total No. Explosions.	Stationary, etc.	Portable.	Saw-mills.	Railway Locomotives.	Steamboats.	Total Killed.	Total Injured.
1884	152	48	18	56	15	15	254	261
1885	155	80	16	33	10	16	220	288
1886	185	88	16	45	22	14	254	314
1887	198	67	20	73	14	14	264	388
1888	246	104	30	69	23	20	331	505
1889	180	85	21	56	15	13	304	433
1890	226	94	16	75	25	16	244	351
1891	257	115	35	68	22	17	263	371
1892	269	122	24	79	33	11	298	442
1893			245				220	151
1894								

The following table gives the total number in Great Britain for the same time:

BOILER EXPLOSIONS IN GREAT BRITAIN.

Years.	Explosions.	Killed.	Years.	Explosions.	Killed.
1882–83	45	35	1889–90	77	21
1883–84	41	18	1890–91	72	32
1884–85	43	40	1891–92	88	23
1885–86	57	33	1892–93	72	20
1886–87	37	24			
1887–88	61	31	Total......	660....	313
1888–89	67	33	Ratio......	....	.482

This table would seem to indicate that the explosions in this country were more disastrous, so far as taking life is concerned, as in this country two people were killed for about every three explosions, whereas in Germany and Great Britain we have about twice as many explosions as deaths. This is probably due to the fact that the statistics in this country classify as boiler explosions only those which are markedly disastrous, whereas in France and Germany every leak or break which appears from this cause is recorded as an explosion.

As showing the disastrous effects often produced by a boiler explosion, the following is abstracted from Thurston's Manual of Steam-boilers. Fig. 170 shows the boiler-room before the explosion. The

FIG. 170.
THE BOILER BEFORE EXPLOSION.

boiler was made of $\frac{5}{16}$ iron, was 3 feet in diameter, and was 7 feet high; the upper tube-head was flush with the

FIG. 171.—PATH TAKEN BY THE BOILER.

top of the shell, the lower forming the crown of the furnace, which was about 2 feet above the grates and the base of the shell, and was flanged upon the inner surface of the furnace. There was a safety-plug in the lower tube-head which was not melted out. The working pressure was 60 pounds per square inch, and the explosion probably took place at or a little below this pressure, throwing the boiler through the roof and high over a group of buildings and a tall tree close by, finally burying itself half its diameter in the frozen ground. There had been a leak in the lower head which had reduced by erosion the thickness of the tubes and the lower head, so that the pressure was sufficient to force the lower head down away from the tubes, opening fifty or more holes 2 inches in diameter from which the fluid contents of the boiler issued at a high velocity, relieving the pressure below and converting the whole boiler into a great rocket weighing about 2000 pounds.

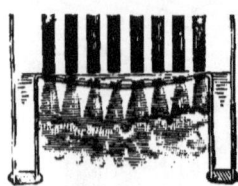

FIG. 172.
SHOWING BEGINNING OF PROCESS OF RUPTURING.

103. Explosions of Hot-water Heaters.—While hot-water heaters provided with an open expansion-tank are to a great extent free from the dangers of explosions, still it is quite possible that extreme carelessness in erection, the freezing up of connections to expansion-tank, or other mishaps, might render the apparatus fully as dangerous as the steam-boiler under its most unfavorable conditions. Some very disastrous explosions have occurred of hot-water heating plants when operated under the Perkins or high-pressure system, and it seems quite probable that such a system, even under the most favorable conditions, is more dangerous than the steam-heating system. The hot-water heating system should be constructed so that the connection between the expansion-tank and heater cannot by any possible means be closed. The placing of a valve in this connection was the cause of a very disastrous explosion in a residence in New York City quite recently, and emphasizes the necessity for caution in this respect.

104. Prevention of Boiler Explosions.—Boiler explosions are probably preventable in every single case by using, first, boilers properly designed, and constructed of excellent material and with good workmanship; and second, by seeing that all appliances, as safety-valves, blow-off cocks, feeding apparatus, etc., are in excellent order; and third, by providing skilled and intelligent attendance.

Disastrous results are usually almost entirely prevented by the use of sectional boilers, and for heating purposes there are at the present time comparatively few of any other kind in use.

As a rule heating-boilers, especially those of small sizes, are not under close supervision, but are attended to and visited only at comparatively long intervals. For this reason automatic appliances for feeding the boiler and for regulating the pressure, opening and closing the dampers, are usually supplied; hence the person erecting the plant should exercise the utmost care to see that such appliances are in excellent order and of such character as are likely to prove durable and reliable. While it is quite certain from our statistics that not one boiler out of ten thousand is likely to explode per year, yet nevertheless the contractor should always bear in mind that a steam-boiler is in every case a magazine of stored energy, and if badly constructed, poorly erected, or carelessly managed may do an immense amount of damage.

CHAPTER IX.

VARIOUS SYSTEMS OF PIPING.

105. Systems employed in Steam-heating.—There are two systems of heating, in the first of which, known as the *Gravity Circulating System*, the water of condensation from the various radiators flows by its own weight into the boiler at a point below the water line; in the second the water of condensation does not flow directly into the boiler, but is returned by some special machinery or, in some cases, wasted. The second system is often called the *High-pressure System*, because steam of any pressure can be produced in the boiler, a portion of which may be employed in operating engines, elevators, etc. It is very seldom, however, that this high-pressure steam is used in radiators, low-pressure steam being obtained directly from the boiler by throttling or passing through a reducing-valve, or, in some instances, indirectly by using the exhaust-steam from engines or pumps.

In this chapter we shall discuss only the systems of piping used with gravity circulating systems of heating, reserving for a later chapter a description of the methods employed in the other system of heating, although there is in the arrangement of pipe lines very little which pertains to either system exclusively.

106. Definitions of Terms used.—Certain terms have been adopted which are always used to describe definite parts in a system of piping, as follows:

The *main or distributing pipe* is the pipe leaving the boiler or heater and conveying the heated products to the radiating surfaces. In steam-heating this is termed the *main steam-pipe*, and in hot-water heating the *main flow-pipe*. It may be carried from the boiler without branches to the top of the building (Fig. 173), where the distributing-pipes are taken off, or it may run in a horizontal or vertical direction from the heater, and branch pipes taken off as required. The pipes in which the flow takes place from the radiating surface toward the boiler are called return-pipes. The pipes which extend in a vertical direction are termed *risers;* when the flow in these pipes is downward they are called *return-risers.*

A *relief* or *drip* is a small pipe run from a steam-main, so as to convey any water of condensation to the return; it must be employed at all points where water is likely to gather. For illustration of use see Fig. 176.

Pitch is the inclination given to any pipe when running in nearly a horizontal direction. In general the inclination or pitch of a supply-pipe should, in steam-heating, be downward from the boiler, and arranged so that the water of condensation will move in the same direction as the current of steam. In hot-water heating the pitch should be upward from the boiler. In all return-pipes the inclination should be downward, toward the heater or boiler.

A *relay* is a term sometimes used to describe a sudden change of alignment, or "jumping up," of a horizontal pipe. This is often necessary in a long line of piping to keep the pipe near the ceiling and preserve the necessary pitch. At such points a drip or relief must permit water of condensation to flow into the return.

Water-line is a term used to denote the height at which the water will stand in the return-pipes. It is usually very nearly the same as the level of the water in the boiler, being higher only in case there is considerable reduction in pressure due to friction. In heating with high-pressure steam it is desirable to have all the relief-pipes discharge into a return filled with water, so that circulation of steam shall be continuously in one direction; this is of less importance with low-pressure steam, provided the water which gathers in returns can move freely and quickly to the boiler.

The term *siphon* is applied to a bend below the horizontal; it is sometimes used in the main return to hold water at a different level from that in the boiler. This is done by admitting steam to the top part of the bend on the boiler side by a relief from the main steam-pipe. It is similar to the siphon-trap, Fig. 159, Article 98. If the relief were not connected to the top of the bend the water would pass over by suction into the boiler.

Steam-traps are vessels designed with valves which open automatically so as to preserve the water-level in the returns at any desired point. Various kinds are described in Chap. VIII, Article 98.

Water-hammer is a term applied to a very severe concussion which often occurs in steam-heating pipes. It is caused by water accumulating to such an extent as to condense some of the steam in the pipe, thus forming a vacuum which is filled by a very violent rush of steam and water. The water strikes the side of the radiators or pipes with great force, and often so as to produce considerable damage. In general a water-hammer may be prevented by arranging the piping in such a manner that the water of condensation will immediately drain out of the radiator or pipes.

A bend in the return of a steam- or water-heating system, when convex upward, will frequently accumulate air to such an extent as to prevent circulation in the system. This is designated as an *air-trap*. When bends of this character must be used a small pipe for the escape of the air should be connected with the highest portion of the bend and led to some pipe which will freely discharge the entrapped air.

An air-valve is not ordinarily to be recommended for such situations.

107. Systems of Piping.—The systems of piping ordinarily employed provide for either a complete or a partial circulating system, each consisting of main and distributing pipes and returns. Several systems of piping are in common use, of which we may mention:

First, the complete-circuit system, often called the one-pipe system, in which the main pipe is led directly to the highest part of the building; from thence distributing-pipes are run to the various return-risers, which in turn connect with the radiating surface and discharge in the main return. The supply for the radiating surface is all taken from the return-risers, and in some cases the entire downward circulation passes through the radiating system.

This system was employed by Perkins in his method of high-pressure hot-water heating, and is mentioned by Péclet as

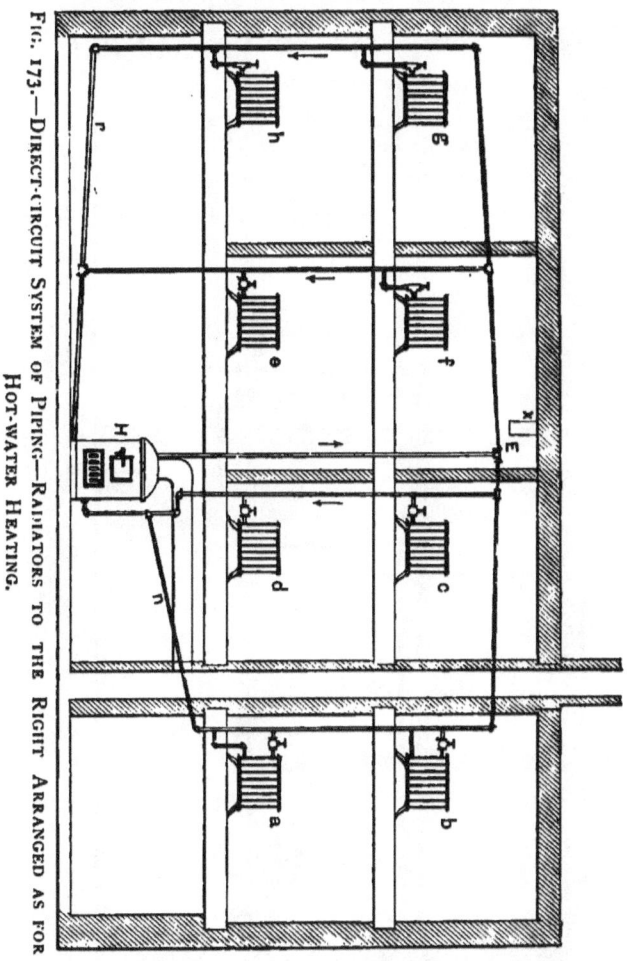

FIG. 173.—DIRECT-CIRCUIT SYSTEM OF PIPING—RADIATORS TO THE RIGHT ARRANGED AS FOR HOT-WATER HEATING.

in use in France in 1830. In this country it seems to have been introduced into use by J. H. Mills, and is often spoken of as the Mills system of piping. The system is equally well adapted for either steam or hot-water heating, and on the score of positiveness of circulation and ease of construction is no doubt to

182 HEATING AND VENTILATING BUILDINGS.

be commended as superior to all others. It is principally objectionable because the horizontal distribution-pipes have to be run in the top story of the building instead of the basement, which may or may not be of serious importance.

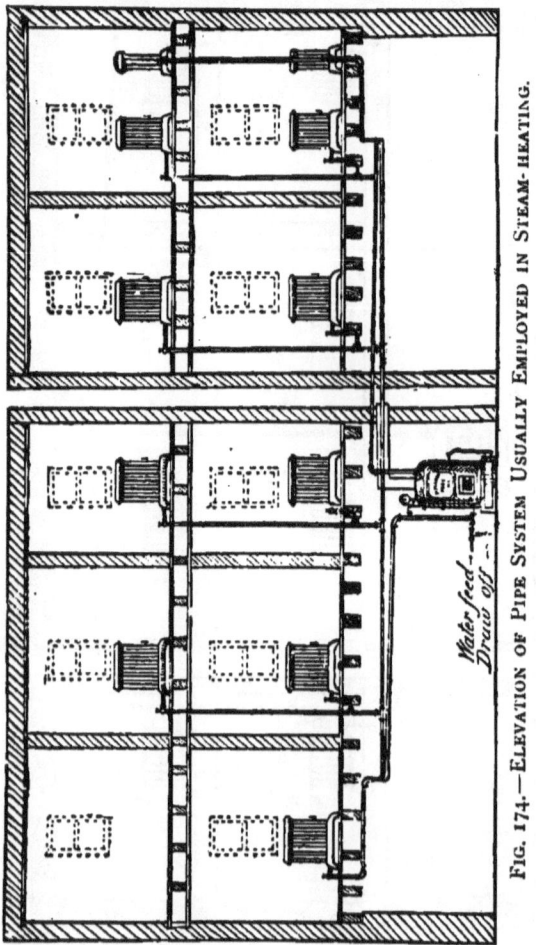

Fig. 174.—Elevation of Pipe System Usually Employed in Steam-heating.

Second, a partial-circuit system, in which the main flow-pipe rises to the highest part of the basement by one or more branches, from whence the distributing-pipes run at a slight incline, often nearly around the basement, and finally connect with the boiler below the water-line. The radiators are connected by risers which carry both flow and return from and to the distributing pipes, as shown in elevation in Fig. 174 and in plan in Fig. 175. This method of piping is employed extensively for steam-heating, and is perhaps less open to objection than any other.

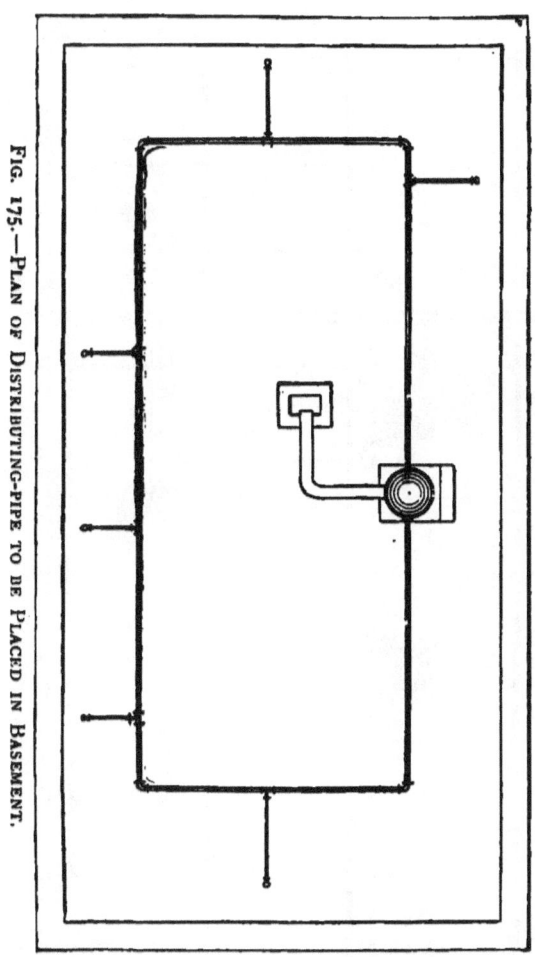

FIG. 175.—PLAN OF DISTRIBUTING-PIPE TO BE PLACED IN BASEMENT.

Third, a system of circulation in which each radiator is provided with separate flow- and return-pipes (Fig. 176). In this case the riser and distributing pipes are run as before, but are connected to the return by a *drip-pipe;* the return is located

184   *HEATING AND VENTILATING BUILDINGS.*

below the water-line of the boiler. The supply-riser from each radiator is taken from the main flow-pipe, and the return-riser is connected to the main return below the water-level. In case two connections are made to a radiator, one for supply and the other for the return, it is quite important that the connection

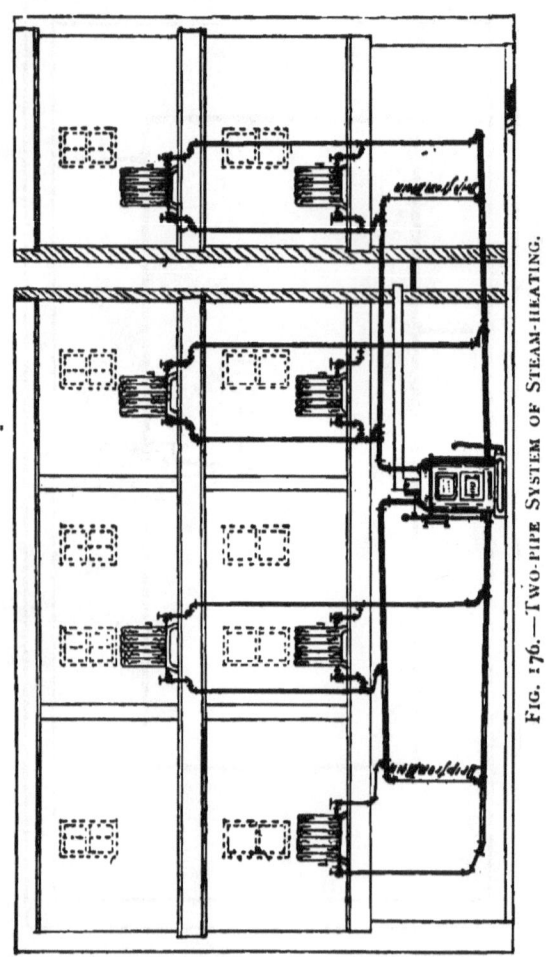

FIG. 176.—Two-pipe System of Steam-heating.

of the return-riser to the main return be made below the water-level of the boiler, in order to prevent steam flowing from two directions to the radiator. Such a condition is certain to cause water-hammer, as the radiator will retain water of condensation.

Various modifications of this third system have been used from time to time with greater or less success. For instance, each radiator has in some cases been connected to a separate flow and return riser, and in other cases simply to a separate return riser. These modifications are unimportant and hardly worthy of notice.

108. Methods of Piping Used in Hot-water Heating.— A system of hot-water heating should present a perfect system of circulation from the heater to the radiating surface and thence back to the heater through the returns; an expansion-tank being provided, as explained, to prevent excessive pressure due to the heating and the consequent expansion of the water. The direct-circuit system, as described for steam-heating, Fig. 173, is well adapted for hot-water heating, and has been used to a limited extent. When this system is employed for hot-water heating two connections are usually taken off from the return riser at different levels for each radiator, as shown in Fig. 103, page 114; although in some cases a single connection is made and a radiator of ordinary form employed, otherwise the method of piping is exactly similar to that described for steam-heating.

The system of piping ordinarily employed for hot-water heating is illustrated in Fig. 177. In this system the mains and distributing pipe have an inclination upward from the heater; the returns are parallel to the main and have an inclination downward toward the heater, connecting at its lowest part. The flow-pipes are taken from the top of the main and supply one or more radiators. The return-risers are connected with the return-pipe in a similar manner. In this system great care must be taken to produce nearly equal resistance to flow in all the branches leading to the different radiators. It will be found that invariably the principal current of heated water will take the path of least resistance, and that a small obstruction, any irregularity in piping, etc., is sufficient to make very great differences in the amount of heat received in different parts of the same system. For instance, two branch pipes connected at opposite ends of a tee, which itself is connected by a centre

186   HEATING AND VENTILATING BUILDINGS.

opening to a riser, are almost certain to have an irregular and uncertain circulation.

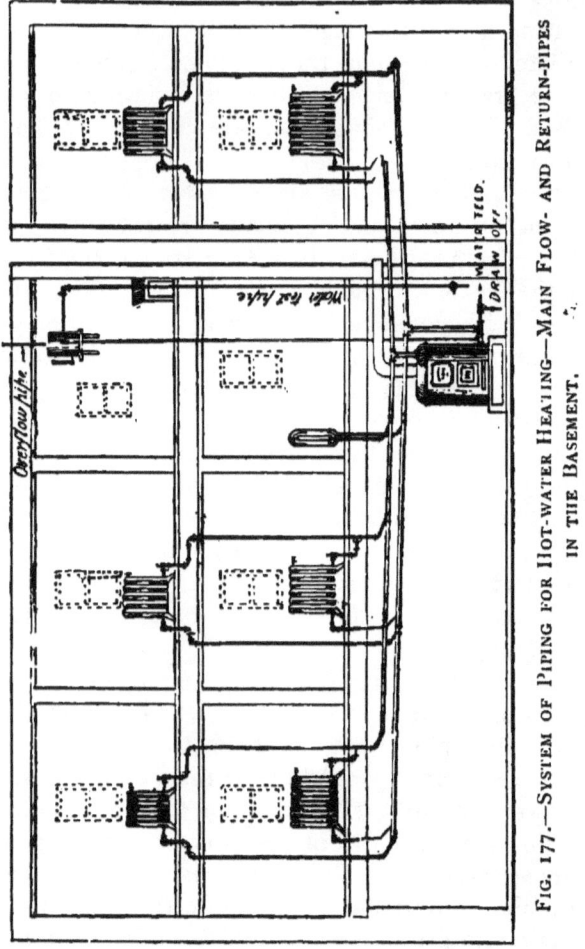

FIG. 177.—SYSTEM OF PIPING FOR HOT-WATER HEATING—MAIN FLOW- AND RETURN-PIPES IN THE BASEMENT.

The method of piping generally adopted for the closed or high-pressure system is that of the complete-circuit or one-pipe system, as illustrated in Fig. 173. .This system when now employed is used only for moderately low pressures, and a safety-valve is provided on the expansion-tank to prevent excessive pressure. In this system, or, in fact, in any of the systems for hot-water heating, the level of the return-pipe can be carried below that of the heater without bad results. The method of applying this system is shown in Fig. 178, which is

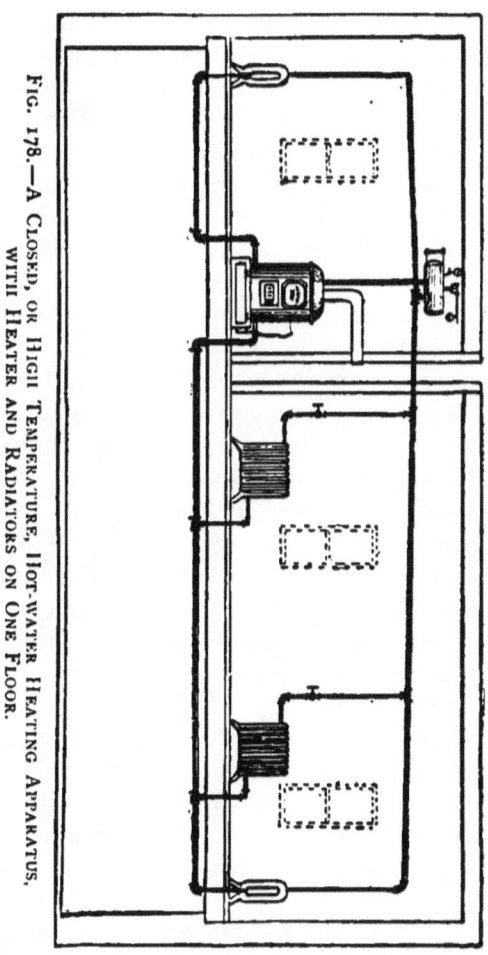

FIG. 178.—A CLOSED, OR HIGH TEMPERATURE, HOT-WATER HEATING APPARATUS, WITH HEATER AND RADIATORS ON ONE FLOOR.

similar in many respects to that used in the Baker system of car-heating.

The expansion-tank must in every case be connected to a line of piping which cannot by any possible means be shut off from the boiler. It does not seem to be a matter of importance whether it is connected with the main flow or with the return. The form of expansion-tank and the different kinds of fittings have been described in Art. 93, page 158.

Single-pipe systems for hot-water heating have been used to some extent. In this case there is a gradual flow of the heated water to the top, and the consequent settlement of the colder water to the bottom. The form of piping would be essentially the same as that shown in Fig. 173 or 174. The writer erected such a system at one time as an experiment, and found that it worked well after the water had once become heated. Where there is no objection to a system which heats slowly, this would probably do well on a small scale, but could not be recommended for an extensive job.

109. Combination Systems of Heating.—Several methods have been devised for using the same system of piping alternately for steam or hot water as the demand for higher or lower temperature might change. The object of this is to secure the advantages which pertain to the hot-water system of heating for moderate temperature and to steam-heating for extremely cold weather. As less radiating surface is required for steam-heating, there is the advantage due to reduction in first cost. This may be of considerable moment, since a heating system must be designed of such dimensions as to be satisfactory in the coldest weather, and this involves the expenditure of a considerable amount for surfaces which are needed only at rare intervals.

The combination system of hot-water and steam heating must require, first, a heater or boiler which will answer for either purpose; second, the construction of a system of piping which will permit the circulation of either steam or hot water; third, the use of radiators which are adapted to both kinds of heating.

These requirements will be met in the best manner by using a steam-boiler provided with all the fittings required for steam-heating, but so arranged that the damper regulator may be closed off from the heater by means of valves when the system is needed for use in hot-water heating. The addition of an expansion-tank is required, which must be arranged so that it can be closed off when the system is required for steam-heating.

Of the different systems of piping, that designated as the complete-circuit or one-pipe system (Fig. 173) is the only one which is equally well adapted for both hot water and steam. In case that system cannot be conveniently installed, the one shown in Fig. 177 for hot water will be found to give fairly good results, it being objectionable in steam-heating only because of the fact that the condensation in the main pipe flows against the current. The radiators and connecting pipes should be of the form required for hot-water heating, but the proportions and dimensions the same as for steam-heating.

While this system has many advantages in the way of cost over the complete hot-water system, yet the labor of changing from steam to hot water will in some cases be troublesome, and should the connections to the expansion-tank not be opened, serious results would certainly follow.

A combination hot-air furnace and hot-water system has been employed to considerable extent. In such a case the water-heating surface is obtained by inserting a coil of pipe or suitable vessel into the hot-air furnace, and certain rooms and portions of the house are warmed by the heated air directly from the furnace, while other parts are heated by the circulation of hot water.

This system is an admirable one from every point of consideration, theoretically; but practically it is a very difficult one to design and construct in such a manner that the supply of heat to the different rooms shall be positive and well distributed. Fig. 179 shows the arrangement of such a system.* In this case the hot-air furnace supplies heat to the lower floors and the hot-water circulating system to the upper floors.

Any system of piping suitable for hot-water heating can be employed for this purpose: the one shown is that of the complete-circuit or one-pipe system, the heated water being taken directly to the top of the building and all radiating surface supplied by the descending current. As the writer knows from experience, it is very difficult indeed to proportion the heating surface in the furnace and the radiating surface in the room so as to give in all cases satisfactory results without an

---

* An admirable series of articles were written on this subject by J. W. Hughes, and appeared in *Metal Worker*, February, 1895.

irregular and uncertain distribution of heat. It will generally be found that the fire maintained in a hot-air furnace is much more intense than that in a steam or hot-water heater; and further, the heating surface which is usually employed is subjected to the full heat of the fire, consequently a smaller amount of heating in proportion to radiating surface must be

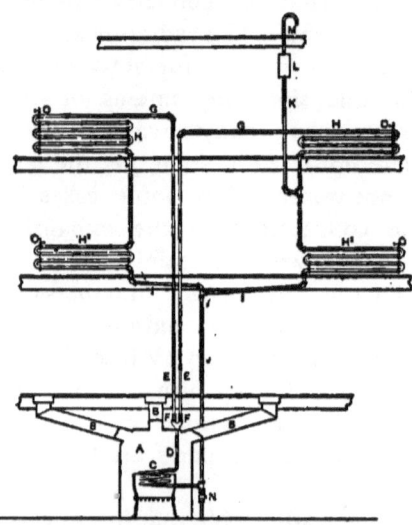

FIG. 179.—COMBINATION SYSTEM, HOT-AIR FURNACE AND HOT WATER.

employed. Whereas in the ordinary hot-water heater one foot of heating surface supplies from 8 to 10 of radiating surface, in this system 1 foot of heating surface will supply 25 to 35 feet of radiating surface in coal-burning furnaces and 50 to 75 in wood-burning furnaces.

Similar combination systems of hot air and steam are also used, but in such cases the heater must be very much like a steam-boiler, and possess all its appliances and also storage capacity for steam. In the case of the hot-water and hot-air system the heater is substantially a hot-air furnace, to which is added a coil of pipe or vessel of suitable form, which serves as the heating surface for the hot water, so that the change in construction is very slight; but for steam-heating the change of construction must be more marked, and is likely to be more expensive and complicated.

VARIOUS SYSTEMS OF PIPING. 191

110. Pipe Connections, Steam-heating Systems.—The manner in which branches are taken off may have great effect on the results obtained in any heating system, since any increase in friction in any part of the system will cause the flow to be sluggish in that portion, and require more pressure to induce circulation. The size of pipes required in order that resistances may not exceed a certain amount are given in the next chapter; but it should be noted that bad workmanship may defeat the operation of a steam-heating plant having the best proportions possible, and that great care is needed, (1) to secure the alignment of every part, (2) the absence of airtraps or any obstructions whatever which would reduce the circulation or make it irregular or uncertain. Some details which are to be considered rather as suggestions than as formal directions are given.

In general, pipe connections should be made so as to afford as little resistance as possible to the flow of steam, and in such a manner as not to interfere with the expansion of the main pipes. The line of piping should present the freest possible channels of circulation for the steam as it leaves the boiler and for the water of condensation as it returns. The expansion, which is not essentially different from $1\tfrac{3}{8}$ inches for each 100 feet in length, can usually be well provided for by the use of two or more right-angled elbows substantially as shown in Fig. 180. No general rule can be laid down for all circumstances and conditions. The following examples and illustrations from *Heating and Ventilation* show the methods of piping commonly employed in setting steam-radiators with one-pipe connections. Fig. 180 illustrates the method where the radiator is set close to the main and no special drip is required.

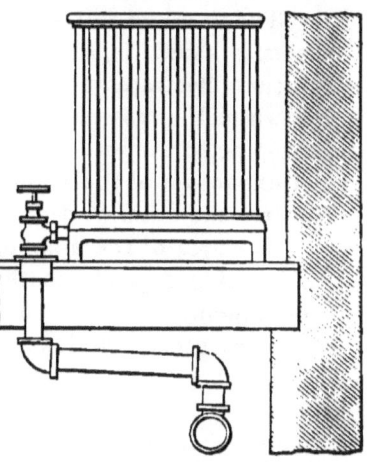

FIG. 180.—CONNECTION TO RADIATOR FROM STEAM MAIN.

The method often employed in connecting a rise horizontal steam main and running a special drip-pipe f densed water to the return main is shown in Fig. 181.

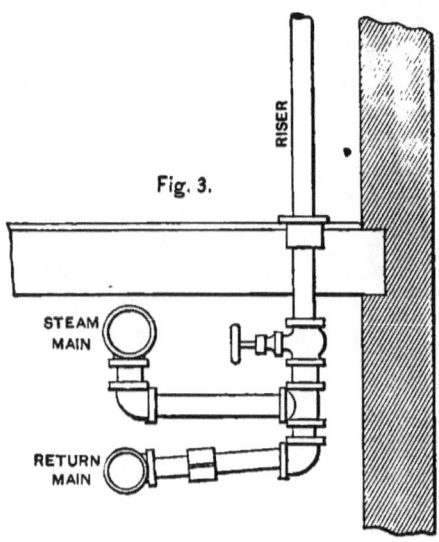

Fig. 181.—Connection to Riser from Main and Return.

The method often employed in connecting radia risers is shown in the upper portion of Fig. 182. The portion illustrates an essentially different method fro shown in Fig. 181 of connecting the riser to the main, a drip-pipe to the return. This method, however, does no for expansion of the steam main; hence this must be pr for in some other portion of its length.

The area of the main pipe must in every case be equ in carrying capacity to that of all the branches taken consequently may be reduced as the distance from the becomes greater and as more branches are supplied.
XVI., Appendix, gives the equivalent capacity of p different diameters, and can be used in determining th tive number of branches of a given size, and also the rec in pipe area which may be made after a certain num branches have been connected. It will, however, in be found, except when large pipes are used, less exper run the main full size than to use reducing fittings.

111. Pipe Connections, Hot-water Heating Systems.—

If the system of circulation adopted is the complete-circuit system, as in Fig. 173, in which the heating main is first taken directly to the top of the building and thence run horizontally

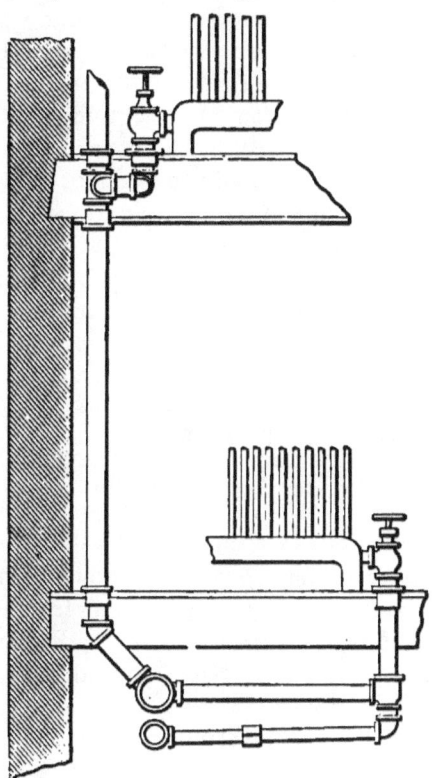

FIG. 182.—CONNECTION OF RADIATOR TO RISER.

to the various lines of return risers, the system of construction would be essentially the same as that described for a steam-heating plant. The main riser should connect into a drum, from the top of which the distributing-pipes leading to the return risers are taken. The size of the distributing-pipes should be proportional to the amount of radiating surface, and the various distributing-pipes should be arranged so that the resistance in each will be substantially equal. The flow connection for each radiator should be taken off at a point about level with the top of the radiator, as in Fig. 103,

page 114, and the return should enter the same
below the radiator. A valve affording as litt
possible is to be put in each connection. Ho
systems have been erected in which the radiato
the riser by one connection only; and while th
to be somewhat slower in heating than that w
tions, it is otherwise quite satisfactory.

In the system commonly employed the mai
ing pipes are erected in the basement, as sho
An offset from the main to the foot of the rise
be made, which should be done as from the ste
180, and in such a manner as to take the flow
part of the pipe; such a connection is also s
Fig. 183. The connection to the main return n

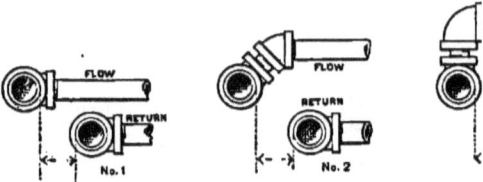

FIG. 183.—CONNECTIONS TO MAINS, HOT-WATER

the side or at the top, as convenient. In some
turned at an angle and a 45-degree elbow ca
good results, as shown at No. 2, Fig. 183.
connecting shown at No. 1 should only be en
the room is not sufficiently high for connectic
No. 3, as its use is attended with doubtful s
cases.

In taking off branches from the top of a ris
seldom or never be employed, since it will b
for any reason the current becomes established
it will be very difficult to induce it to flow
When branches running in opposite directions l
from the main riser, long-radius tees, as sho
page 95, should be employed; but unless the ris
in general be better to erect a separate line fo
Precautions should be taken in every case that
two currents shall not exert an opposing force
pede the circulation.

The connections to radiators for this system need to be made in such a way that the horizontal branches which are taken off from the risers will receive a strong current of water. There is a tendency for water to flow directly in the line of motion, and to the highest radiators in the system. This renders it necessary to increase the resistance in the riser beyond the branch a greater or less amount in order to induce circulation into the side connections. This may be done in several ways, as shown in Fig. 184: (1) by connecting

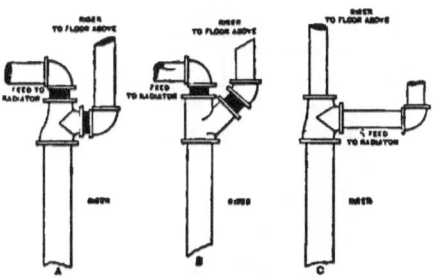

FIG. 184.—CONNECTION TO RADIATORS, HOT-WATER HEATING.

the radiator to an elbow placed on the main pipe and continuing the main pipe from the side opening of a tee or Y, as shown at *A* and *B*; or (2) by using a reducing fitting, as shown at *C*, and continuing the riser with a reduced diameter. The return connections can be made in a similar manner, but they will in every case work well if the return riser be run in a direct line and the connection be made into the side opening of a Y.

112. Position of Valves in Pipes.—If a valve has to be used on a horizontal pipe it should be located so as to afford the least possible obstruction to the flow of water in the required direction. If a globe valve be used with the stem set vertically, Fig. 185, it will form an obstruction sufficient to fill the pipe very nearly full of water; if the stem be placed in a horizontal direction the flow of water will be less impeded. Globe valves form a great obstruction to the flow in water-heating pipes, and under no circumstances should they be used for that work. In the case of steam-heating they are less objectionable, provided they are located in such a manner as to permit free drainage of the pipes. In general, angle or gate valves can be used, however, in every place with better satisfaction.

For hot-water heating special valves have been designed,

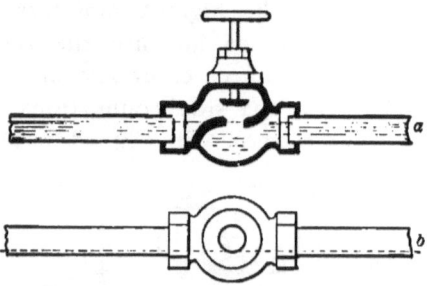

FIG. 185.—ILLUSTRATION OF WATER HELD BY GLOBE VALVE.

which when open offer no special impediment to the flow, and which close sufficiently tight to prevent circulation, although not sufficient to prevent leaks. See page 88.

113. Piping for Indirect Heaters.—Indirect radiators have been described and methods of setting them illustrated in Article 69, page 116. These radiators are generally set in a case or box which is suspended from the basement ceiling and made of matched boards lined with tin, Fig. 186. The sides of the casing should be removable for repair of the radiator. The system of pipes which supply the indirect radiators are generally most conveniently erected, like those shown in Fig. 175 or 177 for steam-heating, and like that shown in Fig. 179 for hot-water heating. The heater should be located above the water-line of the boiler a sufficient distance to afford ready means of draining off the water of condensation. In case this is impossible, a style of radiator should

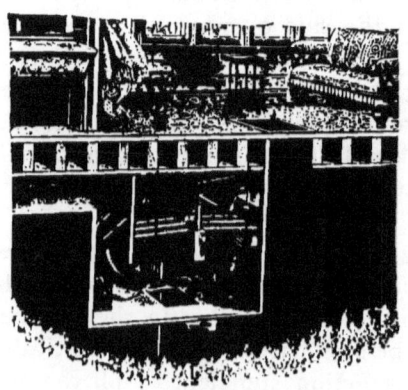

FIG. 186.—INDIRECT SURFACE.

be adopted which can be heated by water circulation. An automatic air-valve should be connected to the heater, and every means should be taken to obtain perfect circulation to and from the boiler. The chamber which surrounds the indirect surface is to be supplied with air from the outside by a properly constructed flue. The air passes up through or over the heater and into the rooms by means of special flues, the sizes of which are given in Chapter X.

114. Comparisons of Pipe Systems.—As to the best system of piping to be adopted little can be said in a general way. The circuit-system, Fig. 173, no doubt gives the freest circulation and is applicable to either hot-water or steam heating. In some respects it is simpler to construct, and it seems quite probable that small errors of alignment, minute obstructions, and error in proportioning the pipes would not be so fatal to the perfect operation of this system as of the others. It requires, however, that distributing pipes be placed in the top story of a building, and this in many cases will be so objectionable that it cannot be used. Regarding other systems there is little to be said. For steam-heating there seems to be little or no use in making more than one connection to any radiator; and this practice, which is now common, will I think become universal.

115. Systems of Piping where Steam does not Return to the Boiler.—For such systems the method of piping and of making connections would be in every case essentially as described; and usually this can be done with less care because of the fact of greater difference of pressure between the supply and the return. Such systems are not often employed except in connection with use of exhaust steam, which is considered in Chapter XI.

116. Protection of Main Pipe from Loss of Heat.—The loss of heat which takes place from an uncovered main steam or hot-water pipe is, because of its isolated position, considerably greater than that which takes place from an equal amount of radiating surface. Unless this heat is actually required it will cause an expenditure of fuel the cost of which is likely to be in a few seasons many times that of a good covering.

The heat lost per square foot of surface from a small uncovered pipe is from 375 to 400 heat-units per square foot per hour in steam-heating, or an amount equal to that required for the evaporation of 0.4 pound of steam. Computing this loss for 100 square feet for a day of 20 hours and for a season of 150 days, it will be found equivalent to the coal required to evaporate 120,000 pounds of steam; this would not be less than 12,000 pounds of coal, which at $5.00 per ton would cost $30.00. The cost per square foot per annum will be found on the above basis to be 30 cents, of which 75 to 80 per cent would have been saved by using the best covering. The loss from hot-water pipes would be about two thirds of the above.

The best insulating substance known is air confined in minute particles or cells, so that heat cannot be removed by convection. No covering can equal or surpass that of perfectly still and stagnant air; and the value of most insulating substances depends upon the power of holding minute quantities in such a manner that circulation cannot take place. The best known insulating substance is a covering of hair felt, wool, or eiderdown, each of which, however, is open to the objection that, if kept a long time in a confined atmosphere and at a temperature of 150 degrees or above, it becomes brittle and partly loses its insulating power.

A covering made by wrapping three or more layers of asbestos paper, each about $\frac{1}{16}$ inch thick, on the pipe, covering with a layer of hair felt $\frac{3}{4}$ inch in thickness, and wrapping the whole with canvas or paper, is much used. This covering has an effective life of about 5 years on high-pressure steam-pipes and 10 to 15 years on low-temperature pipes. There are a large number of coverings regularly manufactured for use, in such a form that they can be easily applied or removed if desired. There is a very great difference in the value of these coverings; some of them are very heavy and contain a large amount of mineral matter with little confined air, and are very poor insulators. Some are composed entirely of incombustible matter and are nearly as good insulators as hair felt. In general the value of a covering is inversely proportional to its weight—the lighter the covering the better its insulating properties; other things being equal, the incombustible mineral substances are to be preferred to combustible material. The following table gives the results of some actual tests of different coverings, which were conducted with great care and on a sufficiently large scale to eliminate slight errors of observation. In general the thickness of the coverings tested was 1 in. Some tests were made with coverings of different thicknesses, from which it would appear that the gain in insulating power obtained by increasing the thickness is very slight compared with the increase in cost. If the material is a good conductor its heat-insulating power is lessened rather than diminished by increasing the thickness beyond a certain point.

PERCENTAGE OF HEAT TRANSMITTED BY VARIOUS PIPE-COVERINGS, FROM TESTS MADE AT SIBLEY COLLEGE, CORNELL UNIVERSITY, AND AT MICHIGAN UNIVERSITY.*

Kind of Covering.	Relative Amount of Heat Transmitted.
Naked pipe	100.
Two layers asbestos paper, 1 in. hair felt, and canvas cover	15.2
Two layers asbestos paper, 1 in. hair felt, canvas cover, wrapped with manilla paper	15.
Two layers asbestos paper, 1 in. hair felt	17.
Hair felt sectional covering, asbestos lined	18.6
One thickness asbestos board	59.4
Four thicknesses asbestos paper	50.3
Two layers asbestos paper	77.7
Wool felt, asbestos lined	23.1
Wool felt with air spaces, asbestos lined	19.7
Wool felt, plaster paris lined	25.9
Asbestos molded, mixed with plaster paris	31.8
Asbestos felted, pure long fibre	20.1
Asbestos and sponge	18.8
Asbestos and wool felt	20.8
Magnesia, molded, applied in plastic condition	22.4
Magnesia, sectional	18.8
Mineral wool, sectional	19.3
Rock wool, fibrous	20.3
Rock wool, felted	20.9
Fossil meal, molded, $\frac{3}{4}$ inch thick	29.7
Pipe painted with black asphaltum	105.5
Pipe painted with light drab lead paint	108.7
Glossy white paint	95.0

---

* These tests agree remarkably well with a series made by Prof M. E. Cooley of Michigan University, and also with some made by G. M. Brill, Syracuse, N. Y., and reported in Transactions of the American Society of Mechanical Engineers, vol. xvi.

The following table translated from Péclet's *Tra Chaleur* gives in a general way the amount of heat tra through coverings of various kinds and of different thic the loss from a naked pipe is taken as 100.

LOSS OF HEAT THROUGH VARIOUS PIPE-COVER

Relative conductivity.	Thickness, in inches.						Kind of Covering.
	0.4	0.8	1.0	1.6	2.0	4.0	6.0	
	Relative Loss of Heat.						
0.04	29	20	18	13	11	7	6	Eider down, loose wool, h
0.08	43	32	29	23	20	13	11	Powdered charcoal.
0.16	56	48	45	38	35	25	22	Wood across fibres.
0.32	66	63	62	58	55	44	41	Sand.
0.64	73	73	73	72	71	70	68	Clayey earth.
1.28	77	83	85	92	96	102	109	Stone, rock.
2.56	78	87	91	103	110	130	150	White marble.
5.12	79	90	95	109	118	149	180	Solid gas carbon.
10.00	100	100	100	100	100	100	100	Naked, or unprotected su

CHAPTER X.

DESIGN OF STEAM AND HOT-WATER SYSTEMS.

117. General Principles.—The general problem of design includes the proportioning of, first, the amount of radiating surface which will be located directly in the rooms to be heated in all systems of direct heating, and in the air-passages or flues leading to the rooms in all cases of indirect heating; second, the size of the pipes which are to convey the heated fluids to the radiating surfaces; and third, the proper size of boiler or heater.

The question of the system or method of heating which is to be adopted will usually depend upon considerations of cost or of personal preference on the part of the proprietor. The various systems of heating, whether by steam, hot water, or hot air, as commonly practised in this country, do not often come in direct competition. Hot-air heating, where the air is moved by natural draft, is adapted only to the smaller sizes of dwelling-houses, and where heat does not need to be carried any considerable distance horizontally. It is generally found that if the horizontal distance exceeds 15 or 20 feet the supply of heat becomes uncertain in amount. With steam and hot-water heating there is no such limitation as to distance; the first cost is, however, considerably greater than that of hot air, but heat can be supplied with certainty to all parts of the system under all atmospheric conditions. Regarding the relative merits of systems of steam and hot-water heating, little can be said. It will generally be found that the first expense of steam-heating is considerably less, and that there is considerable difference of opinion regarding the relative economy of operation of steam and hot-water heating plants. The tests which have been made have generally shown somewhat in favor of water.* The difference, however, is not great, and may be due to local conditions, but is probably due to the fact that the temperature of the discharged gases may be somewhat lower for the hot-water heater than for the steam-boiler, and also to the fact that in comparatively mild weather the fire in the hot-water heater may be regulated somewhat closer, to meet the demand for heat. The hot-water system in general requires rather better workmanship in the erection of pipe lines than steam-heating, and more care must be taken in proportioning the various pipes and fittings. The heat from hot-water radiators is somewhat less in intensity and more pleasant than that from steam-radiators, and the temperature can be regulated by simply throttling the supply-pipe of the radiators, which is not the case with steam.

Whether direct or indirect heating shall be used will depend also on circumstances. It will be found that in general the surface required for indirect heating is one third to one half greater than that for direct, and it will give off 50 per cent more heat per square foot, so that the operating expense is practically twice that of direct heating. Indirect heating assures excellent ventilation, and it is advisable to use it for certain rooms of residences because of that fact.

118. Amount of Heat and Radiating Surface required for Warming.—The amount of heat required for buildings of various constructions has been considered quite fully in Chapter III. From which it may be seen (page 59) that in ordinary building construction the amount required in heat-units, for each degree difference between inside and outside temperature, is approximately equal to the area of the glass surface plus one fourth the area of the exposed wall surface plus one fifty-fifth of the number of cubic feet of air required for ventilation.

The air required for ventilation will vary with the conditions ; but in direct heating it seems necessary to allow for three changes per hour in halls, two in rooms on first floor, and one in rooms on upper floors. (See page 59.)

---

* See Transactions American Society Mechanical Engineers, vol. x, paper by the author. See also Report Massachusetts Experimental Station No. 8, 1872.

The amount of heat given off by one square foot of radiating surface, as determined by a great number of experiments, is given in Chapter IV, from which it is seen (pages 66 and 80) that for the ordinary radiating surface, with a temperature of 150 degrees above the surrounding air, 1.8 heat-units will be given off per square foot of surface per degree difference of temperature per hour, and when the temperature is 110 above the surrounding air about 1.7 heat-units are emitted.

The total heat emitted from radiating surfaces of different characters, corresponding to the average results of experiments is shown on the diagram, Fig. 187, in which the horizontal distances correspond to the mean difference of temperature between the air in the room and the radiator, while vertical distances, the value of which is read on the scale at the left, correspond to the total heat-units transmitted per square foot per hour.

To use the diagram assume the difference of temperature between the air of the room and the radiator, then look on vertical line until intersection with the line representing the desired condition is found, thence read results on the left. Thus, for instance, if the difference of temperature is 150 degrees the intersection of the line from this point with that representing direct ordinary radiation corresponds to 275 heat-units, and with that representing 1-inch horizontal pipe, 375 heat-units, as read on the scale at the left. The dotted lines in the diagram give the heat transmitted from various indirect surfaces for different velocities of the moving air. The results are to be found as for direct radiation, but the difference of temperature is that estimated from the mean of the surrounding air and the radiator.

Having the total heat required for warming and that which is given off from one square foot of radiating surface, it is quite evident that the surface required may be computed by the process of dividing the former by the latter.

Expressing results algebraically we can produce a formula from which the radiating surface may be calculated quickly and easily as follows:

Let R equal the total radiating surface required, t the required temperature of the room, t' the temperature of the outside air, T the tem

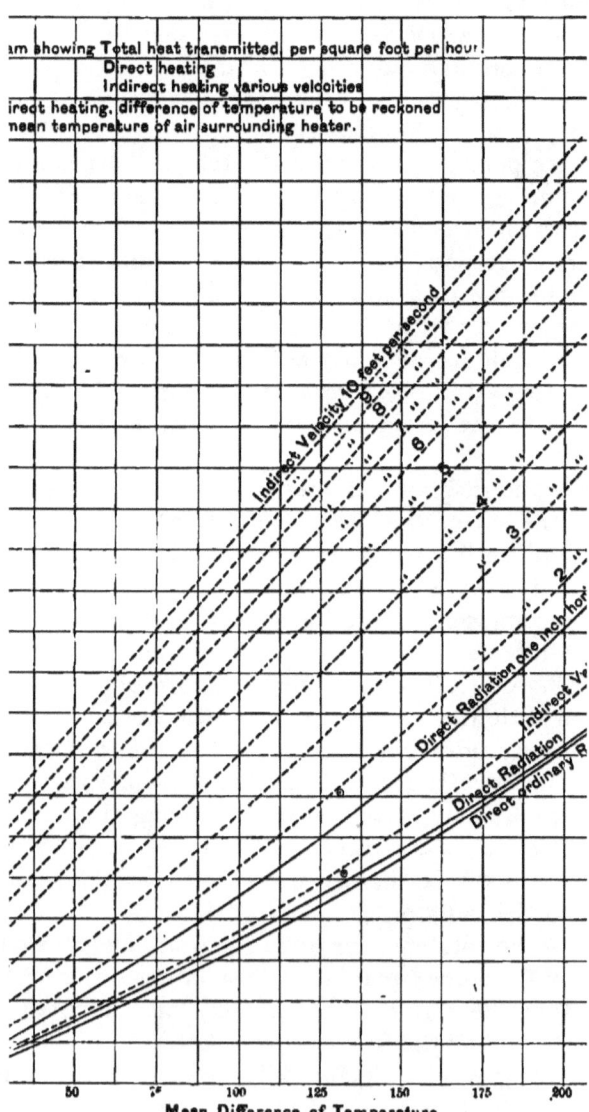

Fig. 187.—Diagram of Heat from Radiating Surfaces.

perature of the radiating surface, C number of cubic feet in the room, G the number of square feet of glass, W the external wall-surface, a the heat given off per square foot of radiating surface per degree difference per hour, r the number of times the air is to be changed per hour.

We have, first, the heat required for one degree difference of temperature as explained, pages 57 and 59, which is approximately

$$H = \frac{r}{55}C + G + \tfrac{1}{4}W. \quad \ldots \quad \ldots \quad (1)$$

Second, the radiating surface is H multiplied by difference of temperature between room and outside air divided by that given off from one square foot. Hence we have

$$R = \frac{t-t'}{(T-t)a} H = \frac{t-t'}{(T-t)a}\left(\frac{r}{55}C + G + \tfrac{1}{4}W\right). \quad \ldots \quad (2)$$

The heat required per degree difference of temperature between room and outside air, as expressed in equation (1), must be computed for every given case. The other quantities which constitute a factor to be multiplied in the above are readily computed and expressed in the table on p. 206, which is calculated for a great variety of conditions.

From this table it is seen that we need to *multiply the area of the glass, plus $\tfrac{1}{4}$ the wall surface, plus $\dfrac{r}{55}$ of the cubic feet of air supplied per hour,* by factors which are approximately as follows: If we are to heat to 70 degrees in zero weather with steam of 10 pounds pressure, multiply by $\tfrac{1}{4}$; if we are to heat to 60 degrees, multiply by $\tfrac{1}{6}$; if we are to heat to 50 degrees, multiply by $\tfrac{1}{8}$. As the steam pressures increase, these factors are reduced. As a method of applying the rule consider a room 20 feet by 12 feet floor surface, and 10 feet high, containing 2400 cubic feet, in which the air is to be changed twice per hour. Suppose that it has 320 square feet of exposed wall surface and 48 square feet of glass. The heating surface required will be found by taking the area of the glass, 48, $\tfrac{1}{4}$ the exposed wall, 80, and $\tfrac{2}{55}$ the cubic contents, which is equal to 87; the total heating surface required would be (48 + 80 + 87) 215, multiplied by the factor given in the table, which is about $\tfrac{1}{4}$, so that the radiating surface required equals 54 square feet. In this case there is about one square foot of heating surface to 44 cubic feet of space.

FACTORS FOR PROPORTIONING DIRECT RADIATORS FOR DIFFERENT TEMPERATURES ROOM AND OUTSIDE AIR.

Number of Column...		1	2	3	4	5
Coefficients for Steam...		1.6	1.7	1.8	1.9	2.4
Temperature Air.	Temperature Room.					
$-10°$	100°	.61	.54	.43	.31	.19
0	100	.55	.49	.40	.28	.17
$+10$	100	.50	.44	.36	.25	.16
-10	80	.42	.38	.31	.23	.145
0	80	.38	.33	.275	.20	.13
$+10$	80	.33	.30	.24	.18	.11
-10	70	.35	.32	.262	.19	.122
0	70	.32	.28	.23	.17	.109
$+10$	70	.26	.24	.20	.14	.092
-10	60	.29	.26	.22	.16	.104
0	60	.25	.22	.19	.14	.089
$+10$	60	.21	.18	.15	.12	.075
-10	50	.23	.23	.18	.15	.087
0	50	.20	.19	.15	.12	.072
$+10$	50	.16	.14	.12	.10	.058

Usual conditions of steam-heating correspond to a mean of columns two and three.

HOT WATER.
(Coefficient 1.6.)

Temperatures water...		140°	160°	180°	200°	212°
$-10°$	80°	.93	.70	.56	.47	.42
0	80	.83	.62	.50	.42	.38
$+10$	80	.73	.54	.435	.36	.33
-10	70	.71	.55	.45	.38	.35
0	70	.62	.47	.40	.333	.32
$+10$	70	.53	.41	.34	.28	.26
-10	60	.54	.44	.41	.31	.28
0	60	.47	.37	.36	.27	.23
$+10$	60	.39	.31	.31	.27	.21
-10	50	.41	.33	.25	.25	.255
0	50	.38	.28	.30	.20	.196
$+10$	50	.275	.225	.20	.175	.156

The radiating surface is in each case found by multiplying heat as required to supply loss from building per degree difference of temperature inside and outside by factor as given in the table. This factor is $\dfrac{t-t'}{(T-t)a}$ in formula (2).

For a room with the same dimensions but on the second floor the quantities will be computed in the same way, except that we will take $\frac{1}{55}$ of the cubic contents to supply that required by ventilation, so that the total heat required for one degree difference of temperature would be $48 + 80 + 44 = 172$. One fourth of this quantity gives the radiating surface for low-pressure steam-heating, which in this case would be 43, or one square foot of heating-surface to 55 cubic feet in the room. For hot-water heating the method of computation would be exactly the same, but the factor would be 0.4 instead of $\frac{1}{4}$. The radiating surface would then be, for the case considered, 0.4 of 216, which is 86, or one to 28 cubic feet for a room on the first floor, and 0.4 of 172 or 69 square feet, which is in ratio of 1 to 35 cubic feet for the second floor.

Many designers of heating apparatus compute the amount of radiating surface required by approximate "rules-of-thumb" which are in current use in their localities. These rules differ in many cases very greatly from each other, and often have to be modified materially in order to give satisfactory results. In the application of the more scientific rules which have been given there will still always be an opportunity for applying judgment and the results of experience and practice, since it is quite impossible that any table of coefficients, no matter how extensive, could be given which would apply to all cases of building construction and to all exposures. Allowance for unusual conditions are given by Mr. Wolff as follows (see page 57):

The amount of radiating surface as given should be increased respectively as follows:

* Ten per cent where the exposure is a northerly one and winds are to be counted on as important factors.

Ten per cent when the building is heated during the daytime only and the location of the building is not an exposed one.

Thirty per cent when the building is heated during the daytime only, and the location of the building is exposed.

Fifty per cent when the building is heated during the winter months intermittently, with long intervals (say days or weeks) of non-heating.

Certain allowances, in addition to the above, the amount of which must be determined by the judgment or experience of

CRUDE ESTIMATE OF SPACE HEATED BY 1 SQ. FT. OF DIR[ECT]
STEAM-HEATING SURFACE.

Authority	A.	B.	C.	D.	E.
DWELLINGS.					
First floor	35 to 60		35 to 50		
Second floor	50 to 80		50 to 75		
Average		60 to 80		50	
Living rooms					50
1 side exposed					50
2 sides "					45
3 " "					40
Halls and bath-rooms					40 to
Sleeping rooms					60 to
PUBLIC BUILDINGS.					
Offices	50 to 80 / 35 to 60	60 to 80		70	50 to
Banks				70	
School-rooms	50 to 80 / 35 to 60	60 to 80			60 to
Factories	75 to 100				80 to
Stores, wholesale	75 to 100	100		150	80 to
" retail		75		125	
" dry-goods				80	
" drugs				70	
Assembly halls	75 to 100	75 to 100			100 to
Auditoriums	125 to 200	75 to 100			
Churches	125 to 200	150 to 200		200	100 to
Large hotels				125	

the engineer, should be made for unusual construction of building, either good or bad.

The rules which have been given for determining the amo[unt] of radiating surface are exceedingly numerous. Some of th[e] rules require the proportioning of radiating surface, as Tredgold's * and Hood's † works, by the amount of gl[ass] others by the amount of glass and exposed wall surface,‡ [but] the great majority by the number of cubic feet of space in [the] room. The discussion which has been given is sufficient [to] show that the amount of heat required is a function of [the] exposed surfaces, so far as the loss from the walls is concern[ed] and of the cubic contents, so far as the supply of air for ve[ntilation]

---

* "Warming and Ventilating Buildings," Tredgold, 1836.
† "Warming Buildings," Hood, 1855.
‡ John J. Hogan in *Metal Worker*, Nov. 10, 1888.

CRUDE ESTIMATE OF SPACE HEATED BY 1 SQ. FT. OF *DIRECT HOT-WATER* HEATING SURFACE.

Authority	F.	A.	B.	C.	G.	H.	E.
DWELLINGS		25 to 50	High Temp. 50 to 70				
First floor	25 to 35	20 to 50 / 20 to 40	Low Temp. 30 to 50	20 to 30	25 to 40		
Second floor	35 to 40	30 to 50		30 to 40			
Average							
Living rooms	20 to 30						
1 side exposed						30	30
2 sides "						28	28
3 " "						25	25
Halls and bath-rooms	15 to 25				30 to 50	20 to 30	20 to 30
Sleeping rooms					30 to 50	30 to 40	30 to 40
PUBLIC BUILDINGS:							
Offices	30 to 60	20 to 50 / 25 to 50	50 to 70 / 30 to 50		35 to 50	30 to 50	30 to 40
Banks							
School-rooms	30 to 60	20 to 50 / 25 to 50	50 to 70 / 30 to 50		35 to 50	30 to 50	40 to 50
Factories	45 to 70	45 to 65 / 35 to 65	65 to 90		40 to 60	50 to 70	50 to 60
Stores, wholesale	45 to 70	45 to 65 / 35 to 65	65 to 90		40 to 60	50 to 70	50 to 60
" retail	45 to 70						
" dry-goods							
" drugs							
Assembly halls	80 to 100	45 to 65 / 35 to 65	65 to 90				75 to 100
Auditoriums	80 to 100	70 to 130 / 80 to 125	130 to 180		75 to 130	80 to 100	
Churches	80 to 100	70 to 130 / 80 to 125	130 to 180		75 to 130	80 to 100	75 to 100

lation, but both of these quantities must be considered in order to give results which are even approximately correct.

In any locality it would seem that the rules which are in common use when modified as to the condition of buildings in which they have been successfully applied would be of considerable value; for that reason the preceding tables are given showing the relation of radiating surface to cubic feet of space to be heated as stated by various authorities; it will be noticed, however, that there is such extreme variation in the amount of heating surface required for the same conditions that the results are almost valueless, and indicate that wide variation is common in the practice of different designers.

119. The Amount of Surface Required for Indirect Heating.—For this case the heat received by the rooms is all supplied by air which passes over the radiating surfaces and is heated by convection. A large number of tests have been quoted of these heaters, both with natural and mechanical draft

(see Article 52, page 79). From these exper
that the amount of heat given off by one squa
varies with the velocity of the air, as shown
page 84 and also in the diagram Fig. 187, the
been explained. From the table on page 84
that with natural circulation the velocity in
will vary from 2.97 for a height of 5 feet to
of 50 feet, and the corresponding convecti
heat-units per degree difference of temperatur
per hour, which in the preceding table is term
varies from 3 to 6.

The entering air is brought into the rc
temperature 20 to 40 degrees above that
this entering air is about 100 degrees, 1 heat-i
cubic feet 1 degree, an amount about 5 per c
when the entering air was 70 (see Table VIII

From these data we can readily compute
cubic feet of air which must be supplied to b
sary heat, and the size of heating-surface
amount of heat to be supplied must be suffi
sate for loss from the room, which is approx
the glass surface $+\frac{1}{4}$ the exposed walled surfa
the difference between the temperature of tl
outside air, or it may be obtained more exa
data, page 57. The number of cubic feet of
be found by dividing this quantity by the exce
of the heated air over that of the air in the ro
ing this result by 58.

The extent of heating surface in squ
obtained by dividing the number of cubic
obtained by the previous calculation by the
feet heated by one square foot of surface. I1
100° F. each heat-unit will warm 58 cubic feet

These results are better expressed in shape of f
tables suited for practical application may be comput
temperature of the room, t' that of the outside air,
temperature of the air surrounding the heating sur
heated air, T that of the radiating surface, H the hea
per degree difference of temperature to supply loss fr

heat given off from 1 sq. ft. radiating surface per degree difference of temperature. We have the following formula:

Loss from the room per hour $(t - t')H = (t - t')(G + \frac{1}{4}W)$ nearly; (1)

Heat brought in by 1 cu. ft. of air $1/58(T' - t)$; . . . . . . (2)

Heat given off from 1 sq. ft. of radiating surface per hour
$$= a(T - t'); \quad (3)$$

Cubic feet of air required per hour $= \dfrac{(t - t')H}{1/58(T' - t)}$; . . . . . (4)

Cubic feet of air heated by 1 sq. ft. of radiating surface per hour
$$= \dfrac{a(T - t')}{1/58(T' - t)} \text{ (see Article 31, page 39); .} \quad (5)$$

Radiating surface $= \dfrac{(t - t')(T' - t')H}{a(T' - t)(T - t')} =$ (Factor as in table) H; . (6)

The table,* page 212, computed from the above formulæ for various conditions gives a series of factors which, multiplied into the building loss H per degree difference of temperature, will give the radiating surface required; it also gives the number of cubic feet of air heated the required amount per square foot of radiating surface per hour.

To use the table, we need simply to know, in addition to temperatures, the probable coefficient of heat transmission, all other conditions being given. For ordinary indirect heating, first floor, the velocity of air can be considered as 2 to 4 feet per second, and the corresponding value of this coefficient as 2. For higher floors the velocity is higher, and coefficients may be taken as 3. (See page 84.) As an example, assume outside temperature zero, inside temperature 70°, and the air leaving the indirect at 100°, the factor with which to multiply the building loss to obtain radiating surface is 0.69. This is practically 3.00 times that for direct heating. Computing the radiating surface required for the same room as that considered in the case of direct heating (page 206), in which there was 48 square feet of glass and 320 square feet of exposed wall surface, and in which the total loss of heat per degree difference of temperature was 128 heat-units, the indirect surface required would be this quantity multiplied by the factor 0.69, which is 88 square feet, or about one half more than required in the calculation for direct heating. For the

---

* In the table the term coefficient is used for the heat transmitted per degree

TABLE OF FACTORS TO OBTAIN INDIRECT HEATING SURFACE AND OF CUBIC FEET OF AIR HEATED PER SQUARE FOOT OF SURFACE PER HOUR.

Temperatures.			B. T. U.—Total Heat per Sq. Ft. Heater.					Factors for Heater Surface.*					Cu. Ft. Air per Sq. Ft. Heat. Surf. per Hour.				
Air Entering Room.	Mean of Air Surrounding Heater.	Mean Difference Air and Radiator.	Coefficient 1.	Coefficient 2.	Coefficient 3.	Coefficient 4.	Coefficient 6.	Coefficient 1.	Coefficient 2.	Coefficient 3.	Coefficient 4.	Coefficient 6.	Coefficient 1.	Coefficient 2.	Coefficient 3.	Coefficient 4.	Coefficient 6.
T'	t''	$T' - t''$	(1)	(2)	(3)	(4)	(6)	(5)	(6)	(7)	(8)	(9)	(10)	(11)	(12)	(13)	(14)
ROOM 70° FAHR., OUTSIDE AIR 0° FAHR., STEAM PRESSURE 0 LBS., STEAM TEMPERATURE 212° FAHR.																	
90	45	167	167	334	501	668	1000	1.92	0.96	0.64	0.48	0.32	108	216	324	432	648
100	50	162	162	324	486	648	972	1.47	0.73	0.49	0.36	0.24	94	188	292	376	564
110	55	157	157	314	471	628	942	1.24	0.62	0.41	0 31	0.21	88	176	264	352	548
120	60	152	152	304	456	608	912	1.10	0.55	0.37	0.28	0.18	73	147	220	394	440
ROOM 70° FAHR., OUTSIDE AIR 0° FAHR., STEAM PRESSURE 5 LBS., STEAM TEMPERATURE 219° FAHR.																	
90	45	174	174	348	522	696	1062	1.72	0.86	0.51	0.43	0.28	112	224	336	448	672
100	50	169	169	338	507	676	1015	1.38	0.69	0.46	0.34	0.23	98	196	294	392	788
110	55	164	164	328	492	656	934	1.18	0.56	0.39	0.29	0.19	86	173	260	346	520
120	60	159	159	318	477	636	954	1.16	0.53	0.35	0.27	0.17	77	154	231	308	462
ROOM 60° FAHR., OUTSIDE AIR 0° FAHR., STEAM PRESSURE 0 LBS., STEAM TEMPERATURE 212° FAHR.																	
80	40	172	172	344	516	688	1032	1.66	0.83	0.55	0.41	0.27	125	250	375	500	750
90	45	167	167	334	501	668	1020	1.16	0.58	0.29	0 29	0.19	108	216	324	332	648
100	50	162	162	326	486	652	972	0.93	0.46	0.31	0.23	0.15	94	188	282	376	564
110	55	157	157	314	461	628	922	0.89	0.42	0.28	0.21	0.14	83	166	249	332	498
ROOM 70° FAHR., OUTSIDE AIR 0° FAHR., HOT WATER AT TEMPERATURE 160° FAHR.																	
90	45	115	115	230	345	460	690	2.8	1.4	0.93	0.7	0.46	74	148	222	296	444
100	50	110	110	220	330	440	660	2.12	1.06	0.70	0.53	0.35	64	128	192	256	384
110	55	105	105	210	315	420	630	1.86	0.93	0.62	0.46	0.32	55	110	165	220	330
120	60	100	100	200	300	400	600	1.68	0.83	0.56	0.42	0.28	48.5	97	145	194	290
ROOM 70° FAHR., OUTSIDE AIR 0° FAHR., HOT WATER AT TEMPERATURE 180° FAHR.																	
90	45	135	135	270	405	540	810	2.36	1.18	0.78	0.57	0.39	87	174	261	348	522
100	50	130	130	260	390	520	780	1.78	0.89	0.59	0.54	0.29	75	150	225	300	450
110	55	125	125	250	375	500	750	1.55	0.72	0.52	0.39	0.26	66	132	198	264	396
120	60	120	120	240	360	480	720	1.4	0.7	0.47	0.35	0.23	58	116	174	232	348

difference of temperature per square foot per hour. Coefficients 1 to 4 correspond to ordinary indirect heating.

* To find surface of heater multiply loss from room for one degree difference of temperature by the factor for the given condition. Results computed by formula (6).

second and third stories the factors are to be found in the column in which the coefficient is 3.

The following table gives the number of cubic feet of air required per hour in indirect heating to maintain the proper temperature, as computed by formulæ (4), for each heat-unit lost from walls and windows of room for a temperature of 60° or 70° above outside air. The total air required will be found by multiplying the values, as given in the table, by the total heat lost per degree difference of temperature from the room. This loss is designated by H in formulæ (4), and is approximately equal to the glass plus ¼ the exposed wall surface expressed in square feet. (See page 59.)

CUBIC FEET OF AIR PER HEAT-UNIT FROM WALLS.

Temperature of Entering Air above that of Room.	Temperature of Room, Degrees Fahr.	
	60°	70°
10	348	406
20	174	203
30	116	135
40	87	103
50	70	81
60	58	68
70	49	58
80	44	51
90	36	45
100	35	41

Thus to find the number of cubic feet of air required to warm a room to 70° in zero weather, in which the glass plus one fourth the exposed wall surface equals 128, and air is introduced 30° above that in the room, multiply 135, as given in the table, by 128.

It is usual to allow 50 per cent more surface for indirect than for direct heating, although some engineers allow only 25 per cent more.

In concluding this subject it may be remarked that the amount of heat which is given off from indirect heating surfaces would seem from the experiments to depend largely on construction. With the surface erected closely together the amount is small. By better arrangement of the surfaces, so that all parts are made hot, and an ample opportunity is provided for circulation of the air, the coefficient of heat transmission may be much increased. If extended-surface radiators are used and the entire surface figured as effective, the coefficient should be taken about 10 per cent less than assumed by the writer in the computation. For forced draft the coefcient may be safely taken as 4 and 6, or about 100 per cent greater than for natural circulation.

The following tables are collected from various authorities, and are of interest as showing character of "rule of thumb" practice in providing indirect heating surface for rooms of various kinds. It will be noted that the amount specified for the same work differs more than 50 per cent, which shows the crudeness of estimates of this character.

CRUDE ESTIMATE OF SPACE HEATED BY 1 SQ. FT. OF INDIRECT STEAM-HEATING SURFACE.

Authority	K.	A.	B.	C.	D.
DWELLINGS:					
First floor............	....	20 to 35	....	25 to 35	
Second floor.........	....	40 to 50	....	40 to 50	
Average	40	....	40 to 50	....	40
Living-rooms........					
One side exposed....					
Two sides exposed...					
Three sides exposed.					
Halls and bath-rooms	....	50 to 70	50 to 70		
Sleeping-rooms......					
PUBLIC BUILDINGS:					
Offices..............	60	{ 20 to 35 } { 40 to 50 }	40 to 50	....	60
Banks..............	60	*See* DWELLINGS	40 to 50	....	60
School-rooms........	....	*See* DWELLINGS	40 to 50		
Factories	....	50 to 70	50 to 70		
Stores, wholesale....	100	70	70	....	100
" retail........	80	50	50 to 70	....	80
" dry-goods....	70	....	....	....	70
" drugs	60	....	....	....	60
Assembly halls......	....	80 to 135	100 to 140		
Auditoriums	....	80 to 135	100 to 140		
Churches	150	80 to 135	100 to 140	....	150
Large hotels.........	100	....	....	....	100

CRUDE ESTIMATE OF SPACE HEATED BY 1 SQ. FT. OF INDIRECT HOT-WATER HEATING SURFACE.

Authority	F.	A.	B.	C.
DWELLINGS:				
First floor	15 to 25	15 to 30 / 15 to 40	30 to 60 / 20 to 40	14 to 20
Second floor	20 to 30			20 to 30
Third "	20 to 30			
Living-rooms	15 to 25			
One side exposed				
Two sides exposed				
Three sides exposed				
Hall and bath-rooms	10 to 20			
Sleeping-rooms				
PUBLIC BUILDINGS:				
Offices	25 to 40	15 to 30 / 15 to 40	30 to 60 / 20 to 40	
Banks				
School-rooms	25 to 40	15 to 30 / 15 to 40	30 to 60 / 20 to 40	
Factories	25 to 40	30 to 45 / 25 to 50	35 to 75 / 25 to 50	
Stores, wholesale	25 to 40	30 to 45 / 25 to 50	35 to 75 / 25 to 50	
" retail	25 to 40			
" dry-goods				
" drugs				
Assembly halls	50 to 80	30 to 45 / 25 to 50	35 to 75 / 25 to 50	
Auditoriums	50 to 80	50 to 100 / 50 to 100	70 to 150 / 50 to 100	
Churches	50 to 80	50 to 100 / 50 to 100	70 to 150 / 50 to 100	
Large hotels				

120. Summary of Approximate Rules for Estimating Radiating Surface.—As the temperature required for buildings of various classes varies but little, and as the heating surface is usually estimated to be sufficient to heat buildings during zero weather to a temperature of 70 degrees, some very simple rules can be given which are founded on a rational basis, and which with certain modifications, as explained (page 57), for those which are especially exposed, will be found to give good results in practice which agree closely with those used by the best heating engineers. They are as follows:

First. The amount of heat required to supply that lost from the room per degree difference of temperature *is approximately equal to the area of the glass in square feet plus 1/4 the exposed wall surface.* (See page 59.)

Second. The heat necessary to supply loss from ventilation *for dwelling-houses, first floor, is 2/55 of the cubic contents per hour for living-rooms;*

3/55 of the cubic contents for halls;

1/55 of the cubic contents for upper stories.

For churches, auditoriums, the loss to supply ventilation should be taken as 3/55 *to* 6/55 *of the cubic contents;* for offices, banks, etc., 1/55 to 2/55 of the cubic contents, depending upon circumstances.

Third. To find the radiating surface for direct steam-heating, multiply the sum of the numbers as given by rules First and Second by 1/4.

Fourth. To obtain the radiating surface for direct hot-water heating, multiply the sum of the numbers as given by rules First and Second by 0.4. These rules may both be summed up in the following concise form:

RULE.—For heating to 70 degrees in zero weather, direct heating: *Radiating surface is equal to the sum of the glass surface plus* 1/4 *the exposed wall surface plus* 1/55 *to* 3/55 *the cubic contents, for rooms as explained, multiplied by* 1/4 *for low-pressure steam-heating or by* 0.4 *for hot-water heating.*

NOTE.—When air is introduced at 100 degrees Fahr., 58 should be used instead of 55. This difference is, however, usually negligible.

For indirect heating the following rules will give quite satisfactory results when the temperature of the room is to be maintained at 70° with outside air at zero and the heated air brought in at a temperature 30° above that in the room. In this calculation the surface of the steam radiator is supposed to be 212°, that of the hot-water radiator 170° Fahr. The coefficients are taken from the preceding table.

RULE.—The radiating surface for indirect heating is equal to the glass surface plus one fourth the exposed wall surface in square feet multiplied by the following factors:

	Steam-heating.	Hot-water Heating.
1st story	0.7	1.05
2d "	0.6	0.9
3d "	0.5	0.8

The total amount of air supplied will be given by the following

RULE.—The air in cubic feet per hour is found by multiplying the radiating surface, computed as in above rule, by the following factors:

	Steam-heating.	Hot-water Heating.
1st story	200	125
2d "	250	160
3d "	300	200

If this is insufficient for ventilating purposes more air must be introduced, which must be heated to 70° F., and this will require approximately an additional foot of surface for each additional 250 cubic feet of air heated by steam, or for each additional 150 cubic feet heated by hot water.

These rules will be found quite simple in application, and they may be easily committed to memory. For rooms which are poorly constructed or especially exposed these results should be increased the same proportional amount as for direct radiating surfaces. For temperatures lower or higher than 70° the table of factors p. 212 may be used with facility.

121. Flow of Water and Steam.—It seems necessary to say a few words respecting the general laws which apply before considering the practical application. The velocity with which water flows in a pipe is computed from the same general laws as those applying to the fall of bodies. The velocity is produced, however, not by actually falling through a given distance, but by a difference of pressure, which must be expressed, not in pounds per square inch, but in feet of head. This head is in every case to be found by multiplying the difference of pressure by the height required for the given fluid to make one pound of pressure. If we denote by h the difference of head as described, by g the force of gravity $= 32.16$, by v the velocity in feet per second, we would have in case of no friction

$$v = \sqrt{2gh}.$$

The quantity discharged per second would be found in every case by multiplying the velocity by the area of the orifice in square feet.

In the flow of water in pipes there is considerable friction, which acts to reduce the velocity and the amount discharged; this increases with the length and decreases with the diameter of the pipe. For the actual flow we depend upon experimental results. An approximate formula, attributed by Robert Briggs* to Prof. Unwin, which is sufficiently accurate for computing the flow of water in pipes is as follows:

Let $v =$ the velocity in feet per second, V the velocity of feet per minute, $q =$ the quantity discharged in cubic feet per second, $Q =$ that discharged per minute, $l =$ the length of pipe in feet, $h =$ the head in feet, $D =$ the diameter in feet, $d =$ the diameter in inches.

$$v = 50\sqrt{\frac{hD}{l}}; \qquad l = 22.3096\frac{hd^5}{Q^2};$$

$$q = 39.27\sqrt{\frac{hD^5}{l}}; \qquad h = 0.0448\frac{Q^2l}{d^5};$$

$$Q = 4.7233\sqrt{\frac{hd^5}{l}}; \qquad d = 0.5374\sqrt[5]{\frac{Q^2l}{h}}.$$

The friction caused by bends and by passing through valves and into entrance of pipes is of considerable amount, and often requires consideration. It can be considered as producing the same resistance to flow as though the pipe had been increased in length certain distances as follows: 90-degree elbow is equivalent to increase in length of the pipe 40 diameters, globe valve 60 diameters, entrance of a pipe in tee or elbow 60 diameters, entrance in straight coupling 20 diameters.

The flow of steam in pipes presents some problems slightly different from that of flow of air (Articles 31 and 32), but in many respects the two cases are similar. There is a tendency for the steam to condense, which changes the volume flowing and affects the results greatly. The effect of condensation and friction is to reduce the pressure in the pipe an amount proportional to the velocity and also to the distance, and these losses are greater as the pipe is smaller. There

---

* Steam-heating for Buildings, p. 75, by Briggs.

seems to be very little exact data regarding the steady flow of steam in pipes, and it has been customary for writers to assume that the same laws which apply to the flow of water hold true in this case, and that the same methods can be used in computing quantities. These results are certainly safe, although no doubt giving sizes somewhat larger than strictly necessary for the purposes required.

In estimating the size of steam-pipe for power purposes it is customary to figure the area of cross-section, such as giving a velocity of flow not exceeding 100 feet per second. This velocity is generally accompanied by a reduction of pressure in a straight pipe of about one pound in 100 feet. For steam-heating purposes the general practice is to use a much larger pipe and lower velocity, so that the total reduction in pressure on the whole system is much less; the effect of a drop in pressure of one pound will cause the water to stand in the return pipe in a gravity system 2.4 ft. above the water-level in the boiler.

The velocity of water and steam in a gravity system of heating is due to a different cause from that in the case just considered, for the reason that the pressure upon the heater acts uniformly in all directions, and exerts the same force to prevent the flow into the boiler from the return, as to produce the flow into the main. For such cases the sole cause of circulation must be the difference in weight of the heated bodies, hot water, or steam in the ascending column and the cooler and heavier body in the descending column. The velocity induced by a given force will be reduced in proportion as the mass moved is greater. In the case of steam-heating the difference between the weight in the ascending and descending column is so great that the velocity will not be essentially different from that of free fall, provided correction is made for loss of head due to friction, etc., as explained, but in case of hot water the theoretical velocity produced will be found very small.

The case is very similar to the well-known problem in mechanics in which two bodies A and B of unequal weights are connected by a cord passing over the frictionless pulley C (Fig. 190).

220 HEATING AND VENTILATING BUILDINGS.

The heavier body B in its descent draws up the lighter body A. In this case the moving force is to the force of gravity as the difference in the weights is to the sum of the weights, and the velocity is the square root of twice the force into the height.

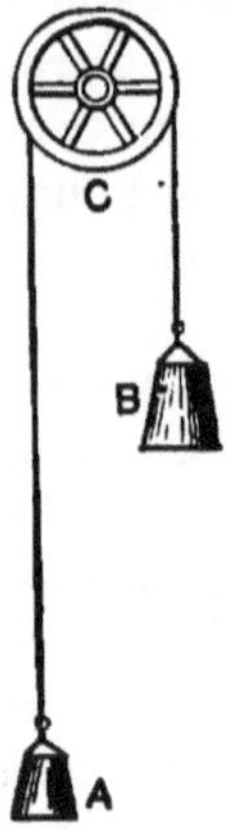

In other words, if f equals the moving force, we have by proportion

$$f : g :: B - A : B + A,$$

from which

$$f = g\frac{B - A}{B + A},$$

which, substituted in place of f in formula $v = \sqrt{2fh}$, gives the following as the velocity:

$$v = \sqrt{\frac{2g(B - A)h}{B + A}},$$

FIG. 190.

h being the height fallen through.

In applying this to the case of hot-water heating we have, instead of the descent and ascent of two solids of different weights, the descent and ascent of columns of water connected as shown in Fig. 191, the heated water rising in the branch AF and the cooler water descending in the branch BC. The force which produces the motion is the difference in weight of water in the two columns; the quantity moved is the sum of the weight of water in both columns. This is equal to the difference in weight of 1 cubic foot of the heated and cooled water divided by the sum, multiplied by the total height of water in the system, so that if W_1 represents the weight of 1 cubic foot in the column BC, and W represents the weight of 1 cubic foot in the column AF, and h represents the total height of the system, then the velocity of circulation will be, in feet per second,

$$v = \sqrt{\frac{2gh(W_1 - W)}{(W_1 + W)}}.$$

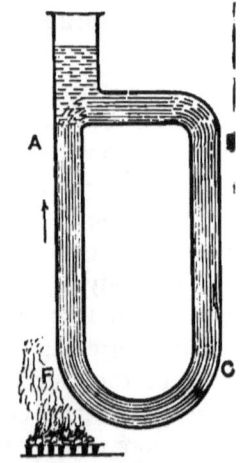

FIG. 191.—CIRCULATION IN HOT-WATER PIPES.

In this formula no allowance whatever is made for friction consequently the results obtained by its use will be much in excess of that actually found in pipes. The amount of friction will depend upon the length of pipe and its diameter As result of experiment the writer found considerable variation in different measurements of velocity, but in no case did he find a velocity greater than that indicated by the formula. The following table is calculated from the formula without allowance for loss by friction. The computation is made with the colder water at 160 degrees F., although little difference would be found in calculations at other temperatures.

VELOCITY IN FEET PER SECOND IN HOT-WATER PIPES.

Height or Head in Feet.	Free Fall in Air.	Difference of Temperature.						
		1°	5°	10°	15°	20°	30°	40°
1	8.03	0.107	0.242	0.335	0.412	0.478	0.593	0.672
5	17.9	0.232	0.541	0.750	0.922	1.09	1.33	1.51
10	25.4	0.328	0.765	1.06	1.32	1.55	1.88	2.14
20	35.9	0.463	1.085	1.5	1.85	2.19	2.66	3.01
30	43.9	0.567	1.33	1.83	2.26	2.68	3.26	3.71
40	50.7	0.656	1.53	2.12	2.61	3.08	3.76	4.26
50	56.7	0.732	1.71	2.37	2.82	3.47	4.22	4.77
60	62.1	0.802	1.88	2.59	3.20	3.79	4.62	5.22
70	67.1	0.866	2.02	2.80	3.45	4.08	4.97	5.65
80	71.8	0.925	2.16	3.0	3.69	4.37	5.32	6.03
90	76.1	0.932	2.27	3.18	3.91	4.64	5.64	6.41
100	80.3	1.037	2.42	3.35	4.13	4.78	5.93	6.72

This table is of interest for the reason that most computations of the velocity of circulation of hot water have entirely neglected the effect that the mass or weight of the water moved has on the velocity, and hence the results as computed have been many times greater than actually found. The method usually employed in computing this velocity has been to consider the denser and lighter fluids occupying the relative positions shown in Fig. 192, the lighter fluid being in one branch of the U tube, the heavier in the other.* If the cock be opened, equilibrium will be established, and the lighter liquid will stand in the branch higher than the heavier a distance sufficient to balance the difference in weight. If we suppose (1) the cock closed and

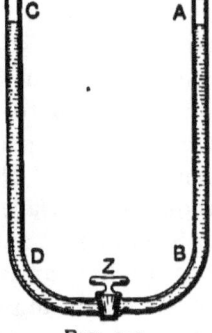

FIG. 192.

---

* See Hood's work on "Warming Buildings," page 27. So far as the writer knows, this theory has not before been questioned.

enough of the heavier material added to the shorter column, so that the heights in each are the same; (2) the cock opened, then the heavier liquid will move downward and drive the lighter liquid upward with a velocity said to be equal to that which a body would acquire in falling through the distance equal to the difference in heights when the columns were in equilibrium. This gives too great results, because it neglects the effect of the mass of the bodies moved. If friction be considered, we should have as a probable expression of velocity, using the same notation as on page 218,

$$v = 50 \sqrt{\frac{(W_1 - W)}{(W_1 + W)}\frac{hD}{l}}.$$

122. Size of Pipes to supply Radiating Surfaces.—The method of computing the size of pipes required for steam heating would be as follows: First find the amount of steam by dividing the total number of heat-units given out by 1 square foot of radiating surface by the latent heat in 1 pound of steam, this will give the weight of steam required per square foot; this multiplied by the number of cubic feet in 1 pound of steam will give the volume which will be required for each square foot of radiating surface. Knowing this quantity the size of pipe may be computed from the considerations already given, either by formulæ of Article 121 or by assuming the velocity of flow as equal that due to the head, corrected for friction; 25 to 50 feet per second can in nearly every case be realized. As an illustration; compute the size of main steampipe required to supply 1000 feet of radiating surface with steam at a temperature of 212 degrees when the surrounding temperature of the air is 70: For this case 1 square foot of radiating surface can be assumed ordinarily as giving off (1.8 times 142) 255 heat-units. To supply 1000 feet of surface 255,000 heat-units per hour would be required; as each pound of steam during condensation (see steam table) will give up 966 heat-units, we will need for this purpose 264 pounds per hour; and as each pound of steam at this temperature makes 26.4 cubic feet, we will require 6970 cubic feet of steam per hour, or 1.94 cubic feet per second.

If we proportion the pipes so that the velocity shall not

exceed 25 feet per second, the area of the pipe must be 0.077 square foot, which equals 11.1 * square inches. For this we would require a pipe 4 inches in diameter. If we had assumed the velocity to be 50 feet per second, the area would have been 5.6 square inches and the diameter 3 inches; if we had assumed a velocity of 100 feet per second, the area required would have been 2.8 square inches and the diameter of the pipe required would have been somewhat less than 2 inches. The friction in a pipe when steam is moving at a velocity of 100 feet per second causes a reduction in pressure of about $1\frac{1}{2}$ pounds in 100 feet, a velocity of 50 feet per second causes about $\frac{1}{4}$ as much, and a velocity of 25 feet about $\frac{1}{16}$ as much. Indirect surfaces of the same extent usually require twice as much steam and a pipe with area twice as great as that needed for direct radiation.

For the *single-pipe system of heating* an additional amount of space must be provided in the steam main to permit the return of the water of condensation. The actual space occupied by the water is small compared with that taken by the steam, but in order to afford room for the free flow of the currents of water and steam in opposite directions, experience indicates that about 50 per cent more area should be provided than is required in the separate return or double pipe system of heating.

By similar computations we obtain the following factors, which are to be multiplied by the radiating surface to obtain areas and diameters of steam-heating mains in inches:

TABLE FOR AREA AND DIAMETER OF STEAM-MAIN.

Velocity of steam, feet per second. (1)	Multiply each 100 sq. ft. radiating surface for area steam main by (2)		Multiply sq. root radiating surface for diam. by (3)		Probable frictional resistance per 100 ft., inches water. (4)	Required steam pressures. Lbs. (5)
	Double-pipe system.	Single-pipe system.	Double-pipe system.	Single-pipe system.		
25	.90	1.35	.107	.131	2.0	0 to 1
37.5	.675	1.01	.092	.113	6.0	2 to 3
50	.45	0.67	.075	.092	8.0	3 to 4
62.5	.375	0.56	.069	.090	12.6	4 to 5
75	.30	0.45	.062	.075	18	5 to 6
100	.225	0.34	.054	.066	32.0	6 to 40

In all cases if the mains are not covered, its surface is to be estimated as a part of the radiating surface.

* This quantity is greater than the area of a $3\frac{1}{2}$-inch pipe, and in such case the safe proceeding is to use the next greater size.

The table on page 223 gives in the first column the velocity of steam, in the second column the corresponding area of pipe in square inches required for each 100 square feet of radiating surface for the double and single pipe systems of heating, in the third column the diameter of pipe for each square foot of radiating surface for both systems of heating, which latter is to be multiplied by the square root of the given radiating surface, to obtain the diameter required. Column 4 gives the approximate back pressure in inches of water per 100 feet in length of the main for steam having the same velocity as in column 1. Column 5 suggests steam-pressures which will render any of these values satisfactory in practice.

As an example showing use of table, suppose that a main pipe to supply 650 square feet of radiating surface is needed in a single-pipe system in which the back pressure shall be about 12 inches of water-column per 100 ft. of length. The assumed resistance is found in column 4 and corresponds to a velocity of about 62.5 feet per second.

Column 2 gives the factor for the area of pipe as 0.56, which, multiplied by 6.50, gives 3.64 sq. in. as the required area. The diameter can be obtained from this result or computed by multiplying the square root of the radiating surface by the number in column 3. The square root of 650 is 25.4. This multiplied by 0.09 gives the diameter required as 2.3 in. For this case a 2½-inch pipe must be used. For the double-pipe system, the factor for area would be 0.375 and that for diameter would be 0.069. The required pipe for the case considered would have a diameter of 1.75 in. The size next largest, viz., 2.0 in. should be used for the steam-main. For calculating return see Article 123.

Most of the rules which have been given for determining sizes of steam-pipe when the radiating surface only is given will be found included in the tabulated values. Thus Mr. George H. Babcock gives a rule for gravity heating-systems with separate returns as follows:* " The diameter of the mains leading from the boiler to the radiating surface should be equal in inches to one tenth the square root of radiating surface, mains included,

---

* Transactions American Society Mechanical Engineers, May, 1885.

in square feet." This rule is also adopted by William J. Baldwin, and given in his book on "Steam-heating."* By consulting the table already given, column 3, this factor would correspond to a velocity of steam slightly exceeding 25 feet per second, and would be adapted for low-pressure steam-heating in small plants.

One authority † gives the following rules for determining the cross-sections of area of pipes: "For steam-mains and returns it will be ample to allow a constant of 0.375 of a square inch for each 100 square feet of heating surface in coils and radiators, 0.375 of a square inch when exhaust steam is used, 0.19 of a sq. inch when live steam is used, and 0.09 of a square inch for the return. Steam-mains should never be less than 1½ inches, nor the returns less than three fourths of an inch, in diameter." Mr. Alfred R. Wolff uses a table which is computed by formulæ similar to those given on page 218 for obtaining the capacity of steam-mains of a given diameter, the capacity being expressed both in heat-units delivered and in radiating surface. This table is given on page 226a and will be found convenient and accurate.

The size of main steam-pipe depends on the consideration already given; the smaller the size the greater the resistance of the steam and the more friction and consequent back pressure on the system; the larger the pipes that are used the less the resistance, and, in general, the more satisfactory the results, but economy, of course, forbids the use of pipes beyond a certain size, and that size should be selected by considerations relating to pressure, velocity of steam, and friction, as explained.

The methods of computing sizes of steam-mains which have been given allow sufficiently for friction for cases in which the pipes are not of considerable length, as in residence heating; but when steam must be carried a long distance more satisfactory results will be obtained by computing the capacity from the formula given in Article 121, page 218. For this computation various cases can be considered respecting both steam-pressure and frictional resistance. The following tables

---

* "Steam-heating for Buildings," Wm. J. Baldwin.
† Van Nostrand's Science Series, No. 68.

INTERNAL DIAMETERS OF STEAM-MAINS FOR A SINGLE-PIPE SYSTEM OF HEATING BY DIRECT RADIATION.*

[Steam-pressure 10 lbs. above atmosphere, frictional resistance 6 in. of water-column.
" " " 0.5 " " " " " " " 12 " " " "]

Radiating Surface, Sq. Ft.	Length of Steam-main in Feet.								
	20	40	80	100	200	300	400	600	1000
	Diameter of Pipe in Inches.								
20	0.5	0.5	0.6	0.6	0.7	0.8	0.8	0.9	1.2
40	0.6	0.7	0.8	0.8	1.0	1.0	1.1	1.2	1.6
60	0.7	0.8	0.9	1.0	1.1	1.2	1.3	1.4	1.8
80	0.8	0.9	1.0	1.1	1.2	1.4	1.5	1.6	2.1
100	0.9	1.0	1.2	1.2	1.4	1.5	1.6	1.7	2.3
200	1.1	1.3	1.5	1.6	1.8	1.9	2.0	2.2	2.9
300	1.3	1.5	1.8	1.8	2.1	2.3	2.4	2.6	3.5
400	1.5	1.7	2.0	2.0	2.4	2.6	2.7	3.0	4.0
500	1.6	1.9	2.2	2.2	2.6	2.8	3.0	3.2	4.2
600	1.8	2.0	2.4	2.5	2.8	3.0	3.2	3.5	4.5
800	2.0	2.3	2.6	2.7	3.2	3.4	3.6	3.9	5.0
1,000	2.2	2.5	2.9	3.0	3.4	3.7	3.9	4.3	5.5
1,400	2.5	2.8	3.3	3.4	3.9	4.2	4.5	4.9	6.5
1,800	2.7	3.2	3.6	3.8	4.4	4.7	5.0	5.4	7.0
2,000	2.9	3.3	3.8	3.9	4.5	4.9	5.2	5.6	7.2
3,000	3.4	3.9	4.4	4.6	5.3	5.8	6.1	6.6	8.5
4,000	3.8	4.3	5.0	5.2	6.0	6.5	6.8	7.5	9.7
6,000	4.1	4.7	5.4	5.7	6.5	7.1	7.4	8.2	10.5
8,000	4.4	5.0	5.8	6.0	7.0	7.5	7.9	8.7	11.3
10,000	4.7	5.3	6.1	6.4	7.4	8.0	8.4	9.2	11.9

* The table is computed by formulæ for d, page 218, in which $h = 318.6$, $Q = 9.2$, cu. ft. of steam per minute for 100 sq. ft. radiating surface. The table is computed for straight pipes with water-level in returns 6 inches above that in boiler. In case there are bends or obstructions consider the length of pipe increased as follows: Right-angle elbow 40 diameters; globe-valve 125 diameters; entrance to tee 60 diameters. For other resistances and steam-pressures multiply the diameters as given above by the following factors:

Water-level in return above boiler......	2 in.	12 in.	18 in.
Multiply by.....................	1.25	0.88	0.80
Steam-pressure above atmosphere......	0.5 lbs.	2 lbs.	5 lbs.
Multiply by.....................	1.22	1.16	1.09

For obtaining the diameter of steam-main to be used in case there is a separate return multiply the above results by 0.82.

For indirect heating without separate return multiply above results by 1.4, with separate return use the results in the form given.

Do not use steam-pipe less than 1¼ inches in diameter.

TABLE FOR THE CAPACITY OF STEAM-PIPES 100 FEET IN LENGTH WITH SEPARATE RETURNS.

By A. R. Wolff.

Diameter of Supply. Inches.	Diameter of Return. Inches.	2 Lbs. Pressure.		5 Lbs. Pressure.	
		Total Heat Transmitted. B. T. U.	Radiating Surface. Square Feet.	Total Heat Transmitted. B. T. U.	Radiating Surface. Square Feet.
1	1	9000	36	15000	60
1¼	1	18000	72	30000	120
1½	1¼	30000	120	50000	200
2	1½	70000	280	120000	480
2½	2	132000	528	220000	880
3	2½	225000	900	375000	1500
3½	2½	330000	1320	550000	2200
4	3	480000	1920	800000	3200
4½	3	690000	2760	1150000	4600
5	3½	930000	3720	1550000	6200
6	3½	1500000	6000	2500000	10000
7	4	2250000	9000	3750000	15000
8	4	3200000	12800	5400000	21600
9	4½	4450000	17800	7500000	30000
10	5	5800000	23200	9750000	39000
12	6	9250000	37000	15500000	62000
14	7	13500000	54000	23000000	92000
16	8	19000000	76000	32500000	130000

In above table each square foot of radiating surface is assumed to transmit 250 heat-units per hour, a safe and conservative estimate, as will be seen by consulting Chapter IV.

For pipes of greater length than 100 feet multiply results in the above table by the square root of 100 divided by the length. In all cases the length is to be taken as the equivalent length in straight pipe of the pipe, elbows, and valves, as given on page 226. For other lengths multiply above results by following factors:

Length of pipe in feet.. 200 300 400 500 600 700 800 900 1000
Factor 0.71 0.58 0.5 0.45 0.41 0.38 0.35 0.33 0.32

For example, the capacity of a pipe 8 inches in diameter and 800 feet long would be 0.35 of 12800 sq. ft. of radiating surface = 4480 sq. ft. It will be noted that the size of return specified by Mr. Wolff is about one pipe-size greater than believed to be necessary by the author, but sizes of main steam-pipe are in substantial agreement with tables on pp. 226 and 226*b*.

The following table will be found convenient for obtaining the size of a steam-main for low-pressure steam-heating, single-pipe system, for various lengths. The table is computed from same formulæ as those on page 226, but for a lower steam-pressure, and results are given in commercial sizes of pipes.

COMMERCIAL SIZES OF STEAM-MAINS FOR A SINGLE PIPE.

(System of heating by direct radiation; pressure 0.5 lbs.; friction resistance 6 inches of water for lengths 100 feet and under, 12 inches of water for greater distances.)

Radiating Surface. Square Feet.	Length of Steam-main in Feet.								
	20	40	80	100	200	300	400	600	1000
	Diameter of Pipe in Inches								
20	1	1	1¼	1¼	1¼	1¼	1¼	1½	1½
40	1¼	1¼	1¼	1¼	1¼	1¼	1¼	1½	1½
60	1¼	1¼	1¼	1¼	1¼	1¼	1¼	1½	1½
80	1¼	1¼	1¼	1¼	1¼	1¼	1¼	1½	2
100	1¼	1½	1½	1½	1½	1½	1½	1½	2
200	1½	1½	2	2	2	2	2	2½	3
300	2	2	2	2	2	2½	2½	3	3½
400	2	2	2½	2½	2½	3	3	3	4
500	2	2½	2½	3	3	3	3½	3½	4
600	2½	2½	3	3	3½	3	3	3½	4½
800	2½	3	3½	3½	3½	3½	4	4	5
1000	3	3½	3½	4	4	4	4	4½	6
1400	3½	3½	4	4	4	4½	4½	5	6
1800	4	4	4	4	4½	5	5	6	7
2000	4	4	4	4½	4½	5	5	6	7
3000	4½	4½	4½	5	5	6	6	7	8
4000	5	5	5	6	6	7	7	7	9
6000	5½	5½	6	7	7	7	7	8	10
8000	5½	5½	6	7	7	8	8	9	11
10000	6	6	6	7	8	8	9	10	12
12000	6	7	7	7	8	8	10	11	12
14000	7	7	7	8	9	9	10	12	14
16000	7	8	8	9	9	10	11	12	14
18000	8	8	8	9	10	11	11	12	14
20000	9	9	9	10	11	11	12	14	16

In using the above table take the equivalent length as explained on page 226.

for capacity of steam-mains are computed for steam 10 pounds above atmospheric pressure, and the frictional resistance 6 inches of water column. Tables computed from the same formula and covering other conditions will be found in "Steam-heating," * by Robert Briggs, and can be consulted when desired.

123. Size of Return-pipes, Steam-heating.—The size of return-pipes, if figured from the actual volume of water to be carried back, would be smaller than is safe to use, largely because of air which is contained in the steam-pipes, and which does not change in volume when the steam is condensed. For this reason it is necessary to use dimensions which have been proved by practical experience to be satisfactory. When the steam-main is large, the diameter of the return-pipe will prove satisfactory if taken one size less than one half that of the steam-pipe; but if the steam-main is small, for instance, 5 inches or less, the return-pipe should be but one or two sizes smaller. The return-pipe should never be less than 1 inch, in order to give satisfactory results. The following table suggests sizes of returns which will prove satisfactory for sizes of main steam-pipes as given:

Diameter Steam-pipe.	Diameter Return-pipe.	Diameter Steam-pipe.	Diameter Return-pipe.
inches.	inches.	inches.	inches.
1½	1¼	5	3
2	1½	6	3½
2½	1¾	8	4
3	2	9	4¼
3½	2¼	10	4½
4	2½	12	5

The size of return-pipes, if computed on basis of reduction in volume due to condensation of the steam, supposing the steam to have a gauge-pressure of 40 pounds and that one half its volume is air, would be, neglecting friction, about one sixth of that of the main steam-pipe, which is much smaller than would be considered safe in practice.

---

* Van Nostrand's Science Series, No. 68.

Main and Return-pipes for Indirect Heating Surfaces.—The indirect heating surfaces require about twice as much heat as the same quantity of direct radiating surface, and hence, for same resistance in the pipe, the area should be twice that required in direct heating. It will usually be sufficiently accurate to use a pipe whose diameter is 1.4 times greater than that for direct heating.

Reliefs and Drip-pipes.—The size of drip-pipes necessary to convey the water of condensation from a main steam to a return cannot be obtained by computation, as there is much uncertainty regarding the amount of water that will flow through.

As the flow through the relief tends to increase the pressure in the return, it may also serve to lessen the velocity of flow beyond the point of junction, provided the size is greater than necessary to carry off the water of condensation from the steam-main. Drip-pipes should be united to the return in such a manner as to re-enforce rather than impede the circulation, which result can usually be attained by joining the pipes with 60 or 45 degree fittings.

The writer would recommend the employment of the following sizes of drip-pipes as ample for usual conditions:

DIAMETER OF DRIP-PIPE FOR STEAM-MAINS OF VARIOUS LENGTHS.

Diameter of Steam-main, Inches.	Length of Steam-main in Feet.			
	0 to 100.	100 to 200.	200 to 400.	400 to 600.
	Diameter of Drip-pipe in Inches.			
0 to 2	$\tfrac{1}{4}$	$\tfrac{1}{2}$	$\tfrac{1}{2}$	$\tfrac{3}{4}$
3	$\tfrac{1}{2}$	$\tfrac{1}{2}$	$\tfrac{3}{4}$	1
4	$\tfrac{3}{4}$	$\tfrac{3}{4}$	1	$1\tfrac{1}{4}$
5	$\tfrac{3}{4}$	1	$1\tfrac{1}{4}$	$1\tfrac{1}{4}$
6	1	$1\tfrac{1}{4}$	$1\tfrac{1}{2}$	$1\tfrac{1}{2}$

124. Size of Pipes for Hot-water Radiators.—Method of computation of the velocity with which circulation will take place in a hot-water heating-system without friction has been considered in Article 121, page 220. In some instances this velocity is increased by bubbles or particles of steam which pass up the main risers and reduce the specific gravity of the water in the ascending pipes to such an extent that the actual velocity produced is much in excess of what would have been possible had no steam formed. This condition is undesirable, as it is usually accompanied with more or less noise and a very high temperature in the boiler, and should not serve as a basis for designing main-pipes to be used in hot-water heating apparatus. It should not be recommended that heaters be run in such a manner as to produce steam in any part of the circulation.

The heat which is given off from radiating surfaces of various kinds has already been considered (page 204), and as each thermal unit given off by the surface is obtained by the cooling of one pound of water one degree in temperature, it is easy to compute from the data already given (1) the weight of water required, and (2) the number of cubic feet needed to heat each square foot of radiating surface.

The following table gives the data necessary for computing the volume of water required to supply radiating surface for various conditions likely to occur in heating:

HOT-WATER HEATING.

Data Used in Computation of Tables.

Temperature outside air........		0	0	0	0	0
Temperature water in radiator..		140	160	180	200	220
Heat-units per degree diff. temperature per square foot per hour......................		1.4	1.45	1.5	1.6	1.8
Weight of cu. ft. water, pounds..		61.37	60.98	60.55	60.07	59.64
Total heat-units per square foot per hour:						
Room 60° per sq. ft........		113	145	180	224	288
" 70° " " "........		98	130	165	208	270
Cubic feet of water required to supply one sq. ft. per hour.						
Radiator cooled 5°—	Room 70°	0.316	0.426	0.546	0.686	0.902
" "	" 60°	0.396	0.472	0.592	0.740	0.970
Radiator cooled 10°—	" 70°	0.158	0.213	0.273	0.343	0.451
" "	" 60°	0.183	0.236	0.296	0.37	0.483
Radiator cooled 15°—	" 70°	0.138	0.142	0.182	0.228	0.339
" "	" 60°	0.132	0.157	0.131	0.247	0.361
Radiator cooled 20°—	" 70°	0.079	0.107	0.137	0.172	0.226
" "	" 60°	0.091	0.118	0.148	0.175	0.241

By dividing the number of cubic feet to be supplied per hour by the velocity with which the water moves per hour we obtain the area of the pipe in square feet.

The general case from which practical tables may be computed can best be considered by the use of formulæ, as follows:

Let w equal the weight of water per cubic foot, let H equal total heat per square foot per hour from radiator, R total radiating surface, Q number of cubic feet of water per hour, A area of pipe in square feet, a area of pipe in square inches, v velocity in feet per second as given in table, page 221, V equal velocity in feet per hour, T loss of temperature of water in radiator. We have the following formulæ:

(1) $a = 144A.$

(2) $V = 3600v.$

(3) $\dfrac{HR}{wT} = Q$ { Total heat divided by heat given off by 1 cu. ft. equals total number of cubic feet.

(4) $\dfrac{Q}{V} = \dfrac{Q}{3600v} = A = \dfrac{a}{144}.$ From which

(5) $Q = 25av.$ Equate (3) and (5), and

(6) $R = \dfrac{25avwT}{H}.$

(7) $a = \dfrac{HR}{25wvT}.$

By taking special values corresponding to temperatures of water and of surrounding air we can reduce these formulæ to simple forms. Thus, if the temperature of the radiator is 180° and of the room 70°, the total heat-units given off per hour, H, will be 165. If we further assume that the water in the radiator cools during the circulation a certain amount, say 10 degrees, T will equal 10, weight of water w will equal 60.5 pounds, and we shall have formulæ 8 and 9:

(8) $R = 92av.$

(9) $a = \dfrac{R}{92v}.$

For the above condition the radiating surface is equal to 92 times the area of the main pipe in square inches times the velocity of the water in feet per second; and further, the area in square inches is equal to the radiating surface divided by 92 times the velocity. The velocity in feet per second will depend upon the height, the difference of temperature, and amount of friction.

The following table gives relations of radiating surfaces to areas of main pipes, friction neglected. For distances less than 200 ft. sufficient allowance for friction will be made by making the main one size larger than required by table.

AREA AND DIAMETER OF HOT-WATER HEATING-MAIN, DIRECT RADIATION.*

DIFFERENCE OF TEMPERATURE, 10 DEGREES.

(1) Height, Feet.	(2) Velocity Water Feet per Second.	(3) Multiply each 100 Square Feet Radiating Surface for Area Main by	(4) Multiply Square Root Radiating Surface for Diameter by	(5) Equivalent Head in Feet.
1	0.335	3.26	0.205	0.0015
5	0.750	1.45	0.133	0.0081
10	1.06	1.03	0.113	0.017
15	1.28	0.85	0.104	0.025
20	1.5	0.723	0.095	0.035
25	1.67	0.65	0.091	0.044
30	1.83	0.595	0.087	0.052
40	2.12	0.513	0.081	0.072
50	2.37	0.46	0.076	0.088
60	2.59	0.42	0.072	0.105
80	3.00	0.362	0.068	0.142
100	3.35	0.324	0.064	0.176

In the above table column (1) gives the height in feet; column (2) the velocity corresponding to the head for a reduction in temperature of 10° F.; column (3) is the area in square inches, neglecting friction, for each 100 square feet of radiating surface; column (4) is the corresponding diameter of pipe required for each square foot of surface, and is to be multiplied by the number of square feet of radiating surface to give the diameter for any given case; the actual diameter should be one pipe size greater; column (5) is the equivalent head which would produce the same velocity if falling freely in the air.

The preceding table is in the same form as that given for diameters of steam-main. If we consider 10 feet as the average height or head producing circulation for the first floor, it will be seen that we shall need, neglecting friction, one square inch in area in our main pipe for each 100 square feet of radiation, or the diameter of our pipe would be found for this case

* As illustrating the use of the table, compute the area of main pipe needed to supply 350 square feet of direct radiation situated 25 feet above the heater. The area is obtained by multiplying 3.5 by 0.65, which will equal 2.28 square inches. The diameter can be found from this, or it may be obtained from column (4), by multiplying the square root of 350 by 0.091. The square root of 350 is 18.7, the product is 1.7. The pipe used, if the distance is about 200 feet, should be 2½ inches in diameter.

as equal approximately to $\frac{1}{4}$ of the square root of the radiating surface in square feet.

If the temperature of the water be supposed to change 20° in passing through the radiators, the required area of the main would be one half of that given by the table; if 15°, two thirds, etc.

In hot-water heating the *return-pipe must have the same diameter as the supply-pipe*, since there is no sensible change in bulk between the hot and cold water.

We may take as a practical rule, applicable when less than 200 feet in length: *The diameter of main supply- or return-pipe in a system of direct hot-water heating should be one pipe-size greater than the square root of the number of square feet of radiating surface divided by 9 for the first story, by 10 for the second story, and by 11 for the third story of a building;* for indirect hot-water multiply above results by 1.5.

125. Size of Ducts and Ventilating-flue for Conveying Air.—The method of computing the sizes of flues would evidently be that of dividing the total amount of air which is required in a given time by that delivered or discharged through a flue one square foot in area. A table has been given for cubic feet of air delivered in ventilating-pipes, see Chapter I, pages 45 to 52. The air required can be found as explained in Article 119, page 211, formula 4, or by consulting the table, page 213, which gives the factors to be multiplied by the area of glass plus $\frac{1}{4}$ the exposed wall surface when the air enters at various temperatures above that in the room.

As an illustration, consider the same problem as in previous cases, viz., that of a room with 48 square feet of glass surface and 320 square feet of exposed wall surface, and from which the heat loss per degree difference of temperature is 128. Supposing air in room to be 70° F. and that supplied by flue to be 100° F., we see by table page 213, that for every heat-unit as above there will be required 135 cubic feet of air per hour, and for this case we will require $135 \times 128 = 17,280$ cubic feet per hour. If excess of temperature of air in flue over that outside be considered as 50°, and height of flue as 10 feet, the discharge per square foot of flue (see table page 45) will be 242 feet per minute, or 14,520 per hour. Hence the required area of the flue will be 17,280 divided by $14,520 = 1.19$ square feet

= 171 square inches. In a similar manner areas of flues may be computed for any given case.

As the velocity of flow increases with difference of temperature between outside air and that in the flue, and is lessened when this difference is small, it is better to assume a mean difference of temperature so low that the computation will certainly afford plenty of air for ventilation.

AREA OF FLUE IN SQUARE INCHES REQUIRED TO SUPPLY GIVEN AMOUNT OF HEAT.

(Excess of temperature is 30°; allowance for friction 50%.)

Building Loss, Total per Hour.*	Building Loss per Deg. Diff. of Temp. per Hr.†	Height or Head of Flue in Feet.									
		5	10	15	20	30	40	50	60	80	100
		Area of Flue in Square Inches.									
B.T.U.	B.T.U.										
700	10	24	17	14	11	9.2	8.2	7.1	6.6	6.1	5.5
1400	20	48	35	28	22	18.4	16.4	14	13.2	12.2	10.9
2100	30	72	52	42	33	28	25	21	19.5	18.3	16.3
2800	40	96	69	56	44	37	33	28	26	24.4	21.8
3500	50	120	87	71	55	46	41	35	32.6	30.5	27.3
5250	75	180	129	116	82	69	61	53	48	45.7	40.8
7000	100	240	173	141	109	93	82	71	66	61	54.5
8750	125	300	216	176	136	115	102	87	82	76.5	68.1
10500	150	360	258	212	164	138	122	105	98	152	81.7
12250	175	420	302	247	191	162	143	123	114	107	95.3
14000	200	480	346	244	218	184	163	141	130	124	109
17500	250	600	432	315	273	231	204	175	163	153	136
21000	300	720	519	423	327	278	245	211	195	183	163
28000	400	960	652	564	436	369	327	281	261	244	218
35000	500	1200	865	715	545	462	408	352	326	306	273
52500	750	1800	1290	1060	825	693	612	527	457	453	408
70000	1000	2400	1730	1410	1090	925	818	705	655	612	545
87500	1250	3000	2160	1760	1360	1150	1018	870	820	765	681
105000	1500	3600	2580	2120	1640	1380	1218	1055	980	1520	817
140000	2000	4800	3460	2440	2180	1840	1630	1410	1300	1240	1090
175000	2500	6000	4320	3150	2730	2310	2040	1750	1630	1530	1360
210000	3000	7200	5190	4230	3270	2780	2450	2110	1950	1830	1630

Table is computed by finding air required to supply heat by formula 4, page 211, when outside air is 0°, inside air 70°, and heated air 100°, and dividing this by the air supplied by a flue one square foot in area for the given height and a difference of temperature of 30°, as obtained in table page 45. Ventilating flues for a given height should be taken one quarter larger than the values given in the table. See note on page 246.

* See page 57.

† Approximately equal to area of glass plus one fourth the exposed wall-surface. See page 59.

The table on p. 233 is computed by the method explained for different heights of flue and for a difference of temperature of the air in the flue over that in the space into which it discharges of 30° F.

For difference of temperature other than 30° multiply results in the table by the following factors to obtain the area of the flue:

Difference Temperature, Degrees.	Factor.	Difference Temperature, Degrees.	Factor.
10	1.74	50	0.775
20	1.22	60	0.71
40	0.87	70	0.655

For usual conditions of residence heating in which the air in the supply-flue is 30° above the temperature of the air in the room, and that in the ventilating-flue 20°, we may compute the approximate area in square inches of the supply- and ventilating-duct, by multiplying each heat-unit per degree difference of temperature lost from the walls by a series of simple factors which are easily memorized.

TABLE OF FACTORS FOR AREA OF AIR-FLUES.

Story of Building.	Supply-duct.			Ventilating-duct.		
	Approximate Head in feet.	Velocity in feet per sec.	Factor for Area, sq. in.	Approximate Distance to Roof.	Velocity in feet per sec.	Factor for Area, sq. in.
	(1)	(2)	(3)	(4)	(5)	(6)
First Floor.....	5	2.8	2.40	47	5.5	0.93
Second "	28	6.8	0.95	32	4.2	1.27
Third "	40	8.1	0.82	20	3.6	1.33
Fourth "	50	9.	0.71	10	2.6	2.17

As an example, find the required area of heat- and ventilating-ducts for a room with 200 square feet of exposed wall-surface and 30 square feet of glass: 30 plus one fourth of 200 is 80, the approximate building loss per degree. This quantity multiplied by factors in columns (3) and (5) gives respective areas of flues in square inches with sufficient exactness for ordinary requirements. The factors afford a ready means of computation in the absence of an extended table, similar to that on page 233.

In some instances the amount of air can be computed as a function of the cubic contents of the room, especially when required for ventilation alone. For ventilation purposes the problem of proportioning the air-passages is solved simply by computing, first, the air required, on the basis of 1800 cubic feet per hour for each person who will occupy the room; second, the number of times the air will be changed per hour, by dividing this result by the volume of the room. This method is considered fully in Article 38, page 53, and a table is given for computing the area of the flue in square inches for different velocities of the moving air.

In applying this method to practical problems, it is best to proportion the ducts so that in no case will the required velocity of the air in the flue exceed 12 feet per second or 43,200 feet per hour, an amount not likely to be reached without a fan or blower, and one which corresponds to a pressure of nearly 0.1 inch of water (pages 42 to 53).

126. **Dimensions of Registers.**—The registers should be so proportioned that the velocity of the entering air will not be sufficient to produce a sensible draft; that is, the area must be such that the velocity shall not exceed 3 to 5 feet per second or 10,800 to 18,000 lineal feet per hour. The writer thinks that very excellent results are obtained by proportioning the registers for first floor so as to give velocity of $2\frac{1}{2}$ feet per second, and those of higher floors and at entrance to ventilating-shafts 3 feet per second.* The results above, except for entrances to ventilating-shafts on the top floor, are less than is usually produced by natural draft, so that the area computed by dividing the total amount of air required by the number which expresses the velocity gives satisfactory results.

The above rules are for effective or clear opening, and this will be found in each case to be about two thirds of the nominal or rated size of the register as shown in the table given in Article 144.

By computing, from the data given, the number of changes of air per hour in room, the table page 53 can be used as explained to determine the effective area in square inches required for each 1000 cubic feet of space.

---

* See page 52, Article 38.

As an example illustrating use of this table, suppose, in a room containing 2500 cubic feet, air to be changed four times per hour, and that velocity in air-flue be 6 feet per second, in ventilating-shaft 4 feet, through fresh-air register 2.5 feet, through ventilating-register 3 feet.

The table on page 53 gives the net area for each 1000 cubic feet of space, so that for above conditions the results as found in the table must be multiplied by 2.5. We should have taking 2.5 times the tabulated values, the following results:

Net area supply-flue 67.5 sq. in.; ventilating-shaft 100 sq. in. fresh-air register 166 sq. in.; ventilating-register 136.5 sq. in.

The nominal area of the register to be used should be about 50 per cent greater than the net area; it may be taken from the table given in Article 144. The velocity corresponding to 2.5 feet per second is taken as the mean of that given in the table for 2 and 3.

It is best to make flue dimensions about one inch greater than obtained by calculation, to allow for surface friction.

127. Summary of Various Methods of Computing Quantities Required for Heating.—The following table gives the required size of steam-pipes and of steam-boiler or hot-water heater, for various amounts of radiating surface. The proportions given will apply to residence heating or where the length of main pipe is not over 200 feet. The value given for the steam-main is that for the single-pipe system when no return is needed. For the system of separate steam- and return-pipe the diameter of the steam-main should be taken ¾ of that given that of the return as in table page 227. The cubic space heated is given if the ratio to radiating surface be known; this is an approximation only, although it may often serve a useful purpose when experience has been gained of heat required in constructions of similar nature in the same locality.

About two thirds as much air is warmed by hot-water as by steam radiators, and flues should be about two thirds as large as given in the table on page 238.

128. Heating of Greenhouses.—Greenhouses and conservatories are heated in some cases by steam and in other cases by hot water, and there is quite a difference of opinion held by florists respecting the relative merits of these two

REQUIRED PROPORTION OF PARTS; DIRECT STEAM-HEATING.

Radiating surface, square feet	100	250	500	750	1,000	1,500	2,000	3,000	4,000	5,000	7,500	10,000
Diameter steam-main, inches*	1.5	2	2.5	2.5	3	3.5	4	4.5	5	5.5	6	7
Heating-surface boiler, square feet	25	55	98	138	178	250	322	447	586	710	833	1,110
Grate-area boiler, square feet	0.9	1.9	3.9	5.4	6	8.9	11.2	15.5	19.5	23.2	32.5	42
Diameter smoke-flue, inches	5	7	9	10	12	13	15	18	20	23	27	30
Cubic feet heated, 40 to 1	4,000	10,000	20,000	30,000	40,000	60,000	80,000	120,000	160,000	200,000	300,000	400,000
" " 50 to 1	5,000	12,500	25,000	37,500	50,000	75,000	100,000	150,000	200,000	250,000	375,000	500,000
" " 75 to 1	7,500	18,750	37,500	56,250	75,000	112,500	150,000	225,000	300,000	375,000	562,500	750,000

DIRECT HOT-WATER HEATING.

Radiating surface, square feet	100	250	500	750	1000	1500	2000	3000	4000	5000	7500	10,000
Diameter pipe, inches, 1st story	1.5	2.5	3.5	4.0	4.5	5.0	5.5	6.5	7.5	8.5	9.5	11
" " 2d story	1.5	2	3.0	3.5	4.5	4.5	5.5	6.0	7.0	7.5	8.5	10
" " 3d story	1.25	1.5	2.5	3.0	3.5	4.5	4.5	5.0	7.0	6.5	7.5	9
Heating-surface heater, square feet	16	36.5	65	92	118	166	215	296	385	470	703	950
Grate-area heater, square feet	0.5	1.25	2.6	3 6	4.3	5.9	7.1	10.3	13.0	15.3	21.5	27.5
Diameter smoke-flue, inches	5	7	9	10	12	13	15	18	20	23	27	30
Cubic feet heated, 20 to 1	2000	5000	10 000	15,000	20,000	30,000	40,000	60,000	80,000	100,000	150,000	200,000
" " 30 to 1	3000	7500	15,000	22,500	30,000	45,000	60,000	90,000	105,000	150,000	225,000	300,000
" " 40 to 1	4000	10,000	20,000	30,000	40,000	60,000	80,000	120,000	160,000	200,000	300,000	400,000

INDIRECT STEAM-HEATING.

Square feet radiation	100	250	500	750	1,000	1,500	2,000	3,000	4,000	5,000	7,500	10,000
Cubic feet air heated per minute	480	1,200	2,400	3,600	4,800	7,200	9,600	14,400	19,100	24,000		48,000
Diameter main steam-pipe*	2.0	3.0	3.5	4.0	4.5	5.0	5.5	6.0	7.0	8.0		9
Heating-surface boiler, square feet	50	110	196	276	356	500	644	894	1,160	1,430		2,220
Grate-area boiler, square feet	1.8	3.8	7.8	10.8	12.8	17.8	22.4	31.0	39	46		84
Diameter smoke-flue, inches	7	10	12	14	17	20	21	25	32	38	4	42
Cubic feet heated, 20 to 1	2,000	5,000	10,000	15,000	20,000	30,000	40,000	60,000	80,000	100,000		200 000
" " 30 to 1	3,000	7,500	15,000	22,500	30,000	45,000	60,000	90,000	120,000	150,000		300,000
" " 40 to 1	4,000	10,000	20,000	30,000	40,000	60,000	80,000	120,000	160,000	200,000		400,000

INDIRECT HOT-WATER HEATING.

Square feet radiation	100	250	500	750	1,000	1,500	2,000	3,000	4,000	5,000		10,000
Cubic feet air heated per minute	374	925	1,850	2,770	3,740	4,600	7,480	11,200	15,000	18,800		37,000
Diameter supply- and return-pipe*	2.5	3.5	4.0	5.0	5.5	6.0	7.0	8.0	9.0	10		12
Heating-surface in boiler, square feet	32	72	130	184	228	332	430	592	730	940		1,900
Grate-area in boiler, square feet	1.0	2.5	4.0	5.2	8.6	11.8	14.2	20.6	26.0	30.6		55
Diameter smoke flue, inches	7	10	12	14	17	20	21	25	32	38		42
Cubic feet heated, 15 to 1	1,500	3,750	7,500	11,250	15,000	22,500	30,000	45,000	60,000	75,000		150,000
" " 25 to 1	2,500	6,250	12,500	18,750	25,000	37,500	50,000	75,000	100,000	125,000		250,000
" " 35 to 1	3,500	8,750	17,500	26,250	35,000	52,500	70,000	105,000	140,000	175,000		350,000

DATA FOR COMPUTATION.—Temperature outside air 0, room 70, entering air 100, temperature steam-surface 220, hot water 180.
* Pipe assumed 100 feet in length. Diameter of main in steam-heating for single-pipe system.

HOT-AIR AND VENTILATING FLUES.

Indirect Radiation Steam Circulation.

Square feet radiation.............	25	50	75	100	125	150	175	200	250
Cubic feet air per minute.........	122	244	362	486	602	729	846	972	1220
Area hot-air flue, square feet :									
1st story........................	0.72	1.45	2.16	2.87	3.57	4.3	5.0	5.7	7.3
2d story........................	0.29	0.59	0.88	1.9	1.47	1.78	2.06	2.35	2.95
3d story........................	0.24	0.49	0.73	0.97	1.22	1.46	1.7	1.95	2.45
Area ventilating flue, square feet :									
1st story........................	0.37	0.74	1.1	1.46	1.81	2.2	2.57	2.95	3.7
2d story........................	0.48	0.87	1.44	1.92	2.37	2.8	3.35	3.84	4.8
3d story........................	0.55	1.1	1.64	2.2	2.71	3.3	3.85	4.4	5.4
Actual area register, square feet :									
1st story........................	1.22	2.4	3.6	4.9	6	7.3	8.4	9.7	12.2
2d and above...................	1.0	2	3	4	5	6	7	8.0	10.0
Ventilating register, square feet...	0.6	1.2	1.8	2.4	3	3.6	4.2	4.8	6.1

methods of heating. The fact, however, that either system when properly proportioned and well constructed gives satisfactory results indicates that the difference is not great, and that the relative value may depend entirely on local conditions.

The methods of piping employed may in a general way be like those described, and the pipes may be located so as to run underneath the beds of growing plants, or in the air above, as bottom or top heat is preferred. In many cases large cast-iron pipes, the method of erection of which is described in Article 58, page 88, are used in hot-water heating of greenhouses. These are generally located beneath the beds of growing plants; the main flow- and return-pipes are laid in parallel lines, with an upward pitch from the boiler to the farthest extremity of the house. Recently small wrought-iron pipes, Article 59, page 89, have been used extensively for greenhouse heating. In this case the main pipe has generally been run near the upper part of the greenhouse and to the farthest extremity in one or more branches, with a pitch upward from the heater for hot-water heating and with a pitch downward for steam-heating. The principal radiating surface is made of parallel lines of $1\frac{1}{2}$-inch, or larger, pipe, placed under the benches and supplied by the return current; this has in all cases a pitch toward the heater. An illustration of the method of piping as designed by A. H. Dudley of the Herendeen Mfg. Co. is shown in Figs. 193, 194, and 195 so clearly as to require no special explanation.

Any system of piping which gives free circulation and which is adapted to the local conditions will give satisfactory results. The directions for erecting and taking off branches are the same as in residence heating (see page 191).

Proportioning Radiating Surface.—The loss of heat from a greenhouse or conservatory is due principally to the extent of glass surface; hence the amount of radiating surface is to be

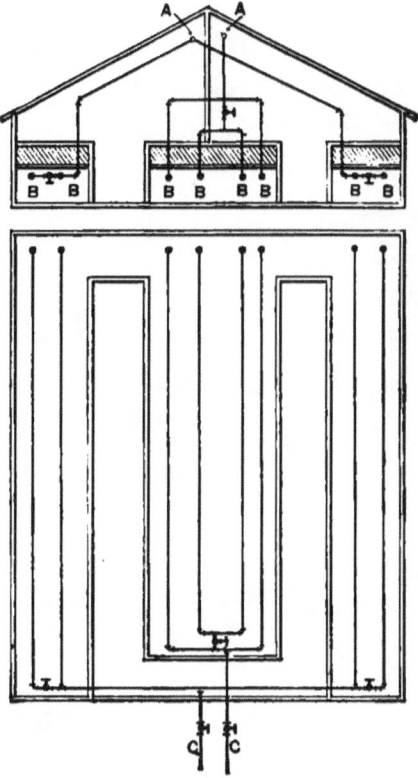

FIG. 193.—PLAN AND ELEVATION OF PIPING.

taken proportional to the equivalent glass surface, which in every case is to be considered as the actual glass surface plus $\frac{1}{4}$ the exposed wall surface. From this surface about 1 heat-unit will be transmitted from each square foot for each degree difference of temperature between that inside and outside per hour; that is, if the difference of temperature is 70 degrees, each square foot of glass surface would transmit 70 heat-units per hour. The radiating surface usually employed for this purpose is horizontal pipe, and hence is of the most efficient kind. From a surface of this nature we can consider without sensible error that 2.2 heat-units are given off from each square foot for each

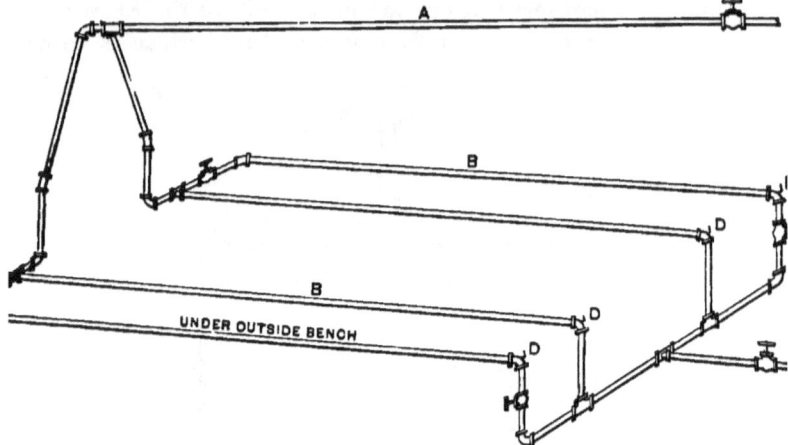

FIG. 194.— PIPING FOR OUTSIDE BENCH.

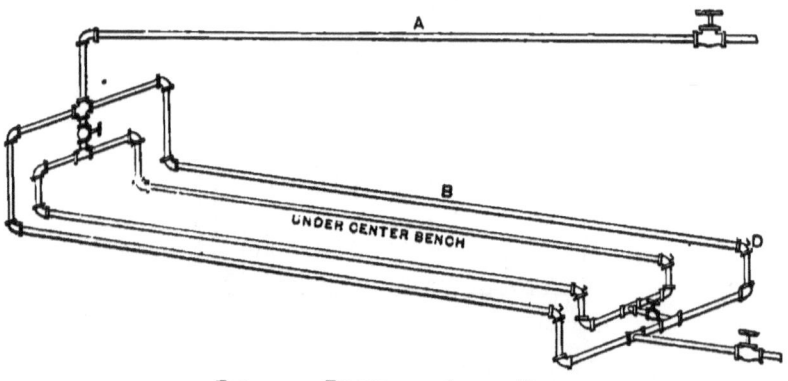

FIG. 195.—PIPING FOR INSIDE BENCH.

degree difference of temperature between the radiator and the air of the room per hour. From this data a table can be computed which gives the ratio of equivalent glass surface to radiating surface, in which the results will be found to agree well with average practice; the results are to be increased or diminished 10 to 20 per cent, according as the circumstances of exposure or the quality of the building vary more or less from the average condition.

TABLE SHOWING AMOUNT OF GLASS SURFACE OR ITS EQUIVALENT WHICH MAY BE HEATED BY 1 SQUARE FOOT OF RADIATING SURFACE IN GOOD BUILDINGS.

	Hot Water.			Steam.	
Temp. of Radiating Surface, Deg. F.	160°	180°	200°	5 lbs. 227°	10 lbs. 240°
	Square Feet of Glass for 1 Square Foot of Radiating Surface				
Temp. of surrounding air, 90° F...	1.9	2.3	2.8	3.3	3.8
" " " " 80° F...	2.3	2.9	3.5	4.0	4.6
" " " " 70° F...	3.0	3.6	4.2	5.0	5.7
" " " " 60° F...	4.0	4.6	5.25	6.0	7.0
" " " " 50° F...	5.0	6.0	6.8	8.0	9.0
" " " " 40° F...	6.9	8.0	8.2	10.0	11.5

From the data above the following table is computed, which gives the radiation in square feet required for greenhouses or conservatories with different amounts of glass surfaces. It also gives divisors from which the heating-surfaces or grate-surfaces in the boilers may be computed by dividing the given amount of radiation. Thus for a greenhouse with 1000 feet of glass surface, which is to be kept at 70 degrees in the coldest weather, we note in the table that 200 square feet of radiation will be required; the heating-surface in the boiler will be 200 divided by 5.6 ($= 36$) square feet, and the area of grate will be (200 divided by $156 =$) 1.28 square feet.

GREENHOUSE HEATING WITH STEAM.

Square feet of glass	100	250	500	750	1000	1500	2000	2500	3000	4000	5000	10,000
Radiation required, sq. ft., tempt. 40.	10	25	50	75	100	150	200	250	300	400	500	1,000
" " " " 50.	13	33	62	82	125	188	250	313	375	500	625	1,250
" " " " 60.	16	43	84	125	167	250	333	416	500	660	830	1,660
" " " " 70.	20	50	100	150	200	300	400	500	600	800	1000	2,000
" " " " 80.	25	64	125	188	250	350	500	625	750	1000	1250	2,500
Divisor of Radiation.												
For heating surface in boiler........	4	4.5	5.1	5.4	5.6	6.0	6.2	6.5	6.7	6.9	7.0	7
For area of grate	25	132	138	144	156	160	180	190	192	204	216	240

GREENHOUSE HEATING WITH HOT WATER.

Square feet of glass	100	250	500	750	1000	1500	2000	2500	3000	4000	5000	10,000
Water 160°.												
Radiation sq. ft. tempt. 40........	15	37	73	110	145	218	290	360	435	580	730	1,450
" " " 50........	20	50	100	110	200	300	400	500	460	800	1000	2,000
" " " 60........	25	62	125	187	250	375	500	625	750	1000	1250	2,500
" " " 70........	33	83	166	250	333	500	666	833	1000	1330	1660	3,333
" " " 80........	37	91	183	333	441	666	888	1000	1330	1660	2100	4,450
Divisors of Radiation.												
For heating surface...............	6.5	6.8	7.6	8.1	8.4	9.0	9.3	10.0	10.4	10.5	10.5	12.0
For grate surface	190	193	207	216	232	252	270	288	306	324	342	360

The sizes of main pipes should be the same as those which are used for direct heating, page 237.

Relative Tests of Hot-water and Steam Heating Plants.— Several tests have been made to determine the relative efficiency and economy of steam and hot-water heating plants. The first test so recorded was made at the Massachusetts Agricultural College by Professor S. T. Maynard, the results of which are given in Bulletins 4, 6, and 8, issued by the Mass. Exp. Station, 1889 and 1890. In this test two houses were used which were located as nearly as possible with equal exposure, and the tests were made with great care and by entirely disinterested observers. The following is a summary of the results and conclusions as taken from the bulletins:

STEAM-HEAT *VERSUS* HOT WATER.
[From Bulletin No. 4.]

In order to get at some facts in regard to this subject, so important to the grower of plants under glass, and gain some positive knowledge as to the relative value of the two systems, two houses were constructed during the summer of 1888, 75 × 18 feet, as nearly alike as possible in every particular. Two boilers of the same pattern and make were put in, one fitted for steam and one for hot water; the steam for heating the east house, and the hot water for the west and most exposed one. The boilers were completed and ready for work in November and were used until January 9, 1889, when these experiments began.

Records of temperature of each house were made at 7.30 and 9 A.M., and 3, 6, and 9 P.M. Sufficient coal was weighed out each morning for the day's consumption and the balance not consumed deducted the next morning. "The two boilers and fittings were put in so as to cost the same sum and were warranted to heat the rooms satisfactorily in the coldest weather."

These experiments were repeated during the months of January and February, 1889, and in summarizing the results it was found that the steam-boiler consumed during the two months referred to 6582 lbs. of coal, while the hot-water boiler consumed in the same time only 5174 lbs., a saving in favor of the latter of nearly 20 per cent. At the same time the temperature of the room heated by hot water averaged 1.7° higher than that heated by steam.

The temperature was more even where heated by hot water, and consequently there was less danger from sudden cold weather. This was strikingly shown on the night of February 22.

The average outside temperature for the day was 34°.

At 9 P.M. it was above 32°, and proper precautions not having been taken for so sudden a change as followed (the average temperature during the 23d of February was 2°), at 6 o'clock on the morning of the 23d the temperature of the room heated by steam was 29°, while in that heated by hot water it was 35°. . . .

[*From Bulletin No. 6.*]

The boilers used were built of cast-iron sections. In the hot-water boiler five sections are used, the area of heating surface exposed to the fire being 74.5 feet.

The steam-boiler consists of eight sections, the total heating surface of which is 85.12 feet.

The experiments reported in the April Bulletin were continued during the two following months of March and April, and from the tables showing the comparative results the following summary is appended:

SUMMARY FOR HOT-WATER BOILER.

Total coal consumed by hot-water boiler from December 23, 1888, to April 24, 1889, 4 tons 1155 lbs. Average daily temperature for the four months, 53.5°.

SUMMARY FOR STEAM-BOILER.

Total coal consumed by steam-boiler from December 23, 1888, to April 24, 1889, 5 tons 1261 lbs. Average daily temperature for the four months, 51.2°.

It will be seen by the above that the average temperature of the house heated by hot water was 2.3° higher than that heated by steam, and that the amount of coal consumed was 2106 lbs. less in the former than in the latter.

[*From Bulletin No. 8, April, 1890.*]

Much discussion having been provoked relative to the accuracy of the results of experiments with steam and hot water for heating greenhouses, reported in Bulletins No. 4 and 6, we have the past winter made a careful repetition of the experiments to correct any errors that might be found and to verify previous results.

The boilers having been run with the greatest care possible from December 1, 1889, to the present date, March 18, 1890, and every precaution having been taken that no error should occur, we give the results in the following table:

| Month. | HOT WATER. | | | | | STEAM. | | | |
| | Lettuce and Carnation Room. | | | | | Lettuce and Carnation Room. | | | |
	Outdoor Average Daily Temperature.	Indoor Minimum Temperature.	Indoor Maximum Temperature.	Indoor Average Daily Temperature.	Lbs. Coal Consumed.	Indoor Minimum Temperature.	Indoor Maximum Temperature.	Indoor Average Daily Temperature.	Lbs. Coal Consumed.
December	34.99	41.52°	57°	47.59°	1505	40.21°	51.69°	46.39°	2350
January	33.27	44.35	62.48	51.41	2304	42.72	61	49.45	3202
February	32.04	43.67	65.96	52.54	1704	42.42	66.32	51.01	2540
March, 17 days	29.75	39.94	58.83	47.44	1085	39.16	58.11	46.73	1692
Averages	32.51°	42.37°	61.06°	49.74°	Total 6598	41.12°	59.28°	48.39°	Total 9784

Summary for Hot-water Boiler.

Total coal consumed from December 1, 1889, to March 18, 1890, 6598 lbs. Average daily temperature for the time, 49.74°.

Summary for Steam-boiler.

Total coal consumed from December 1, 1889, to March 18, 1890, 9784 lbs. Average daily temperature for the time, 48.39°.

A saving of fuel in favor of hot water of about 33 per cent.

Similar tests were made under the general direction of Prof. L. R. Taft at the Michigan Experiment Station and are to be found reported in full in a paper by the writer, read before the American Society of Mechanical Engineers, Volume XI. For this test two houses were used, each of the same size and of the same grade of construction. The houses were equally exposed to the heat of the sun, but the hot-water house was rather more exposed to the wind. The general method of testing was essentially the same as that described, and the results show substantially the same difference. The heaters used were cast-iron of the drop-tube form, quite different from those used in Massachusetts, but well adapted for the work.

The following table gives a summary of the results:

SUMMARY OF RESULTS OF TEST OF HOT WATER AND STEAM.

Year.	1889		1890						
Months.	December.		January.		February.		March.		April.
Days of Experiment.	10		31		28		20		30
	Steam.	H. W.	Steam.	H. W.	Steam.	H. W.	Steam.	H. W.	Steam.	H. W.
Total coal	1025	825	3475	2799	3400	2775	2714	2288	1800	1800
Average coal per day	93.2	75	112.1	90.3	121.4	99.1	114.4	135.7	60	60
Average outside temperature, 6 A.M.	31.8	31.8	27.7	27.7	22	22	19.2	19.2	36.2	36.2
Average outside temperature, 4 P.M.	38.5	38.5	38	38	33.8	33.8	29.2	29.2	42	42
Average outside temperature, 9 P.M.	35.1	35.1	27.2	27.2	27	27	22.0	22.0	38	38
Average inside temperature, 6 A.M.	53.9	54.9	52.5	54.1	54.1	54.4	53.3	54.3	51.8	58.4
Average inside temperature, 9 P.M.	54.9	60.3	53.8	54.8	53.5	56	55.7	57	54.9	60.2
Extreme variation	....	13	4.4	4	4.3	4.2	....	....	5.9	4.3

During the month of April, 1890, the same amount of coal was burned in both heaters in order to see what the effect would be on the resulting temperature of the two houses. The results gave a temperature which averaged 8.5 degrees higher in the hot-water house than in the steam-heated house.

Experiments were made by Prof. L. H. Bailey, of Cornell University, in 1891 with houses which were not similar either as to exposure or methods of piping, the results of which were in general somewhat more favorable to steam than to hot water. In 1892 Prof. Bailey arranged the same room so that it could be alternately heated with steam and hot water. The results of this last test so far as economy is concerned were also somewhat in favor of the steam-heat. The general conclusions which Prof. Bailey drew from this test were as follows:

CONCLUSIONS.

Under the present conditions the following results can be deduced. It will be observed that they confirm several of the conclusions of last year.

1. Hot water maintained a slightly greater average difference between the minimum inside and outside night temperature than steam.
2. There was practically no difference in the coal consumption under the two systems.
3. With a small plant like this the fluctuations under both systems are much greater than in larger ones, and neither proved very satisfactory.
4. The utility of slight pressure in enabling steam to overcome unfavorable conditions is fully demonstrated.
5. The addition of crooks and angles is decidedly disadvantageous to the circulation of hot water and of steam without pressure, but the effect is scarcely perceptible with steam under low pressure.
6. In starting a new fire with cold water, circulation commences with hot water sooner than with steam, but it requires a much longer time for the water to reach a point where the temperature of the house will be materially affected.
7. The length of pipe to be traversed is a much more important consideration with water than with steam.
8. A satisfactory fall towards the boiler is of much greater importance with steam than the manner of placing the pipes.

129. Heating of Workshops and Factories.—Workshops or factories where counter-shafts and belting are running which keeps the air in agitation can be heated satisfactorily by erecting coils of pipe for radiating surface near the ceiling of the room. Coils made with branch-tees, as described in Article 64, page 107, may be used, with the pipes placed in a horizontal plane and parallel to each other. In such a position the radiating surface is very efficient, and the heat given off as shown by experiment is a maximum. In a coil located near the ceiling the temperature of the room in the upper portion will become very high and will not be evenly distributed unless the air is mechanically agitated, so that the overhead system of piping is only satisfactory in shops and places where there are moving belts or other means for agitating the air. The method of proportioning supply-pipes and radiating surface for this case has already been considered. Mr. C. J. H. Woodbury gives, in Vol. VI, page 861, "Transactions of American Society Mechanical Engineers," considerable useful data relating to this method of heating. It is the favorite method for heating cotton-mills, about one foot in length of $1\frac{1}{4}$-inch pipe being used for 90 cubic feet of space.

NOTE.—In the preceding discussion a loss of 50 per cent due to friction has been assumed as probable in flues and registers through which air discharges into a room. This is a reasonable allowance under many conditions, but on the other hand is fully double the loss which will be experienced in flues which are smooth and well aligned, and provided with long and easy turns. When the conditions are favorable, flues which have an area of two thirds those specified in the table on page 233 will be perfectly satisfactory.

CHAPTER XI.

HEATING WITH EXHAUST STEAM. NON-GRAVITY RETURN-SYSTEMS.

130. General Remarks.—Steam after being employed in an engine contains the greater portion of its heat, and if not condensed or utilized for other purposes it can usually be employed for heating without materially affecting the power of the engine. The systems of steam-heating which have been described are those in which the water of condensation flows directly into the boiler by gravity. In other systems in use high-pressure steam is carried in the boilers, high- or low-pressure steam in the heating-mains and radiators, and the return-water of condensation is received by a trap and delivered either into a tank from which it is pumped into the the boiler or in some instances wasted. The exhaust steam may need to be supplemented by live steam taken directly from the boiler, which may be reduced in pressure either by passing, through a valve partly open, or a reducing-valve, as described in Article 137.

It will often be found that little attempt is made to utilize the heat escaping in the exhaust steam from non-condensing engines, and consequently a good opportunity exists for construction of systems which will save annually many times their first cost.

131. Systems of Exhaust Heating.—The exhaust steam discharged from non-condensing engines contains from 20 to 30 per cent of water, and considerable oil or greasy matter which has been employed in lubricating. When the engine is freely exhausting into the air, the pressure in the exhaust-pipe is, or should be, but slightly in excess of that due to the atmosphere. The effect of passing exhaust steam through heating-pipes is likely to increase the resistance and cause back pressure which will reduce the effective work of the engine. The engine delivers steam discontinuously, but at regular intervals at the end of each stroke. The amount is likely to vary with the work done by the engine, since the engine-governor is always adjusted to admit steam in such amount as is required to preserve uniform speed; if the work is light very little steam will be admitted to the engine. For this reason the supply available for heating varies within wide limits.

The general requirements for a successful system of exhaust-steam heating must be, first, the arrangement of a system of piping having such proportions as will make little or no increase in back pressure on the engine and will provide for using an intermittent supply of steam; second, provision for removing the oil from the exhaust, since this will interfere materially with the heating capacity of the radiating surfaces; third, provision against accidents by use of a safety or *back-pressure valve* so arranged as to prevent damage to the engine by sudden increase in back pressure.

These requirements can be met in various ways. To prevent sudden change in back pressure due to irregular supply of steam the exhaust-pipe from the engine should be carried directly to a closed tank whose cubic contents should be at least 30 times that of the engine and as much larger as practicable. This tank can be provided with diaphragms or baffle-plates arranged so as to throw all or nearly all the grease and oil in the steam into a drip-pipe, from which it is removed by means of a steam-trap, as described in Article 98, page 164. To this tank may be connected a relief-pipe leading to the back-pressure valve, and also a supplementary pipe for supplying live steam. The supply of steam for heating should be drawn from the top of the tank.

Any system of piping may be adopted, but extreme care should be taken that as little resistance as possible is introduced at bends or fittings. The radiating surface employed should be such as will give the freest possible circulation. In general, that system will be preferable in which the main steam-pipe is carried directly to the top of the building, the distributing-pipes run from that point, and the radiating surface is supplied by the down-flowing current of steam (Fig. 173). It is desirable to have a closed tank at the highest point of the system, from which the distributing-pipes are taken, and provided with drips leading to a trap so as to remove, before it can reach the radiating surface, any water of condensation or oil which has been carried to the top of the building.

132. Proportions of Radiating Surface and Main Pipes Required in Exhaust Heating.—The size of exhaust pipe required for an engine of given power, in order that the back pressure shall not exceed a certain amount, may be computed, the only data required in addition to that already given for heating with live steam, being that relating to the steam required by engines. The amount of steam used by engines will depend upon the workmanship and class to which they belong, but we can assume with little error that non-condensing engines will require the following weights of steam per horse-power per hour: simple with throttling-governor 40 pounds, with automatic governor 35 pounds, with Corliss valves 30 pounds; compound using high-pressure steam 25 pounds. In order that the pipes may be sufficiently large it is better to proportion the systems for the more uneconomical type.

TABLE OF DATA FOR COMPUTATION.

	0	1	2	3	10	−2	−5
Steam-pressure from Atmosphere. Absolute	14.7	15.7	16.7	18.7	24.7	12.7	9.7
Temperature of steam, F	212	216	219	222	239	204	192
Temperature of air	70	70	70	70	70	70	70
Difference	142	146	149	152	169	134	122
Heat per min. from 100 sq. ft. radiation in B. T. U. equal 3 times difference	426	438	447	456	507	402	366
Total heat of steam above 212°	966	967	967	967	973	962	958
Latent heat steam B. T. U	966	963	960	957	946	970	978
Cubic feet steam per lb	26.4	24.6	23.3	21.0	16.2	30.3	39.0
Cubic feet steam to weigh ¾ lb	17.6	16.4	15.5	14.0	10.8	20.2	26.0
Cubic feet steam required each min. to supply 100 ft. rad. sur., air 70°	11.6	11.25	10.85	10.1	8.8	12.6	11.5
Weight of 1 cubic foot steam lbs.	0.0379	.0403	.0427	.0475	.0640	.0326	.0257
Radiating surface per H. P. Throttling	152	146	143	139	126	162	179
Automatic	134	129	127	122	112	146	158
Corliss	114	110	107	104	95	122	134
Compound	95	91	90	87	79	102	112
Head of steam in feet equal 1 foot water of water column	1669	1585	1455	1317	1010	1902	2440

In the following discussion the dimensions of piping are computed for an engine using 40 pounds of steam per horse-power per hour (⅔ pound per minute), and exhausting

against a back pressure above or below atmosphere as stated.*
The preceding table gives properties of steam, also radiating surface supplied per horse-power by engines of various classes.

The computation of the size of exhaust-pipes can be made by the following algebraic process:

Let V equal velocity of the steam in feet per second; v, velocity in feet per minute; l, length of pipe in feet; D, diameter of pipe in feet; d, diameter in inches; A, area of pipe in square feet; Q, cubic feet of steam discharged per minute; h, back pressure above atmosphere expressed in feet of steam; p, back pressure expressed in pounds per square inch; HP, horse-power of engine; c, number of cubic feet in one pound of steam.

From the formulæ, page 218, we have, for velocity in feet per second

$$V = 50\sqrt{\frac{h}{l}D}; \quad \quad (1)$$

from which by reduction the velocity in feet per minute

$$v = 3000\sqrt{\frac{h}{l}D} = 866\sqrt{\frac{h}{l}d}. \quad \quad (2)$$

The discharge in cubic feet per minute

$$Q = Av = 3000A\sqrt{\frac{h}{l}D} = 4.723\sqrt{\frac{h}{l}d^5}. \quad \quad (3)$$

Since $\frac{2}{3}$ pound of steam is used per horse-power per minute,

$$Q = \tfrac{2}{3}cHP. \quad \quad (4)$$

From above by reduction

$$d = 0.537\sqrt[5]{\frac{Q^2l}{h}} = 0.457\sqrt[5]{\frac{c^2l}{h}HP^2}; \quad \quad (5)$$

$$HP = 7.135\sqrt{\frac{d^5h}{c^2l}}. \quad \quad (6)$$

In case the back pressure is equal to one foot of water column (0.433 pound per square inch) above atmosphere, $h = 1598$, $c = 25.7$, and we have

$$HP = 1.11\sqrt{d^5}.$$

For one pound back pressure

$$HP = 1.18\sqrt{d^5}.$$

It is advisable to make the diameter one inch greater to overcome additional resistances. (See table.)

---

* Radiating surface 25 per cent less. See Article 121, page 218.

RADIATING SURFACE AND HORSE-POWER OF ENGINE FOR A GIVEN DIAMETER OF EXHAUST-PIPE.

Diam. Exhaust-steam Pipe 100 Ft. Long. Back Pressure not to Exceed 0.4 Lb.	Corresponding H. P. of Engine.	Radiating Surface in Sq. Ft. Supplied by AutomaticType of Engine.	Diam. Exhaust-steam Pipe 100 Ft. Long. Back Pressure not to Exceed 0.4 Lb.	Corresponding H. P. of Engine.	Radiating Surface in Sq. Ft. Supplied by AutomaticType of Engine.
Inches.			Inches.		
2	1.12	110	6	63	6,200
2½	3.1	300	7	99.3	9,500
3	6.4	605	9	304	19,500
3½	11.1	1,050	12	356	34,000
4	17.5	1,650	14	562	54,000
4½	22.9	2,200	16	825	89,000
5	36.6	3,400	18	1,150	110,000

The foregoing table is computed for steam having a pressure of 0.43 pound above the atmosphere. For other pressures of exhaust multiply the results given in the table by the following factors (for other distances multiply by $0.1\sqrt{l}$):

Pressure.	Factor.
Atmospheric	1.05
2 pounds below	1.125
5 pounds below	1.27
2 pounds above	0.98
3 pounds above	0.895
10 pounds above	0.79

As an example; find the size of exhaust-pipe and amount of radiating surface supplied by the exhaust of a 50 horse-power engine of the automatic type, working against a back pressure of 0.43 pound. For this condition, the exhaust from one horse-power will supply 25 per cent less than 131 square feet of radiation (see table page 249), or 4900 square feet. From the table at top of page we see that a 6-inch pipe will be somewhat larger than required, but should be used. The amount of radiating surface needed to warm a given building will depend on pressure of the steam, exposure, and class of building, as explained on page 55.

133. Systems of Exhaust-heating with Less than Atmospheric Pressure.—If a system of exhaust-heating discharge the water of condensation directly into the atmosphere, the pressure must be slightly above atmospheric; but systems have been used with success in which the back-pressure was less than atmospheric, and in the table of proportions which has been given such cases are considered.

Such a system can be constructed by connecting the discharge from the system to an air-pump which will remove the water of condensation and to a great extent the atmospheric pressure; the heating surface will act as a condenser for the engine, and in case it is insufficient for this purpose a jet or surface-condenser, supplied with cold water may be used to supplement it. Instead of an air-pump and condenser, a siphon condenser, Fig. 196, may be used. This latter instrument is regularly on the market, and consists of a chamber above a convergent tube which receives the exhaust steam and a jet of water. This condenser depends for its action upon the fact that a column of water 34 feet in height will balance and overcome the atmospheric pressure. For its successful use it must be set so that the top of the condenser is at least 34 feet higher than the end of the discharge-tube, the bottom of which is to be submerged.

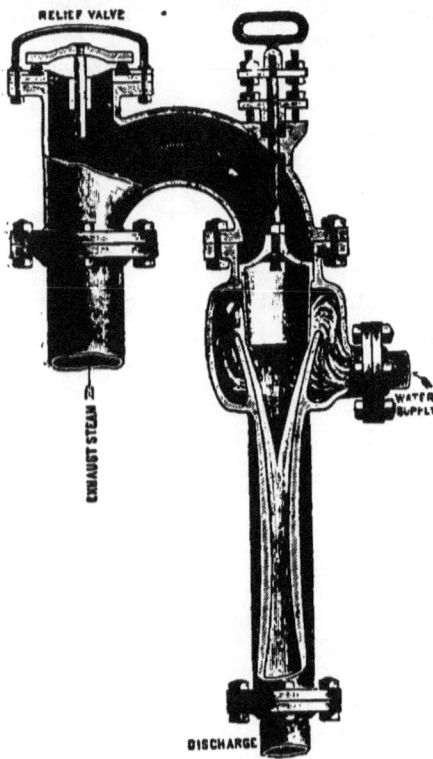

FIG. 196.—SIPHON CONDENSER.

In a system of exhaust heating by-pass connections to the condenser should be provided, so that the heating surface would not need to be used in warm weather.

Besides the general system which has been described, other systems of great merit have been devised and put on the market with many special and patented features. Of these we may mention first the Willames system, which is shown in Fig. 197, with details of construction. It will be seen that the exhaust from the engine is received into a large upright stand-pipe with back-pressure valve at top, and that the steam is drawn from near the top, and after passing through the

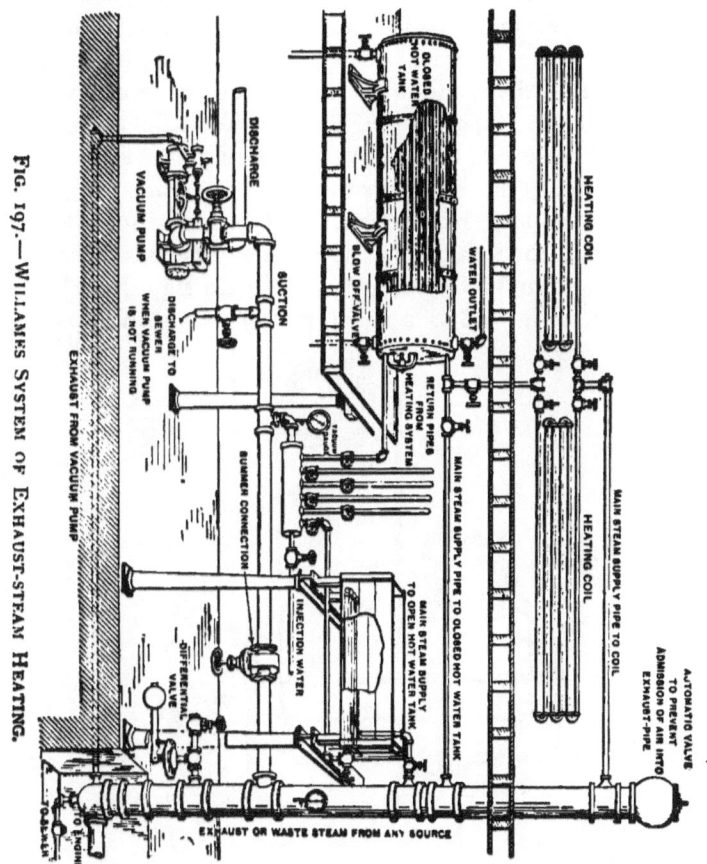

FIG. 197.—WILLAMES SYSTEM OF EXHAUST-STEAM HEATING.

radiating system, is received into a large branch-tee, which is supplied with injection-water and serves as a condenser. The suction-pipe of the air-pump is connected to the branch-tee and acts to remove the atmospheric pressure from the entire system. A by-pass for summer use is shown. Water is heated in the closed hot-water tank by a portion of the return, and may be used for any purpose needed, as, for instance, feed-water for boilers, heating by hot-water circulation, etc.

Another system of this kind which, by increasing the efficiency of surface, has met with much favor is that invented by Andrew G. Paul. This differs in construction and principle of operation from that described, in that instead of using an air-pump which receives all the exhaust, a small tank is connected with an induction condenser called an *exhauster*, which is connected to all the drips and to the air-valves of the radiators. An automatic device stops the operation of the exhauster as soon as the air is removed. The advantages of this system depend principally upon the quick removal of air from the various radiators and pipes, which constitutes the principal obstruction to circulation; the inductive action in many cases is sufficient to cause the system to operate at a pressure slightly below the atmosphere. Fig. 198 is a diagram * showing an application

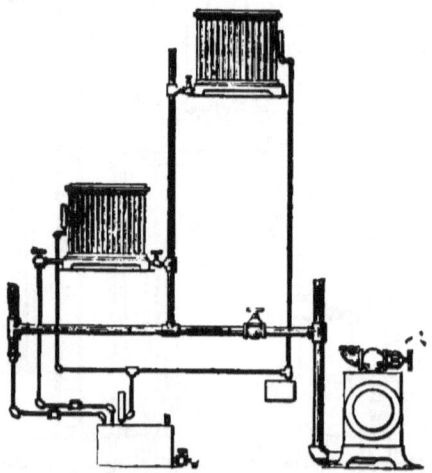

FIG. 198.—PAUL SYSTEM.

of the Paul system to the exhaust-piping of a steam-engine. The connections of two radiators are shown, one of which is of the single-pipe, the other of the two-pipe, system. The exhauster, shown in the lower left-hand corner, receives all the

---

* *Heating and Ventilation*, November 15, 1894.

drips from the piping and radiators, and is connected with the air-valve of each radiator.

134. Combined High- and Low-pressure Heating-systems.—In nearly all systems of heating with exhaust steam it is

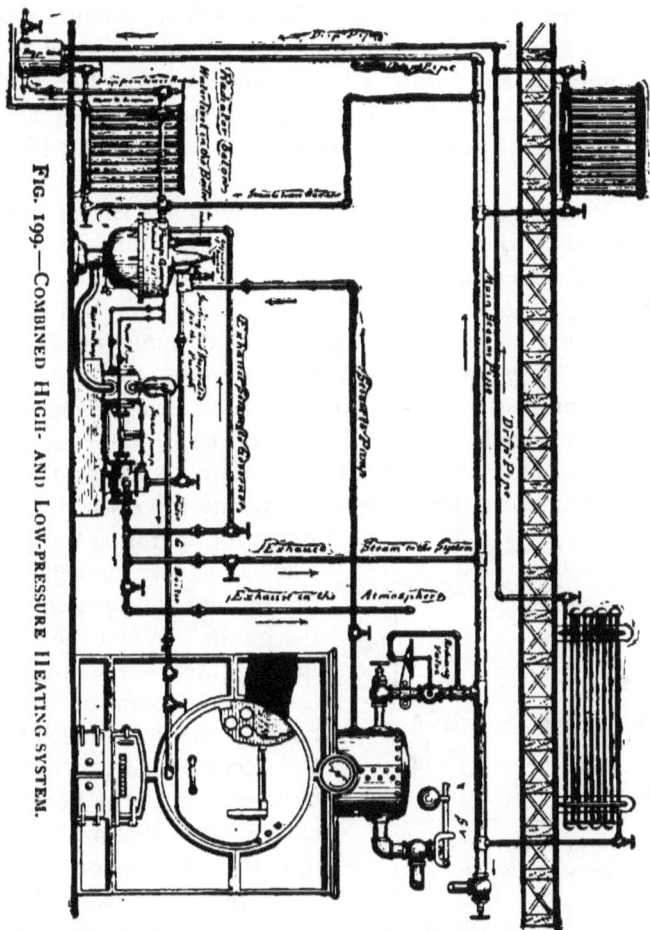

FIG. 199.—COMBINED HIGH- AND LOW-PRESSURE HEATING SYSTEM.

necessary to arrange the piping so that at times live steam may be admitted in any amount required, as substantially described in Article 130.

In some instances high-pressure steam is carried in the boiler and may possibly be used in a few radiators, while the principal part of the building is heated with low-pressure steam which is drawn directly from the boiler, and is reduced in pressure by passing through a reducing-valve. In this case the return-water of condensation passes to a tank or chamber at the lowest portion of the system, and is fed into the boiler by means of a return-trap or steam-pump. The principal elements of such a system is shown in Fig. 199, as designed by the Albany Steam Trap Company, and forms a useful illustration of the method of piping essential. To start the pump automatically and to keep it moving at the proper speed a pump-governor (Article 135) is used.

135. Pump-governors.—In non-gravity systems of heating the water of condensation is returned to the boiler by return-traps, as described in Article 99, page 167, or by steam-pumps. The trap is automatic, and when in good order will operate without attention, but the ordinary steam-pump needs to be started and stopped, as required, to remove the water. To render the pump automatic a device termed a *pump-governor* is often employed. Many forms are used, but they consist in nearly every case of a tank containing a float or equivalent device, connecting with levers to the valve which admits steam for operating the pump. The tank is connected to the suction and located above the pump. When the tank is full of water, the steam-pump is put in operation by the rising of the float, which opens the steam-valve. When the tank is empty, the float falls, closing the steam-valve and thus stopping the pump.

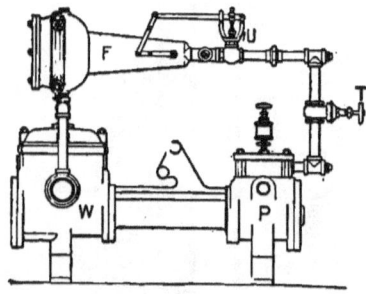

FIG. 200 —PUMP-GOVERNOR WITH OUTSIDE LEVERS.

A pump governor consisting of a float-trap with outside connections to a steam-valve, as described by F. Barron,* is shown in Fig. 200.

A steam-pump with attached governor is shown partly in section Fig. 201. In this case the float is of the bucket form, the valve for supplying steam to the pump is flat with a single

---

* *Heating and Ventilation*, March, 1894.

port, and is connected by an internal lever to the bucket in such a manner that when the tank is filled the valve will be opened and the pump will operate, and when the tank is empty the valve will be closed, and the pump will stop.

The pump-governors are frequently set some little distance

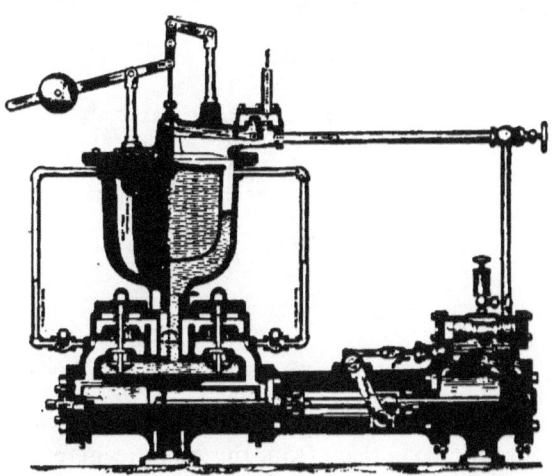

FIG. 201.—INTERNAL CONNECTED PUMP-GOVERNOR.

from the pump, but attached in every case so as to produce the results described.

136. The Steam-loop.—A device which has been used quite extensively for returning water of condensation to the boiler when the pressure has been reduced only a few pounds is called a steam-loop, the construction and principle of operation of which, as described by Walter C. Kerr, is as follows:

The figure shows the loop returning the water, from a separator attached to an engine-main, to a boiler above the separator level. "From the separator drain leads the pipe called the 'riser,' which at a suitable height empties into the horizontal. This runs back to the drop-leg, connecting to the boiler anywhere under the water-line. The riser, horizontal, and drop-leg form the loop, and usually consist of pipes varying in size from three quarters of an inch to two inches, and are wholly free from valves, the loop being simply an open pipe, giving free communication from separator to boiler. (Stop- and check-valves are inserted for convenience, but take no part in the loop's action.)" Supposing, for example, the boiler-pressure to be 100 pounds and the pressure at the separator reduced to 95. "The pressure of 95 pounds at the separator extends (with even further reduction) back through the loop,

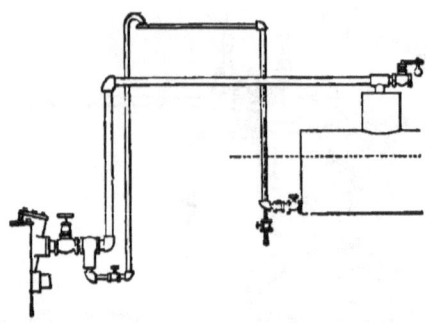

FIG. 202.—THE STEAM-LOOP.

but in the drop-leg meets a column of water (indicated by the broken line) which has risen from the boiler, where the pressure is 100 pounds, to a height of about 19 feet, that is, to the hydrostatic head equivalent to the 5 pounds difference in pressure. Thus the system is placed in equilibrium. Now the steam in the horizontal condenses, lowering slightly the pressure to 94 pounds, and the column in the drop-leg rises two feet to balance it; but meanwhile the riser contains a column of mixed vapor, spray, and water, which also tends to rise to supply the horizontal, as its steam condenses, and being lighter than the solid water of the drop-leg it rises much faster. By this process the riser will empty its contents into the horizontal, whence there is a free run to the drop-leg and thence to the boiler."

137. Reducing-valves.—The reducing-valve is a throttling-valve arranged to be operated automatically so as to reduce the pressure and also to maintain a constant pressure on the steam-mains. A great many forms of these valves are in common use. In one a diaphragm of metal or rubber is employed, as in Fig. 203. The low-pressure steam acts on one side of the diaphragm, a weight or spring which may be set at any desired pressure on the other side. This diaphragm is

The three important requisites in the construction of such plants are, first, a removal of all surface-water so that it cannot possibly come in contact with the steam-pipe; second, provision for taking up expansion of pipe and keeping it in proper alignment; and, third, insulation of the pipe from heat losses.

The first condition, which is the most important of all, is also the most likely to be overlooked, and many failures to secure economic transmission have been caused by allowing the surface-water to come in contact with the heated pipes. This water can be removed by the construction of a drain beneath or by the side of the pipe-system, provided with proper outlets. A perfect drainage-system for the soil is in every case an essential requisite for success.

Provision for expansion may be made by the use of expansion-joints, as already described in Article 62, page 105, or by the use of elbows and right-angled offsets arranged to partly turn as the line expands. The writer has had experience with various forms of these joints, and found nothing equal to the straight expansion-joint, Fig. 90, which should, however, be constructed so that it cannot by any possible accident be pulled apart; this may be done either by use of an internal lug or external brace. These joints should be thoroughly anchored, so that they will stay in position, and should be placed sufficiently close together to take up all expansion without strain on the pipe-line. If the ordinary slip-joints are used, they will need to be placed at distances of about 120 feet apart. The pipe between the joints should rest on rollers or connecting hangers which permit its free motion. If elbows and offsets are employed to take up expansion, there will be an abrupt change in grade, and if any part dips below the main steam-line it should be drained by a pipe connecting to a trap or to the return. If bends convex upward are necessary, means must be provided for removing the air.

In general, in systems where the steam is transmited long distances the best results will be possible only when the boiler-plant can be located on lower ground than the buildings to be heated, so that the water of condensation may be returned by gravity. This cannot always be done, and in many cases it will only be possible to return the water of condensation by

a pump located in one of the buildings to be heated, and regulated by a pump-governor. This in some cases may involve more expense than will be warranted by the saving due to returning the water of condensation.

For the insulation of the pipe many methods have been adopted, of which we may mention first the wooden tube and concentric air-space surrounding the pipe, Fig. 205. The

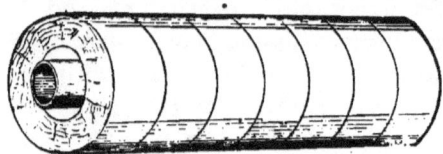

FIG. 205.—PIPE WITH WOODEN-TUBE INSULATION.

tube is usually made by sawing out the interior portion of a log, leaving a shell or wall about two inches thick. Each length is provided with a mortise and tenon joint, and the different lengths are joined together by driving. These wooden tubes are slipped over the steam-pipe as it is laid, the pipe being held in a central position by collars, so as to leave an airspace about one inch thick surrounding the pipe. This pipe is usually strongly banded with hoop-iron, and the joints can be made water-tight when laid, but checks soon form in the wood-pipe and make crevices through which the soil-water can reach the steam-pipe. Recently a form of tube made of two layers of inch board separated by tarred felting has come into use and is in general to be preferred to the solid tube as having superior insulating qualities. A view of such tubing partly in section is shown in Fig. 206.

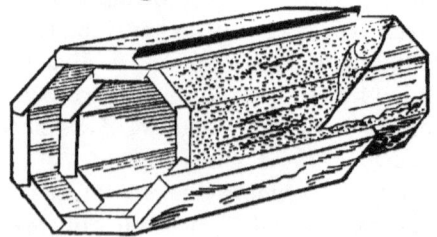

FIG. 206.—WYCKOFF BUILT-UP WOOD TUBING.

The wooden-tube system of insulation is objectionable, principally because it does not protect the pipe from groundwater, its durability, as proved by experience, is not great, and leaks in the steam-pipe are very difficult to locate and repair. A modified plan of the construction described has been employed, in which both steam- and return-pipes were covered with asbestos and hair-felt and placed in a box made of 2-inch plank; the box was laid on a concrete bottom three inches thick, and after the pipes were laid it was completely surrounded with concrete. This was arranged so that the steam-pipes would not be disturbed by decay of the wood. The concrete would in that event support the steam-pipes and constitute a protecting tube. The heat insulation proved on trial to be much superior to that of the solid wooden tube, while its cost was somewhat less. Similar constructions in which the wooden tube has been replaced by sewer-pipe are in use and are of superior durability. In one case familiar to the writer a wooden tube lined with sewer-pipe was laid outside the steam-pipe, the whole being covered with earth; such a construction replaced one shown in Fig. 205, but in practice its heat-insulation properties have not proved to be better.

The best system of transmitting steam long distances, but probably also the most expensive, is to be obtained by building a conduit lined with brick or masonry laid in cement and sufficiently large for inspection and repairs. The pipe should be carried in it on proper hangers and thoroughly wrapped with insulating material, as described in Article 116, page 200. Every required condition can be easily met in this construction.

The loss of heat from systems protected by a simple wooden tube is considerable, rising in many cases to from 30 to 40 per cent of that from the bare pipe. This is, however, due to the poor system of insulation used, since it should not exceed in any case 20 per cent of that from naked pipe (see page 199). The loss from the underground system of piping at Cornell University, which is somewhat over one half mile in length, and in which the steam-pipes are laid inside of sewer-pipe, with a wooden tube outside the sewer pipe, the whole covered with about 4 feet of earth, causes the consumption of about two and one half tons of coal per day, which is about 10 per cent of the total coal consumption when the plant is working at normal capacity. This heat loss is very nearly a constant amount and cannot be expressed as a fixed percentage of the total steam used, for the reason that when the steam consumption is large this percentage of loss is small and *vice versa*.

High-pressure steam for power purposes is also sometimes transmitted in this manner and engines operated at a great distance from the boiler-plant. The losses from such a system of transmission are often serious, especially if a long pipe-line has to be kept hot, and if the engine is operated only a part of the time or only at partial capacity. Where the engine is worked to its full capacity, the loss is usually less than by any other system of transmission. The following paragraph gives a careful estimate, based on actual experiment, of the loss experienced in transmitting constant power by various methods a distance of 1000 feet.

The loss in transmitting power by any system is principally constant, and hence when the power is greatly increased the percentage is correspondingly reduced. The following estimate is based on the transmission of 100 horse-power 1000 feet:

Method of Transmission.	Percentage of Loss.
Line shafting:	
Loss by friction..................(average 32)	25 to 40
Electricity:	
Loss in transforming from mechanical to electrical, and *vice versa*..........................	20 to 30
Line loss...................................	2 to 5
Total loss, electrical transmission..............	22 to 35
Conveying steam:	
Naked steam-pipe (still air)...................	37 to 45
Pipe covered with solid wood and earth.........	11 to 13

For operating machinery which is required occasionally or at intervals electricity is no doubt the most economical medium, since when the demand for power ceases the expenditure on account of transmission also becomes nothing, which is rarely the case either with line-shafting or steam.

The diagram, Fig. 207, gives the summary of the results of a test of the Lehigh Coal-storage Plant, South Plainfield, N. J., made by the writer to determine the heat lost in supplying an engine situated 740 feet from a boiler-house, the connecting

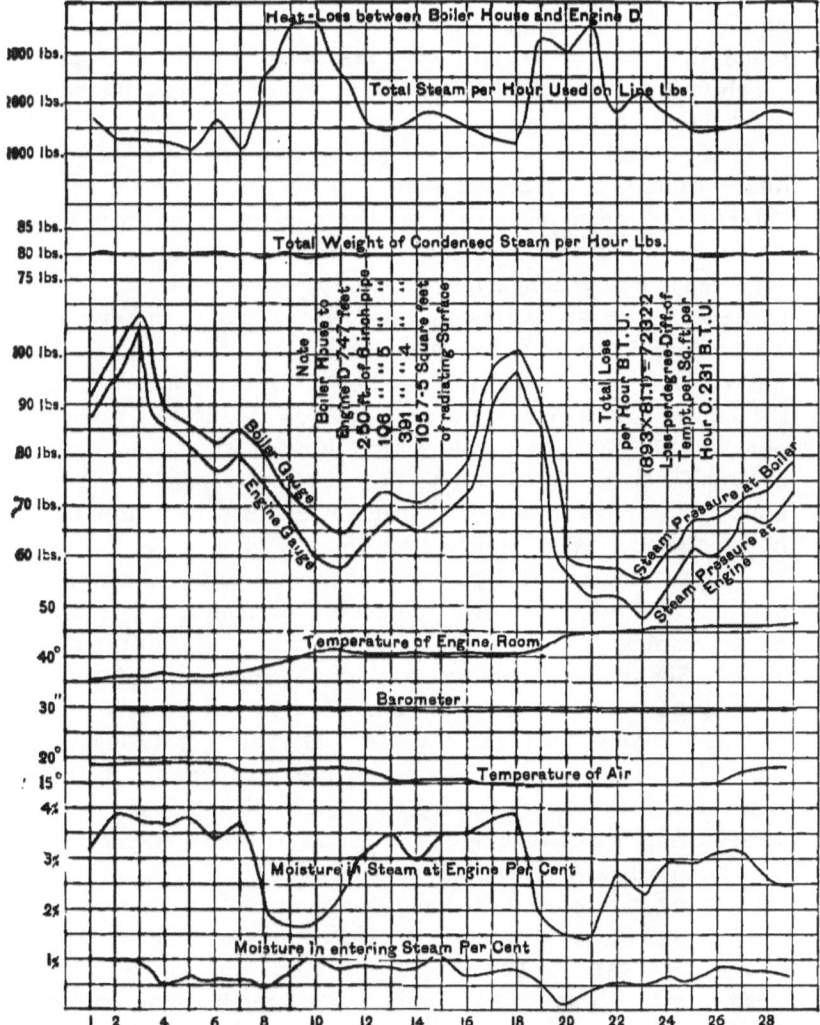

FIG. 207.—DIAGRAM SHOWING RESULTS OF TEST TO DETERMINE HEAT LOSSES IN UNDERGROUND PIPE.

pipe-line consisting of 250 feet of 6-inch, 106 feet of 5-inch, and 391 feet of 4-inch pipe, having a total radiating surface of 1057.5 square feet.

The engine was 12-inch diameter, 16-inch stroke, running with a piston speed of about 600 feet a minute, thus producing, when cutting off at one third stroke, a velocity of steam of about 60 feet per second in the 4-inch supply-pipe. As this pipe was 391 feet long, more reduction in pressure was anticipated than was actually found. As shown by the summary which follows, the actual reduction varied from 5 to 7 pounds, averaging 6 pounds.

The general method of testing adopted was such as to give information, first, of the amount of water in the steam as it entered the steam-pipe; second, the amount of water in the steam as it reached the engine; third, the amount of water collected at intervening drips; fourth, the total amount of steam used; fifth, the fall in pressure between the boilers and engine. These determinations were made as follows: The amount of water in the steam was determined by a throttling calorimeter, the sample of steam being drawn in each case from a vertical pipe located close to a bend from a horizontal, and collected by a half-inch nipple extending past the centre of the vertical pipe. The drip was caught at places which had been provided in the pipe, and was weighed from time to time.

The barometer readings were taken with an aneroid which had been compared with a mercurial barometer. The corrected readings are given in the summary as well as in the diagram. Simultaneous observations of the quantities given in the summary were taken every ten minutes. A study of the summary shows that the loss was sensibly constant during the run. This is clearly shown by noting the fact that any increase in the amount of steam flowing through the line had the effect of decreasing the percentage of moisture at the engine.

The total heat loss per hour was equivalent to that required to evaporate $(36 + 45.1 =) 81.1$ pounds of water from a temperature of $314°$ F., to a pressure of 70.1 pounds by gauge. This is equal to $(81.1 \times 893 =) 72,322$ B. T. U. The average steam-pressure was 70.1 pounds by gauge, its temperature $313.6°$ F., the average outside temperature $16.6°$ F.; hence the difference of temperature was $297°$. The loss for each degree difference of temperature between that of outside air and that of steam becomes $(78,342 \div 297 =) 243.7$ B. T. U. per hour. The total radiation surface was 1057.5 square feet; hence the loss in B. T. U. per square foot per hour was 0.229 per degree difference of temperature.

This for a difference of temperature of 150° corresponds to 0.17 B. T. U. per degree difference per square foot per hour, an amount about 10 per cent of that which would have been given off from a naked pipe. (See page 66.)

The loss by condensation varied from 3 to 8 per cent, the loss of pressure and consequent ability to do work about 6 per cent. The total loss was not far from 10 per cent from both these causes; if this had been proportional to length, it would have been 13.5 per cent for a line 1000 feet in length.

The diagram shows variations in the observed quantities as they occurred from time to time. It is to be noted that as the demand for steam at the engine was large the moisture in the steam delivered was correspondingly reduced.

CHAPTER XII.

HEATING WITH HOT AIR

139. General Principles.—The general laws which apply to hot-air heating have already been considered in the articles relating to Ventilation and to the Methods of Indirect Heating with Steam or Hot Water.* The method of heating with hot air, as usually practised, consists in first enclosing a suitable heater, termed a *furnace*, in a small chamber with brick or metallic walls, which is connected to the external air by a flue leading to its lower portion and to the various rooms to be heated by smaller flues leading from the upper part. In operation the cold air is drawn from the outside, is warmed by coming in contact with the heated surfaces of the furnace, and is discharged through the proper flues or pipes to the various rooms. The rapidity of circulation depends entirely upon the temperature to which the air is heated and the height of the flue through which it passes; the velocity will be in every case essentially as given in the table on page 45. In order that a system of circulation may be complete flues must be provided for the escape of the cooler air from the room to be heated, otherwise the *circulation* will be very uncertain and the heating quite unsatisfactory. Registers and flues for the escape of the air from the room are often neglected, although fully equal in importance to those leading to the furnace.

Regarding the relative merits of hot-air heating by furnace as described and of the various systems of steam or hot-water heating, little can be said in a general way, since so much depends on circumstances and local conditions. It is rarely that these systems come in direct competition. The force which

---

* See pages 52 and 211.

causes the circulation of the heated air is a comparatively feeble one and may be entirely overcome by a heavy wind; consequently it is generally found that the horizontal distance to which heated air will travel under all conditions is short; hence the system is in general not well adapted for large buildings. When properly erected and well proportioned, this system gives, in buildings of moderate size, very satisfactory results.

It may be said, however, that, in erecting a hot-air system of heating, competition has been in many cases so sharp as to induce cheap, rather than good, construction. Small furnaces have been used in which the temperature of the exterior shell had to be kept so high, in order to meet the demands for heat, that the heated air absorbed noxious gases from the furnace and entered the room in such condition as to impair, rather than to improve, the ventilation. Ventilation-ducts for removing the air from the rooms have often been neglected, and hence the results obtained have been far from satisfactory. Such faults are to be considered, however, as those of design and construction rather than as pertaining to the system itself.

In order that the hot-air system should be satisfactory in every respect, the furnace should be sufficiently large, and the ratio of heating surface to grate such that a large quantity of air may be heated a comparatively small amount rather than that a small quantity shall be heated a great amount. As air takes up heat very much more slowly than steam or water, it would seem that the relative ratio of heating surface to grate surface should be more than that commonly employed in steam-heating. By studying the proportions which have already been given for steam-heating boilers (page 125) it will be seen that the ratio of heating surface to grate surface for the steam-boiler varies between 20 and 45, averaging about 32. From a study of the results in catalogues of manufacturers of furnaces the ratio of air-heating surface to grate surface in hot-air furnaces seems to vary from 20 to 50 as extremes. These proportions are essentially the same as used in steam-heating and are much too small for the best results in hot-air heating. It is quite evident that since air cannot be heated by radiation, and is warmed only by the contact of its particles against the heated surface, that the exterior form of the furnace should be such as will induce a current of air to impinge in some portion of its course directly against the surface.

Regarding the economy of this or any other system of indirect heating, it is simply a question of perfect combustion and relative wastes of heat. If the fuel is perfectly burned and all the heat which is given off is usefully applied, the system is perfect. The waste of heat in any system of combustion is that due to loss in the ashes, to radiation, and to escape of hot gases into the chimney. If the furnace is properly encased and if the hot-air pipes are well covered, there is no reason why losses from imperfect combustion and from radiation should not be a minimum. The chimney loss depends largely upon the temperature of the surface of the heater: if this is high, the loss will be large. In general, it may be said that the larger the heating surface provided the lower may be its temperature, and the greater the economy. It should be noted, however, that this or any system of indirect heating requires the consumption of more fuel than when the heating surfaces are placed directly in the room, and for that reason the operating expense must be considerably greater than that of direct systems of hot-water and steam heating. (See page 202.)

Furnaces, or in fact heating-boilers of any kind, are uneconomical if operated with a deficient supply of air. In this case the product of combustion will contain carbon monoxide,* an extremely poisonous and inflammable gas, which is quite likely to take fire and burn, on coming in contact with air, at the base or top of the chimney.

140. General Form of a Furnace.—The principles which apply in furnace construction are not essentially different from those already given in Chapter VII for steam and hot-water boilers. In the case of a hot-air furnace the fire and heated products of combustion are on one side of the shell and the air to be warmed on the other. In the case of steam or hot-water boilers the water and steam occupy the same relative

---

* See Article 24, page 26.

positions as the air in the case of the hot-air furnaces. The types and forms of furnaces which are in use may be classified exactly the same as heating-boilers, Articles 77 and 82, as having plain or extended surface, and as being horizontal or vertical, tubular or sectional; it may be said that the forms which are in use are fully as numerous as those described for steam-heating and hot-water heating.

The material which is employed in construction is usually cast iron or steel, and there is a very great difference of opinion as to the relative merits of the two. It seems quite probable that cast iron, because of its rough surface, may be a better medium for giving off heat than wrought iron or steel, but it is quite certain that at a very high temperature, some carbon from the cast iron will unite with the oxygen from the air forming carbonic acid. When very hot it may be slightly permeable to the furnace gases. Such objections are, however, of little practical importance, since the temperature of a furnace never should, and never does if properly proportioned, exceed 300 or 400 degrees Fahr., and for this condition the difference in heating power of cast iron and steel is very slight. It is of great importance that the shell of the furnace be tight, so that smoke and the products of combustion cannot enter the air-passages.

Furnaces can be purchased with or without magazine feed, but the demand of late years is principally for those without the magazine, since it has not been proved to present any special advantages.

Furnaces are often set in a chamber surrounded with brick walls, as explained for steam-boilers, but they are more frequently set inside a metallic casing, this latter being termed a portable setting; this casing varies somewhat as constructed by different makers, but usually consists of two sheets of metal, the outer of galvanized iron, with intervening air-space empty or filled with asbestos. The casing is placed at such a distance from the furnace as to provide ample room for the passage of air.

Some form of dumping or shaking grate which can be readily and quickly cleaned is almost invariably employed. The draft-doors which admit air below the grate and check-dampers in the stovepipe are usually arranged so they can be opened or closed from some convenient place on the first floor of the house by means of chains passing over guide-pulleys.

A pan in which water may be kept is added to every furnace for the purpose of increasing the moisture in the air; this is of importance, since the heated air requires more moisture than cold to maintain a comfortable degree of saturation, as explained in Article 28, page 30.

141. Proportions Required for Furnace Heating.—The proportion of the area of heating surface in the furnace to that of the grate cannot be computed from any data accessible to the writer, and the proportions given are assumed to be twice those which have been found to give best results in steam-heating; these apparently agree well with the best practice. The tables which are given are computed for a maximum temperature of 120° F. for the air leaving the furnace, which is 50 degrees in excess of the ordinary temperature in the house. No doubt better practice might require the introduction of more air at a lower temperature, but considering the fact that this high temperature only has to be maintained when the outside weather is extremely cold, it seems quite doubtful if the expense of a furnace large enough for this additional duty, would be warranted.

The ratio which the grate surface of the furnace should bear to the glass and exposed wall surface of the room can be computed with sufficient accuracy from known data relating to the heat contained in coal and to the probable efficiency of combustion. The heat given off from the walls of a room for each degree difference of temperature between the inside and outside has been shown on page 59 to be approximately equal to the area of the glass plus one quarter the area of the exposed wall surface, which we will in this place denominate as the *equivalent glass surface.* One pound of good anthracite coal will give off about 13,000 heat-units in combustion. One pound of soft or bituminous coal will give off in combustion from 10,000 to 15,000 heat-units, depending on the kind and quality. Of this amount a good furnace should utilize 70 per cent.* The amount of coal which is burned per square foot

---

* It is quite probable that the efficiency of combustion in an ordinary furnace is much less than the above, often as low as 50 per cent.

of grate surface per hour will depend very much upon the character of attendance; in ordinary furnaces used in house heating, and where it is expected to replenish the fires only two or three times per day, this amount is low, being not greatly in excess of 3 pounds. If the air is 120 degrees in temperature, nearly 60 cubic feet will be required, when heated one degree, to absorb one heat-unit (see Table VIII), and if such air is delivered 50 degrees above that of the air in the room, each cubic foot will bring in $\frac{5}{6}$ of one heat-unit.

The velocity of air in feet per minute with ample allowance for friction is given in a table on page 45, from which it is seen that it will be safe to assume velocities of 4, 5, and 6 feet respectively, per second in the flues or stacks leading to the various floors. The velocity of the air passing the register may be assumed as 3 feet per second in every case; this lower velocity is obtained by making the area of the register somewhat larger than that of the pipe leading to it.

The following mathematical discussion gives these various considerations in general and algebraic terms, as follows:

Let F = square feet in grate, C = weight of coal burned per square foot of grate per hour, r = heat-units per pound of coal, E = efficiency of furnace, h = heat-units per hour, T = temperature of air leaving furnace, t' = temperature outside air, t = temperature of room, G = area of glass in room, W = area of exposed wall surface, H = heat lost by room for one degree difference of temperature, K = cubic feet of air heated by furnace per hour, K' = cubic feet air required to warm room.

We have, as explained,

$h = CFEr$ = total heat given off by furnace, equal to that required for all the rooms. . . . . (1)

$K = \dfrac{60CFEr}{T - t'}$ = cubic feet of air heated per hour by furnace. . . (2)

$h' = (G + \frac{1}{4}W)(t - t')$ = total heat-units to warm the room. . . (3)

$K' = \dfrac{60(G + \frac{1}{4}W)(t - t')}{T - t}$ = cubic feet of air to warm the room. . (4)

For average conditions substitute in above, as explained, $T = 120$, $t = 70$, $t' = 0$, $C = .70$, $r = 13{,}000$, $Cr = 9100$, and we have

$h = 9100 CF = 2K$. . . . . (5)

$K = 4550 CF = 0.5 h$. . . . . (6)

$K' = 84(G + \frac{1}{4}W)$. . . . . (7)

When $K = K'$, $CF = \dfrac{G + \frac{1}{4}W}{54.2}$; when $C = 3$, $F = \dfrac{G + \frac{1}{4}W}{162.6}$. . . (8)

$h' = 70(G + \frac{1}{4}W)$. . . . . (9)

For computing areas of leader-pipes and stacks, for residence heating, assume velocities which can safely be taken as follows: First floor, 4 feet per second or 240 per minute; second floor, 5 feet per second or 300 per minute; third floor, 6 feet per second or 360 per minute. (See table, page 45.)

Through a cross-section of the flue equal to one square inch 100 cubic feet will pass in one hour when the velocity is 4 feet per second, 125 when the velocity is 5 feet per second, 150 when the velocity is 6 feet per second, $25v$ when the velocity in feet per second is represented by v.

Denote area of flue in square inches by L; then from equation (7)

$$L = \frac{K'}{25v} = \frac{84(G + \frac{1}{4}W)}{25v} = \frac{3.36(G + \frac{1}{4}W)}{v} \quad . \quad . \quad (10)$$

From this, by transposition, we have

$$(G + \tfrac{1}{4}W) = \frac{v L}{3.36}. \quad . \quad . \quad . \quad . \quad (11)$$

If for first-floor rooms $v = 4$.

$$G + \tfrac{1}{4}W = 1.19L.$$

If for second-floor rooms $v = 5$.

$$G + \tfrac{1}{4}W = 1.53L.$$

If for third-floor rooms $v = 6$.

$$G + \tfrac{1}{4}W = 1.78L.$$

(Also see table on page 53.)

The following table gives the relative values of these various quantities, computed for the conditions as explained:

PROPORTIONS REQUIRED IN FURNACE HEATING.

Equivalent glass surface * ...	25	50	75	100	125	150	200	250	500	750	1000
Cu. ft. air to be heated per hr.	2100	4200	6300	8400	10,500	12,600	16,800	21,000	42,000	63,000	84,000
Grate area, square inches ..	22	43	64	85	107	127	170	212	425	640	850
Equivalent diameter round grate, inches............	7.5	8.5	9.5	11.5	12.5	13.5	15	17	24	29	33
Heating surface, square feet.	11	21	32	42	53	62	85	106.	212	320	425
Diameter smoke-pipe, inches	...	....	....	....	......	......	7	7	8	10	11
Approximate cubic feet space	420 525	840 1050	1260 1570	1680 2100	2100 2625	2520 3150	336 4200	4200 5250	8400 2100	12,600 15,750	16,800 21,000
Area stack—											
1st floor (vel. 4) sq. in.	21	42	63	84	105	126	168	210	420	630	840
2d " (vel. 5) "	17	33	51	68	85	102	135	170	345	500	670
3d " (vel. 6) "	14	28	42	55	70	84	112	140	280	420	560
Diameter leader-pipe— †											
1st floor..	7	7.5	9	10.5	11.6	12.7	14.7	16.5	19	23.2	26.7
2d " ..	7	7	8.2	9.5	10,5	11.5	13.2	14.7	21	25.2	29.2
3d " ..	7	7	7.5	8.5	9.5	10.4	12	13.4	19	23.2	26.7
Net area register, sq. in.—											
1st floor (vel. 3)...	28	56	84	110	210	168	224	280	560	840	1120
2d " and above	21	42	63	84	105	126	168	210	420	630	840
Area ventilating flue	21	42	63	84	105	126	168	210	420	630	840
Net area ventilating register .	17	33	51	68	85	102	135	170	345	500	670

* This quantity is (defined page 272).
† For pitch of one inch per foot. Use larger pipe for less pitch.

NOTE.—The proportions in the above table agree very well with those given by the Excelsior Steel Furnace Co. for the condition of changing the air in each room four times per hour, which can be taken as representing the average amount required to bring in the heat.

The grate surface is computed for combustion of 3 pounds per square foot per hour, with an efficiency of 70 per cent. or a greater amount at less efficiency. The heating surface given in above table is much larger than ordinarily found in furnaces, but not too large for best results.

142. Air-supply for the Furnace.—The air-supply for the furnace is usually obtained by the construction of a passage-way or duct of wood, metal, or masonry leading from a point beneath the furnace casing or near its bottom to the outside

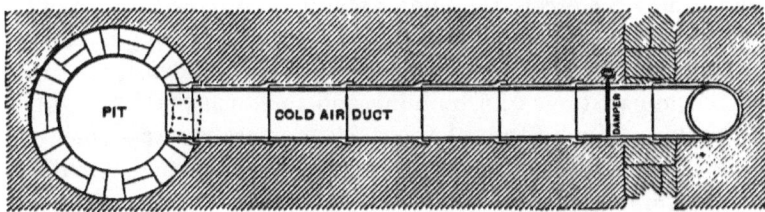

FIG. 208.—HOT-AIR FURNACE WITH COLD-AIR BOX BELOW CELLAR BOTTOM.

air, essentially as shown in section Fig. 208. This duct or pipe is usually termed the *cold-air* box and is often constructed of wood. In all cases there should be a screen over the outer end to keep out vegetable matter or vermin, and doors should be arranged so that it can be cleaned periodically. A damper is usually desirable, arranged so that it can be partly or entirely opened to regulate the admission of the cold air. The cold-air box should be made perfectly tight and in a workmanlike manner, so that air cannot escape into or be drawn from the cellar or basement. This should join onto the furnace casing at as low a point as the character of the cellar bottom will permit. In some instances it is desirable to erect two cold-air boxes, opening to the air on opposite sides of the house, so that the supply may be drawn from either direction as required to obtain the help of wind-pressure, to aid in the circulation of the air over the furnace.

The cross-sectional area of the cold-air box is proportioned, by different authorities, from 66 to 100 per cent of the sum, of the areas of all pipes taken from the furnace. If this were proportioned so that its area should be in ratio to the respective volume of cold and heated air, the sectional area of the cold-air box should be about 80 per cent, of the sum of the areas of the various stacks. To avoid frictional resistances it would seem to be advisable when practicable to make its area equal to that of the sum of the areas of the stacks.

143. Pipes for Heated Air.—The pipes for heated air are of two classes: first, those which are nearly horizontal and are taken from near the top of the furnace casing—these are usually round and made of a single thickness of bright tin, and if possible erected with an ascending pitch of one inch to one foot, and are termed *leader-pipes;* second, rectangular vertical pipes or risers, termed *stacks*, made in such dimensions as will fit in the partitions of a building and to which the leader-pipe connects. The bottom of the stack is enlarged into a chamber termed a *boot*, which is made in various forms and provided with a round collar for connection to the leader-pipe. The top part of the stack may be provided with a similar boot from which horizontal rectangular stacks are taken, or it may be connected to a rectangular chamber into which the register may be fitted and which is known as the *register box*. The stacks usually pass up or near the woodwork of partitions, and for lessening the fire risk as well as preventing loss of heat should be made with double walls separated by an intervening air-space. The register boxes should also in every case have double walls. The general form of a stack in position in a partition, with boot attached at bottom for leader-pipe and with round connection for register box, is shown in Fig. 209.

The leader-pipes and stacks, boots, and register boxes are now a standard article of manufacture by several firms. I am indebted to the Excelsior Steel Furnace Company of Chicago for the table of capacity and dimensions of various forms of stacks and leader-pipes, given on page 278.

It will be found profitable in nearly every case to wrap the leader-pipes with two or more thicknesses of asbestos paper

FIG. 209.—REGULAR STACK WITH COLLAR AT TOP AND FLAT BACK BOOT AT BOTTOM.

FIG. 210. — FIRST-FLOOR OUTSIDE REGISTER BOX WITH COLLAR ATTACHED.

and mineral wool in order to prevent loss of heat. It is desirable to locate the stacks in the inside partition-walls of the building, or where they will be protected as much as possible from loss of heat, since any loss affects the rapidity of circulation. It is generally necessary to have the leader-pipes not over 15 feet in length, otherwise the circulation will be uncertain in amount and character.

144. The Areas of Registers or Openings into Various Rooms.—Registers are made regularly in various forms, square or round, and arranged for use either in the floor or side walls

* TABLE OF SIZES AND DIMENSIONS OF SAFETY DOUBLE HOT-AIR STACKS.

Size of Stack as Listed. (In Inches.)	Actual Size of Outside Stack.	Actual Size of Inside Stack.	Area of Inside Stack in Inches	Capacity as compared with that of Hot-air Pipe with Pitch of 1 Inch to 1 Foot.	Equivalent in Round Pipe with Pitch of 1 Inch to 1 Foot.	Sizes of Round Pipe which should be used with each Stack.	Area of said Round Pipes in Inches.	Size of Registers and Register Boxes which should be used with each Stack.	Cubic Feet of Space (approximate) that can be Heated with each Stack with Pipe and Registers of size given.	Equivalent of said Space on Floor of Rooms 10 Ft. high	Area in Inches of Registers with Space occupied by Bars deducted.
4 × 8	3⅝ × 7⅝	3¼ × 7	23	35	6½	7	38	6 × 8	500	6 × 8	35
4 × 10	3⅝ × 9⅝	3¼ × 9	29	43	7½	8	50	8 × 10	850	8 × 10	45
4 × 11	3⅝ × 10⅝	3¼ × 10	32½	48	8	8	50	8 × 12	1000	9 × 11	55
4 × 12	3⅝ × 11⅝	3¼ × 11	35	53	8¼	9	63	9 × 12	1250	10 × 12½	60
4 × 14	3⅝ × 13⅝	3¼ × 13	41	63	9	9	63	10 × 12	1650	12 × 14	70
6 × 10	5⅝ × 9⅝	5¼ × 9	47	71	10	10	78	10 × 14	2000	12 × 17	80
6 × 12	5⅝ × 11⅝	5¼ × 11	58	87	11	12	113	12 × 15	2300	14 × 17	115
6 × 14	5⅝ × 13⅝	5¼ × 13	68	102	12	12	113	12 × 17	2600	15 × 18	120
6 × 16	5⅝ × 15⅝	5¼ × 15	79	119	12½	14	154	14 × 20	3000	15 × 20	156
8 × 18	7⅝ × 17⅝	7¼ × 17	124	186	15	16	201	16 × 24	4000	20 × 20	210
10 × 20	9⅝ × 19⅝	9¼ × 19	176	264	18	18	254	20 × 24	5400	20 × 27	270
10 × 24	9⅝ × 23⅝	9¼ × 23	213	330	20½	20	314	21 × 29	7000	20 × 35	340

Stacks for 4 inch studs carried in stock. Other sizes made to order.

* This table is copyrighted by Excelsior Steel Furnace Co.

as required. These registers are usually supplied with a series of valves which may be readily opened or closed. The

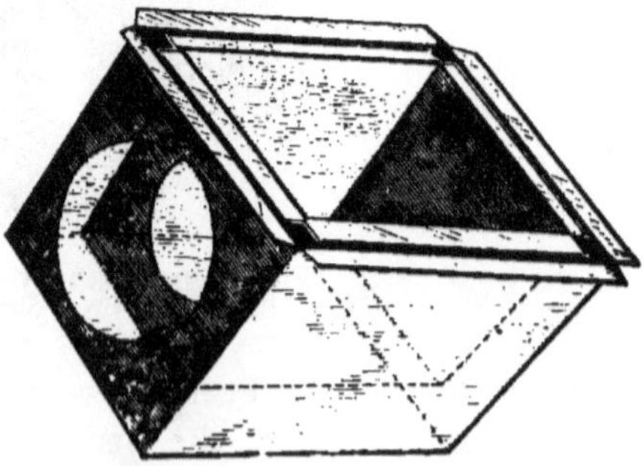

FIG. 211.—REGISTER BOXES SHOWN IN POSITION.

space taken by the screen and valves is usually about $\frac{1}{3}$ of that of the register, so that the effective or net area is about $\frac{2}{3}$ of the nominal size of opening. These registers may be obtained finished in black or white japan, or electroplated with nickel, brass, bronze, or copper. The table on page 280 gives the various sizes of registers which are regularly on the market, their effective area in square inches, and diameters of round pipe having the same capacity.

The areas of stacks may be considerably less than those of the registers, since it is generally required that the velocity of air entering the room shall not exceed 3 or 4 feet per second, while that passing through pipes and stacks may have the highest velocity possible, which for the different floors will not differ greatly from 4 to 6 feet per second, as already explained. For methods of proportioning ventilating flues see page 233.

FIG. 212.—SIDE-WALL REGISTER HEAD OR FLANGE.

Considerable difference of opinion exists as to the relative merit of floor and wall registers for heating purposes. It is the common practice to use floor registers for most rooms on the first floor, and wall registers for rooms on the second and higher floors. The floor register, from its general form and position, can be supplied with hot air somewhat more readily than the wall register, and for that reason may induce somewhat stronger circulation, but it is a receptacle for dust and sweepings of the room and in a position to materially interfere with the carpets. It will be found that the experiments made by Briggs (see page 46) as to diffusion of air hold in the case of furnace heating the same as in that of any other system. From these experiments it would seem that the highest efficiency would be attained when the inlet for the heated air was at the side near the top of the room and the outlet for ventilation near the floor. This distribution is one that, so far as the writer knows, has never been practised in furnace heating of residences, although it is the commonly accepted method in school-house heating, whether with a furnace or an indirect system of steam or hot-water heating.

TABLE OF REGISTERS.

Size of Opening. Inches.	Effective Area. Square Inches.	Diameter Round Pipe. Inches.	Size of Opening. Inches.	Effective Area. Square Inches	Diameter Round Pipe. Inches.
4½ × 6½	20	5.1	10 × 20	132	13.0
4 × 8	21	5.2	12 × 12	96	11.1
4 × 10	26	5.8	12 × 14	112	11.9
4 × 13	34	6.6	12 × 15	120	12.4
4 × 15	40	7.2	12 × 16	128	12.8
4 × 18	48	7.8	12 × 17	136	13.2
6 × 6	24	5.6	12 × 18	144	13.5
6 × 8	32	6.4	12 × 19	152	13.9
6 × 9	36	6.7	12 × 20	160	14.3
6 × 10	40	7.2	12 × 24	192	15.6
6 × 14	56	8.5	14 × 14	130	12.8
6 × 16	64	9.1	14 × 16	149	14.8
6 × 18	72	9.6	14 × 18	168	14.7
6 × 24	96	11.1	14 × 20	186	15.5
7 × 7	32	6.4	14 × 22	205	16.2
7 × 10	52	8.2	15 × 25	250	17.8
8 × 8	42	7.4	16 × 16	170	14.7
8 × 10	53	8.2	16 × 20	213	16.5
8 × 12	64	9.6	16 × 24	256	18.1
8 × 15	80	10.1	18 × 24	288	19.2
8 × 18	96	11.2	20 × 20	267	18.5
9 × 9	54	8.2	20 × 24	320	20.2
9 × 12	72	9.6	20 × 26	347	21.0
9 × 13	78	10.0	21 × 29	406	22.7
9 × 14	84	10.3	24 × 24	384	22.1
10 × 10	66	9.2	24 × 32	512	25.5
10 × 12	80	9.1	27 × 27	486	25.0
10 × 14	93	10.9	27 × 38	684	29.5
10 × 16	107	11.7	30 × 30	600	27.7
10 × 18	120	12.4			

145. Circulating Systems of Hot Air.—By connecting the cold-air box with the hall floor or the lower portion of a passage communicating with all rooms of the building and closing outside connections a downward current of air will pass from the rooms to the furnace, which, being warmer than the outside air, will aid materially in heating. Such a connection if properly made and used with judgment may be of great service in reducing the cost of operation without seriously affecting the ventilation. Such a system if erected, however, should be supplied with devices to prevent overheating and arranged so that cold air can be drawn from outside of the building whenever desired. There is so much danger that ventilation will be poor with this system that it is not recommended.

146. Combination Heaters.—A combination heater consisting of a hot-air furnace, with the addition of a boiler for hot water or steam, is meeting with somewhat extensive use and has been described on page 189 so far as relates to the construction of the steam and hot-water appliances. In case a combination heater is used the area of the grate and heating surfaces will need to be proportioned for both systems. A combination heater is better suited to large buildings than a hot-air furnace. In practice, however it will be found, difficult to so proportion the amount of heating and radiating surface, as to give a perfect distribution of heat in rooms some of which are heated with hot water or steam, and some with hot air, but this difficulty will no doubt be largely overcome by experience.

147. Heating with Stoves and Fireplaces.—The manufacture of stoves for heating purposes is a very great industry in the United States and they are extensively used in the cheaper classes of dwellings. In every case the stove is located directly in the room to be heated and is connected with a chimney by means of several lengths of sheet-iron pipe. Stoves are built in many forms, some of which are very elaborate and highly ornamented, and in many cases they are provided with magazines from which the coal feeds itself automatically as required. The heat, given off from a stove, is generally nearly all utilized in warming, perhaps not over 10 or 15 per cent being carried off by the chimney. Stoves do not, however, present an economical mode of heating, largely because the wastes which occur from the operation of small fires are very great and cannot be avoided. It is doubtful if the efficiency averages much above 25 per cent. In addition, the stove occupies useful room, is the source of very much dirt and litter, and requires a great deal of attention.

Open fireplaces which were used at one time extensively are very wasteful, as little more than the direct radiant heat from the fire is absorbed in warming. They are also subject to all the wastes which pertain to stoves, and their probable efficiency cannot be considered as over 15 or 20 per cent. They are, however, valuable adjuncts of a system of ventilation, since large quantities of air are drawn from the room and discharged into the chimney. In the use, of a stove called a fireplace heater, the heated gases from an open fire pass through a drum or radiating surface in the room above, and the heat which otherwise would be discharged from the chimney and wasted is partly utilized in heating.

148. General Directions for Operating a Furnace.—The general directions for operating a furnace so far as regards the care of the fire are the same as those which have been previously given for the operation of steam-heating furnaces, page 169; there are, however, no steam-gauges or safety appliances needed. In regulating the temperature of the house the drafts of the furnace should be operated rather than the valves of registers leading to various rooms. In some instances if the circulation is strong in certain directions and weak in others so that certain rooms cannot be heated, it may be a good plan to shut all registers except the one to the room where heat is required until circulation is established, after which circulation will usually continue without further attention. In the operation of a furnace great care should be taken that the metal never becomes red hot or even cherry-red. If it will not warm the building without being excessively hot, the furnace is too small, or else has too little radiating surface in proportion to the fire-pot. The water-pan should be kept filled with water. Thermostats arranged to open or close the drafts when desired are in use in many systems of furnace heating with success.

For protection of the furnace during summer months some makers recommend that the fire-pot be filled with lime. For burning soft coal, furnaces of special construction only should be employed.

NOTE.—*Rules for Furnace Heating:*

First. To find area of grate in square inches: *Divide total window surface plus $\frac{1}{4}$ total exposed wall surface in square feet by 200.*

Second. To find area of flue for any room in square inches: Divide window surface plus $\frac{1}{4}$ wall surface in square feet by 1.2 for first floor, by 1.5 for second floor, by 1.8 for third floor.

CHAPTER XIII.

FORCED-BLAST SYSTEMS OF HEATING AND VENTILATING.

149. General Remarks.—In the systems of hot-air heating which have been described the circulation of air is caused by expansion due to heating, which is a feeble force and is likely to be overcome by adverse wind currents, by badly proportioned pipes, or by friction ; by employing a fan or blower of some character for moving the air the circulation will be rendered positive and so strong as to be unaffected by these causes.

This system can be employed where power is available, and in many cases will be found to present an economical and satisfactory system of heating, comparing well with any that has been devised, especially when the amount of ventilation provided is considered. The cost of heating a large quantity of air is, however, in every case one of considerable amount, so that it is quite probable that in expense of operation no system of indirect heating, whether by furnace or steam-pipes, can compare with that of direct hot-water or steam radiation. The systems of forced-blast heating are in almost every case employed in connection with steam-heated surfaces, but in some instances the system has been applied successfully with furnace heated surfaces.*

150. Form of Steam-heated Surface.—The heating surface is generally built of inch pipe, set vertically into a square cast-iron base, connected at top with return-bends, although the box coil, Fig. 94, page 109, or any form of indirect radiating surface could be used. The fan or blower is placed either

---

* *The Metal Worker*, May 25, 1895, gives an interesting example showing the successful use of a blower and furnace for heating a church.

so as to draw the air by suction over the heated surface and then deliver by pressure into the rooms, or it is placed so as to force the air by pressure over the heating surface and thence into the conduits leading to the various rooms. The heating surface is usually surrounded with metallic walls forming a chamber through which the air is discharged. Fig. 213 shows the arrangement often adopted, in which a pressure fan is directly connected to an engine, and arranged to take air from

FIG. 213.—BLOWER CONNECTED TO ENGINE.

the atmosphere and force it into the chamber in which the heating surface is placed.

151. Ducts or Flues—Registers.*—The dimensions of the ducts or flues leading from the heater should be such that the required amount of air may be delivered with a low pressure and velocity, so as to avoid excessive resistances due to friction. The velocity which will be produced by various pressures in

---

* General formulæ for the motion of air in long pipes is to be found in Weisbach's Mechanics, and in article Hydrodynamics in Encyclopædia Britannica, by Prof. W. C. Unwin. The formula given by Weisbach is elaborated in an article by Carl S. Fogh in *Engineering Record*, Feb. 16, 1895, and a graphical diagram given for practical application. The uncertainty which relates to the application of these elaborate formulæ is well shown by the fact that a factor of safety of 4 is used by Mr. Fogh, and serves in the writer's opinion to render such estimates as crude as those obtained by the approximate formulæ given here. The article is of great value, however, to those desiring to study the theory of motion.

excess of that of the atmosphere is given in table on page 42, from which it is seen that a pressure sufficient to balance $\frac{1}{2}$ inch of water (0.29 ounce per square inch) will produce a velocity of 30 feet per second in a pipe 100 feet long and 1 foot in diameter; this is generally considered to be the maximum velocity which should be pemitted in any of the pipes or passages. In proportioning apparatus in this system of heating it is generally required that sufficient air shall be brought in to change the cubic contents of the room four times per hour. By consulting the table on page 53, it will be seen that for this condition, and without allowance for friction, it will require a flue with 5.7 square inches of area for each 1000 cubic feet of space in the room. By adding two inches to the diameter obtained as above, a fair allowance for friction will be made.

The pipes are usually made of galvanized iron or bright tin and should have tight joints and be protected from loss of heat by some good covering (see page 197). Flues of brick or masonry cause more friction than those of galvanized iron, and if used should generally be about two inches larger in diameter than provided for by this table. As branch pipes for various apartments are taken off, the main pipe can be reduced in size; this should never be done abruptly, but only by the use of tapering tubes, the angle of whose sides measured from the line of the main pipe should rarely be greater than 15 degrees. The fan can be located in a chamber which is connected with the external air, as in Fig. 214, or it may be placed in a tube or passageway leading from the heating surface to the outside.

The area of the cold-air duct or passageway leading to the fan should be as great as possible in order to keep the velocity of entering air low; if the area of cross-section is equal to the sum of the areas of all the ducts leading from the heating surface, the velocity will probably be about three quarters of that in the hot-air pipes, and may draw in considerable dust and dirt from outside. The flues which convey air to the rooms should discharge near the upper part of the room substantially as described on page 49 and shown in Fig. 21. The friction in small pipes is greater than in large ones, being relatively proportional to the circumference or perimeter; hence the sum of the areas of the branch pipes should be considerably greater than that of the main.*

The table on opposite page gives the number of small pipes which provide an area equivalent to that of one large pipe of similar cross-section; in case no table is at hand the same results may be obtained by dividing the larger diameter by the smaller one and taking the square root of the fifth power of the quotient.

The following table gives the actual amount discharged with constant resistance, and with pressure equal to one half inch of water column in round pipes, as computed from Unwin's formulæ, page 41:

VELOCITY AND QUANTITY OF AIR DELIVERED IN PIPES OF DIFFERENT DIAMETERS, EACH 100 FEET LONG, WITH AN AIR-PRESSURE EQUAL TO ½ INCH OF WATER COLUMN.

Diameter of Pipe. In.	Velocity of Air. Ft. per Sec.	Cubic Feet of Air per Min.	Diameter of Pipe. In.	Velocity of Air. Ft. per Sec.	Cubic Feet of Air per Min.
1	8.7	2.6	16	35.6	3,024
2	12.4	16	18	36.8	4,032
3	15.0	45	20	38.8	5,184
4	17.3	90	22	40.6	6,480
5	19.4	160	24	42.4	8,208
6	21.3	253	26	44.2	9,936
7	23.0	380	28	46.0	11,952
8	24.5	515	30	47.4	14,256
9	26.1	720	36	52.0	23,040
10	27.4	900	42	56.1	33,120
11	28.6	1190	48	61.0	46,080
12	30.5	1440	54	63.6	61,920
13	31.3	1620	60	67.0	80,640
14	32.4	2160			

* The velocity of flow of air is given in formulæ on page 41; the amount discharged is equal to the area of the pipe multiplied by the velocity, and will be equal in every case to the square root of the fifth power of a constant multiplied by the diameter of the pipe. If we denote diameter of larger pipe by D, of smaller pipe by d, and the number of smaller pipes required to make one of area equivalent to that of larger by n,

$$n = \left(\frac{D}{d}\right)^{\frac{5}{2}}.$$

To find diameter of round pipe, d, which shall be equivalent in carrying capacity to a rectangular pipe with dimensions a and b, we would have

$$d = \sqrt[5]{\frac{32a^3b^3}{\pi^2(a+b)}}.$$

FORCED-BLAST SYSTEMS. 287

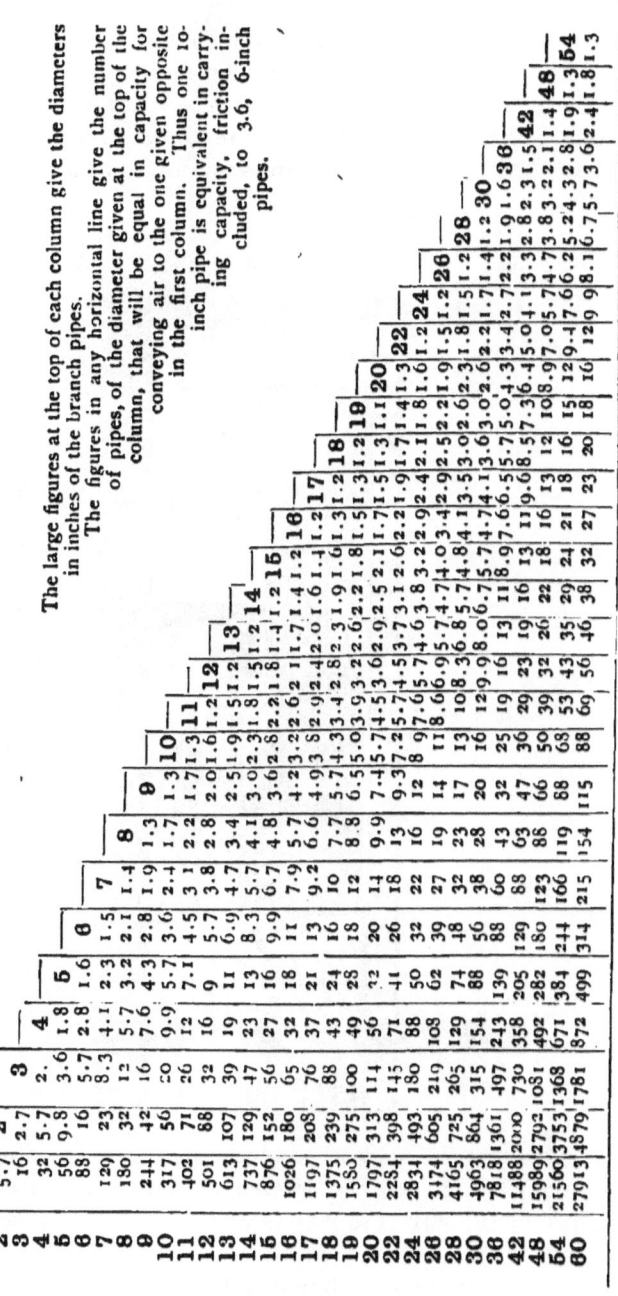

TABLE FOR EQUALIZING THE DIAMETER OF PIPES.

The large figures at the top of each column give the diameters in inches of the branch pipes. The figures in any horizontal line give the number of pipes, of the diameter given at the top of the column, that will be equal in capacity for conveying air to the one given opposite in the first column. Thus one 10-inch pipe is equivalent in carrying capacity, friction included, to 3, 6, 6-inch pipes.

288   HEATING AND VENTILATING BUILDINGS.

Air which is drawn in from outside at high velocity is often loaded with dust, and for this reason filters made of some textile material, or baffle-plates which discharge the dust into vessels of water, are sometimes required in the passageway leading to the fan. Where fans are required for ventilation as

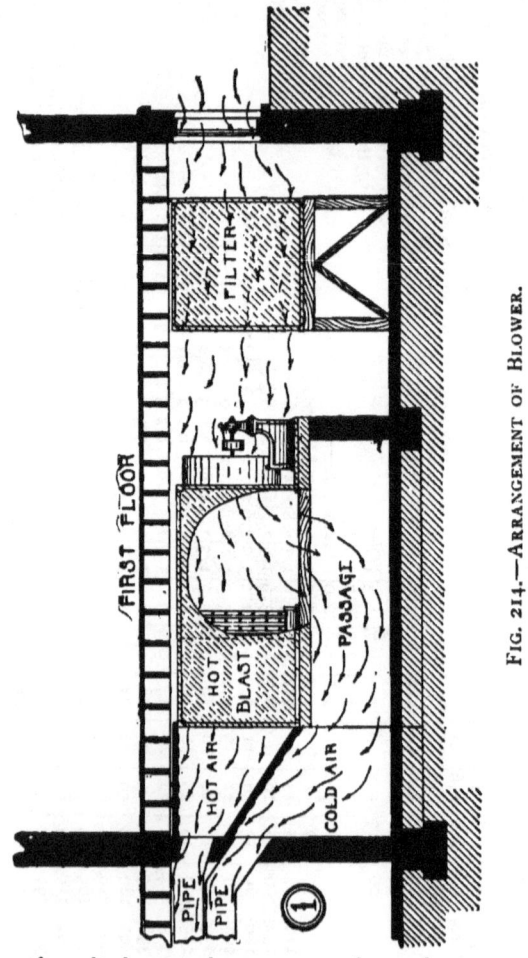

FIG. 214.—ARRANGEMENT OF BLOWER.

well as heating, it is an advantage to have by-pass pipes leading from the fan around as well as over the heating surface and provided with proper dampers so that the air can be delivered into the room fully or partly heated as required. Such an arrangement is shown in Fig. 214.

The net areas of registers should be sufficiently great to prevent the velocity in the entering air becoming so great as to produce a sensible draft. Taking this limiting value at 5 feet per second, the area of the register can be obtained from table on page 53. If the air is to be changed four times per hour, there should be 34 square inches in the register for each 1000 cubic feet of space. The nominal area of the register should be about 50 per cent greater than given by this computation; the actual areas of commercial registers is given in table page 280.

152. Blowers or Fans.—Blowers or fans are made in a great variety of forms, and there is little reliable data as to the best shape of fan-blades for practical use. It is quite certain that in the centre of the fan there is very little useful work done, and in some cases a back current is produced which reduces the capacity of the fan, although probably not affecting to any great extent the power required to drive it. It is quite probable that the workmanship and character of bearings have more to do with the efficiency than any theoretical form of blades. The limits of this book do not permit a discussion of the various forms of blowers. The reader is referred for some experiments on this subject to the work on "Warming and Ventilating of Buildings," by J. H. Mills, vol. II, page 559.

The motive power employed to drive fans may be obtained from a running countershaft, from an engine either directly connected or belted, or from an electric or water motor. Where the fan is to be used only at intervals, the electric motor will be found more desirable and fully as economical as the engine.

The fan should be located in a position where the noise caused by its operation is likely to be of little importance, and it should be arranged so that a portion or all of the blast can be deflected from the heating surface and sent to the rooms without being warmed if so required. This can be done by proper construction of ducts and dampers.

The actual power required to drive fans cannot, for the reasons mentioned, be determined from theoretical considerations, but must be obtained by actual test for each given make of fan.

The following table gives the sizes, capacity, and power required for various dimensions of the Sturtevant pressure fan-wheels, which are built to be set in wood or brick housing:

TABLE OF CAPACITIES, REVOLUTIONS PER MINUTE, AND HORSE-POWER REQUIRED.

PRESSURE IN OUNCES PER SQUARE INCH, AND INCHES OF WATER.

Size. In.	Diameter Fan in inches.	Diam. of Inlet in inches.	Width of housing in inches.	Pulley.		¼ oz. Pressure. 0.43 in.			½ oz. Pressure. 0.86 in.			¾ oz. Pressure. 1.3 in.			1 oz. Pressure. 1.73 in.		
				Diam.	Face.	Revol's per min.	Cub. ft. per min.	H. P.	Revol's per min.	Cub. ft. per min.	H. P.	Revol's per min.	Cub. ft per min.	H. P.	Revol's per min.	Cub. ft. per min.	H. P.
5½ × 3	66	47	36	22	9½	150	24734	1.2	212	34992	3.4	259	42892	5.6	300	49524	9.7
6 × 3½	72	52½	42	24	10½	138	31694	1.5	194	44857	4.4	238	54966	7.1	275	63463	12.4
7 × 4	84	60½	48	28	12½	118	42167	2.1	166	59681	5.8	204	73130	9.5	235	84436	16.5
8 × 4	96	68½	48	32	12½	103	47486	2.3	145	67218	6.6	178	82355	10.8	206	95086	18.6
9 × 4½	108	77	54	36	12½	91	60992	3.0	129	86180	8.4	159	105602	13.7	183	121938	23.8
10 × 5	120	85½	60	42	12½	82	75816	3.7	116	107313	10.5	143	131475	17.1	165	151800	29.6
12 × 6	144	102½	72	48	14½	69	108703	5.3	97	153850	15.0	119	188520	24.5	137	217340	42.4
14 × 7	168	120	84	54	14½	59	149840	7.3	83	212070	20.7	102	259765	33.8	118	300000	58.4
15 × 7½	180	128½	90	60	16½	55	172740	8.4	78	244487	23.8	95	299584	39.0	110	345900	67.4

The following table gives the capacity and power required for Sturtevant steel-plate exhaust fans:

TABLE OF CAPACITIES, REVOLUTIONS PER MINUTE, AND HORSE-POWER REQUIRED.

PRESSURE IN OUNCES PER SQUARE INCH AND INCHES OF WATER.

Size.	¼ oz. Pressure. 0.43 in.			½ oz. Pressure. 0.86 in.			¾ oz. Pressure. 1.3 in.			1 oz. Pressure. 1.73 in.		
	Rev. per min.	Cub. ft. per min.	H. P.	Rev. per min.	Cub. ft. per min.	H. P.	Rev. per min.	Cub. ft. per min.	H. P.	Rev. per min.	Cub. ft. per. min.	H. P.
40 in.	412	2,388	.11	582	3,380	.32	714	4,141	.54	824	4,782	.93
50 in.	329	4,396	.20	465	6,220	.60	571	7,623	.98	659	8,802	1.70
60 in.	274	6,458	.31	388	9,140	.89	476	11,200	1.49	549	12,932	2.52
70 in.	235	8,412	.41	333	11,906	1.16	407	14,588	1.90	470	16,848	3.29
80 in.	206	11,234	.54	291	15,900	1.55	366	19,483	2.53	412	22,495	4.38
90 in.	183	15,195	.74	258	21,507	2.10	317	26,354	3.43	366	30,427	5.92
100 in.	165	19,646	.95	233	27,804	2.71	286	34,070	4.44	329	39,338	7.67

An examination of this table will show the superior economy of moving a given volume of air under low pressure with a large fan as compared with the movement of the same volume under high pressure by a small fan. Thus to move 8400 cubic feet of air we can use (see above table) a fan 70 inches in diameter revolving at 235 revolutions, and requiring 0.41 horse-power to drive it, or we can use a 50-inch fan moving at 659 revolutions and requiring 1.7 horse-power. It is therefore evident that true economy can be best attained by purchasing a large fan, and thereby saving the running expense necessary for additional power to drive a smaller fan up to the same capacity.

An exhaust fan in the ventilating shaft has been used, in some instances with good results, for removing air from a building and producing circulation over the heater, but there is liability of leakage or infiltration of air into the flues from the outside. In case air enters this without passing over the heating surface, which is likely to reduce its efficiency, so that in practice it has not proved as satisfactory as the pressure-system. For purposes of ventilation only, or for the removal of foul and noxious gases where the ventilating ducts are tight, or as an accessory to the pressure-system, the exhaust fan is very efficient and often invaluable.

153. Heating Surface Required.—The methods of proportioning the heating surface, will be the same in every particular as those previously described for indirect heaters, page 211, and for hot-air furnaces, page 278. In this case, however, as the air passes over the heating surfaces with considerable velocity, the amount of heat which is given off is many times more than that from ordinary radiating surfaces in direct heating. Experiments have already been quoted on page 83 which show that the number of heat-units given off per degree difference of temperature per square foot of surface per hour is approximately equal to twice the square root of the velocity of the air in feet per second; for a velocity of 36 feet per second this would amount to 12 heat-units. For very cold weather the difference of temperature between heating surface and air will be from 160 to 170 degrees, and in this case the total heat given off per square foot will be about 2000 heat-units, or the equivalent of that given off in the condensation of somewhat more than 2 pounds of steam.

The following general formula will apply to this case:

Let T = temperature of heating surface, t that of the air of the room, t' that of outside air, t'' that of air leaving heating surface, t_1 the mean temperature of air surrounding heating surface = $\frac{1}{2}(t'' - t')$, $n =$

number of times air is to be changed per hour in the room, C cubic contents of room, $a =$ coefficient giving number of heat-units per degree difference of temperature per square foot per hour from heating surface. We have, since one heat-unit is capable of heating 56 cubic feet of air one degree: *

Cubic feet of air heated per hour $= nC$;

Heat-units required for warming this air $= \dfrac{nC}{56}(t'' - t')$;

Square feet of radiation $= \dfrac{nC(t'' - t')}{56a(T - t_1)}$.

If in the above equations outside air is zero and air leaving heating surface is at 120 degrees, and the air in the room is changed 4 times per hour and maintained at 70 degrees, we shall have $T = 220$, $t' = 0$, $t_1 = 60$, $n = 4$, $a(T - t_1) = 2000$, from which is deduced the following simple rules: First, the heat required expressed in heat-units per hour is equal to 8.6 times the number of cubic feet in the room; second, the number of square feet of heating surface will be equal to the number of cubic feet in the room multiplied by 0.0041. The amount of heat given off per square foot of surface is about 6 times that in direct heating; hence the areas of main steam- and return-pipes should be 6 times greater than those given by the table on page 237.

154. Size of Boiler Required.—From the preceding statement and by reference to page 124 it is seen that one square foot of heating surface in hot-blast heating will condense from 0.7 to 0.9 the amount of steam that can be produced by one square foot of heating surface in the boiler. Hence there should be from 0.7 to 0.9 as much area of heating surface in the boiler as in the indirect heater, or in other words there should be one boiler horse-power for every 15 to 18 square feet in the heater. The proportions of grate surface, chimney, etc., will be found by consulting Article 74, page 124.

155. Practical Construction of the Hot-blast System of Heating.—The following matter regarding the construction of hot-blast heating plants has been kindly furnished for this book by Mr. F. R. Still of Detroit, who has had an extensive engineering experience in this particular kind of work:

---

* See Table VIII, temp. at 70° F.

"*Air Required.*—The following is intended to give the basis of calculation for different parts of a plant of the so-called *hot-blast* system. The first thing to consider with this system usually is the amount of air to be delivered and warmed per minute. Experience has proved that the delivery of an amount of air into a building or apartment equal to its cubic contents every 15 minutes, will warm it under average conditions of construction to 70 degrees F. when the outside temperature is zero. This amount of air will accomplish like results in some buildings when the outside temperature is 10 or even 20 degrees below zero, and in other cases this amount will be found insufficient; the variation being due to construction, glass surface, and conditions which have been previously mentioned on page 58. In some classes of buildings, for instance, churches, school-houses, theatres, and hospitals, a change of air may be required every 10 minutes.

Amount of Heating Surface. — Having determined the amount of air required, the next consideration is the amount of heating surface to be used in the indirect heater. This can be treated better by taking a specific example, for instance, suppose that 20,000 cubic feet of air to be delivered into the building every minute (1,200,000 cubic feet per hour) at a temperature of 120 degrees, when air outside is zero, that the steam-pressure on the coils or heating surface is 10 pounds per square inch, and that the temperature of the water of condensation is 213 degrees. In one pound of steam at a pressure of 10 pounds above the atmosphere there is 1186.5 units of heat, while in one pound of water of condensation there is 213 units, leaving 973.5 units, which is given off by the heating surface. By consulting Table VIII it will be seen that at temperature of 70° F. one heat-unit will warm 56 cubic feet of air one degree, and hence to heat one cubic foot 120 degrees will require 2.15 heat-units: each pound of steam gives off 973.5 heat-units and will heat 452 cubic feet of air from zero to 120 degree. To heat 1,200,000 cubic feet of air to 120 degrees will require 2660 pounds of steam. The indirect heater provided with blower will condense under average conditions 2 pounds of water per square foot of surface per hour, and hence we should require as many square feet of surface as the quotient of 2660 divided by 2, or 1330 square feet.*

Size of Boiler.—To find the size of boiler needed, divide the total steam required per hour, in the example 2660, by that required for one boiler horse-power; this, when water of condensation is all returned to boiler, is 34.5 pounds (see page 122), and we obtain 77 horse-power. This computation gives a larger boiler than would generally be installed for work of this magnitude. The rated horse-power of a boiler is capable of considerable increase in times of necessity and for short periods. It can hardly be considered good practice to overwork a boiler, but as extremely severe weather is usually of very short duration and the balance of the season mild, there is good reason, on the score of economy in first cost, for this practice. The boiler is usually rated on the supposition that it will need to supply 1.5 pounds of steam for each square foot of surface in the radiator per hour, in which case 23 square feet of surface would be supplied by one boiler horse-power. This estimate would require the normal rating of the boiler to be developed during the average stress of weather; this method would require a boiler of about 60 horse-power for the plant considered in the example. Such a method of proportioning has proved quite satisfactory in actual practice, although greater economy could, no doubt, be obtained by using a larger boiler.

Size of Blower.—We are next to determine the size of blower required, which in some respects is the most difficult part relating to the design, as much depends on the location of the fan and the various uses to which the building is to be put. Noise is an objection in any kind of a building except perhaps one devoted to manufacturing. A good basis from which to determine the velocity of the air is that relating to the highest speed at which the blower can be driven without making a serious noise. This limit of speed is found to be about 250 revolutions per minute, but except in rare cases the blower should run at from 180 to 200 revolutions. We do not advo-

---

* Some manufacturers claim 5 pounds of condensation at zero weather; highest results obtained by Mr. Still were 3.5 pounds.

cate a linear velocity of the air through the discharge of a blower in excess of 2400 feet per minute, and it will be found to give better economy and more satisfactory results if the velocity does not exceed 1500 or 1800 feet; though in some instances this low velocity may require a large and unsightly fan.

Assuming, as in the example, that the blower is to deliver 20,000 cubic feet of air per minute at a velocity not exceeding 2000 feet per minute, the following considerations must receive attention: A blower standing in an open room and having a free inlet and outlet will discharge air at a velocity nearly 10 per cent greater than the peripheral velocity of the fan-blades, but attach this blower to a bank of heating coils and a system of conduits and the resistance due to friction becomes so great that it reduces this velocity nearly 50 per cent. To allow for this loss and retain a factor of safety it is customary to call the peripheral velocity of the fan-blades equal to the linear velocity of the air, and to figure on the efficiency of delivery in actual work as 50 per cent of this amount. On this basis a velocity of 2000 feet through the discharge of the blower will be maintained when the peripheral velocity of the fan is about 4000 feet. Having determined that the blower is not to run over 200 revolutions per minute, it will be necessary to have fan-wheels for this peripheral speed 6.4 feet in diameter. A fan 6 feet in diameter running 200 revolutions has a peripheral velocity of 3770 feet per minute, so that the air delivered, with 50 per cent allowance for friction will move at the rate of 85 feet per minute. A blower of any standard make, having a wheel 6 feet in diameter, would be provided with a discharge-opening at least 11.5 square feet in area. The product of this area by 1885 gives a discharge of 21,677 cubic feet per minute, which is slightly more than is required in the example considered.

Power Required.—The next consideration is the amount of power required to drive the blower, regarding which we will say that we know of no formula which has sufficient elasticity to apply alike to large and small blowers at high and low speeds. The following tables give the results of the actual power, as obtained by testing, required to operate various sizes of Smith fans when delivering a specified amount of air:

TABLE OF CAPACITY AND POWER FOR STEEL-PLATE BLOWERS OF VARIOUS SIZES.

Size. In.	Diameter of Wheel.	¼ Oz. Pres.			½ Oz. Pres.			¾ Oz Pres.			1 Oz. Pres.			2 Oz. Pres.		
		Rev.	Cub. Ft. per Min.	H.P.	Rev.	Cub. Ft. per Min.	H.P.	Rev.	Cub. Ft. per Min.	H.P.	Rev.	Cub. Ft. per Min.	H.P.	Rev.	Cub. Ft. per Min.	H.P.
70	42	214	10336	.3	312	14628	1.3	377	17928	1.6	428	20700	3.7	607	29352	10.5
80	48	188	12584	.5	265	17809	1.6	325	21827	2.4	367	25202	4.5	539	35736	12.8
90	54	167	16150	.7	236	22856	2.0	289	28012	3.7	333	32343	5.7	472	45860	16.4
100	60	150	20723	.9	212	29329	2.6	260	35945	4.8	300	41503	7.4	425	58850	21.1
110	66	137	24548	1.1	193	34741	3.1	236	42579	5.7	273	49162	8.8	387	69711	25.0
120	72	125	30165	1.3	177	42678	3.8	217	52304	7.0	250	60392	10.7	354	85634	30.7
140	84	107	40465	1.8	152	57268	5.1	186	70188	9.4	214	81040	14.4	304	114913	41.3
160	96	94	51344	2.3	133	72264	6.4	163	89057	11.5	152	102807	18.3	260	145806	52.4

TABLE OF DIMENSIONS.*

Fan with Steel-plate Housing.

Size of Fan. Inches.	Diameter of Wheel.	Diameter of Shaft.	Size of Pulley. In.	Inlet.		Outlet.		Size of Engine Cylinder In.		Weights.		
				Diameter.	Area in Sq. In.	Size. In.	Area in Sq. In.	Single.	Double.	Fan Only.	Single Engine.	Double Engine.
70	42	2 7/16	14 × 8½	26	530	24 × 24	576	4 × 4	3 × 3	1000	1290	1330
80	48	2 7/16	16 × 8½	30	706	26½ × 26½	702	4 × 4	3 × 3	1300	1590	1630
90	54	2 7/16	18 × 10½	34	907	30 × 30	900	5 × 5	4 × 4	1650	2150	2190
100	60	2 7/16	20 × 10½	38	1134	34 × 34	1156	6 × 6	5 × 5	2000	2640	2850
110	66	2 7/16	22 × 12½	42	1385	37 × 37	1369	6 × 6	5 × 5	2500	3140	3350
120	72	2 7/8	24 × 12½	46	1661	41 × 41	1681	7 × 7	6 × 6	3000	3870	4300
140	84	3 1/4	28 × 12½	53	2206	47½ × 47½	2256	....	7 × 7	4000	5600	5700
160	96	3 1/4	32 × 12½	60	2827	53½ × 53½	2862	....	7 × 7	5200	6800	6900

* Catalogue American Blower Co.

Capacity of Blower.—The 120-inch fan has a wheel 6 feet in diameter, as shown by the table of dimensions. By consulting the table of powers it will be seen that this fan, running with a speed of 200 revolutions per minute, requires less than 7 horse-power to drive the blower. The capacity, as given under the same head, is that for a fan working with free inlet and outlet, and, as before remarked, is about 10 per cent greater than the capacity when delivering into conduits. To totally close off either inlet or discharge of the blower causes the air to move around with the fan; this removes so much load from the engine that unless it is provided with an excellent governor it will speed up to a very great rate and may run away. This fact that an increase of resistance diminishes the power required at different speeds is not considered in the tables given; consequently these powers are somewhat in excess of those actually required. The excess of power would depend upon friction and other resistances; consequently no allowance can be made which would be accurate for all conditions.

Dimensions of Horizontal Conduits.—We now come to the question regarding dimensions of horizontal conduits that convey the air from the blower to various parts of the building. There is a great difference of opinion as to the proper velocity of the air through such conduits, and circumstances have a great deal to do with this question. In my opinion the easier you make it for the air to travel the more successful will be the plant. In no plants, in public-buildings, do we advocate a velocity of air that exceeds 15 feet per second, or 900 feet per minute; 600 feet, or even 400 feet, is better, although in an extensive plant the conduits might be so large as to be unsightly and interfere with the convenience of the building. Vertical flues in the walls leading to the various apartments should be so large that the velocity of the air will not exceed 10 feet per second, or 600 feet per minute.

Maximum Velocity of Air.—From an economical and efficient standpoint air should never enter a room through a register, screen, or grille at a velocity exceeding 400 feet per minute (6.6 feet per second). A greater velocity is liable to create such a rapid movement of the air as will stir up the dust in the room and create serious throat affections. Again, air coming in contact with the screen at a very high velocity will cause a low whirr or whistle often proving very annoying. Better ventilation, or perhaps we should say better circulation of the air, takes place when introduced at a moderate velocity than at a high velocity, because in that case the air enters gently and is distributed by gravitation, due to the cooling of the air in contact with cold walls, and the whole body of air is thus kept in slight motion and the entering air is more evenly distributed. If the air is forced in at high velocity, it creates swift currents and counter-currents, which will completely prevent the equitable distribution of the fresh air.

Introduction of Air.—My method of introducing air into a room is from a register about 8 feet above the floor, connected with a flue located in an inside wall, and discharging the current of air in the direction of an outside wall. The vent register should be located in the same wall as the fresh-air register, but at the opposite side and in the warmest corner of the room.*

General Remarks.—Architects very often combat such arrangements on the ground of interfering with their plans or of taking up too much room, and very often seriously object to making even the slightest alteration. This often leads to sorry arrangements for heating and ventilating plants, which will probably always continue so long as competing manufacturers design those to be installed in certain buildings.

It may be said generally that while the method of designing, followed by different manufacturers, may be essentially different from that given here, yet the experience of the writer has shown that the quantities, as computed by various manufacturers when submitting plans in competition for the same building, are essentially the same as those stated here.

156. Systems of Ventilation without Heating.—Where large quantities of air are required, especially in seasons when heat is not needed, systems of ventilation may be constructed which are independent from the systems of heating. The circulation of the air through the building may be produced either by exhausting or rarefying the air in the discharge-ducts, or by delivering fresh air to the rooms under pressure, as described for hot-blast heating.

The air may be rarefied in the discharge-flue by heating either with steam or hot-water radiators, with an open fireplace or a stove. When circulation is produced by heat, the amount of air moved will depend upon the height of the chimney or discharge-duct and its temperature, and will be essentially as that given in the table on page 45. The air may also

---

* The above opinion gives the practice of Mr. Still, and is different from that of many engineers. See a full discussion of the matter on pages 44 to 49.

be exhausted from the building by induction, for which may be used a jet of steam, water, or compressed air which is delivered from a nozzle into a convergent pipe of somewhat larger diameter and with both ends open. A very strong draft can be produced in this way, although at the expense of more energy than that required to operate exhaust fans or blowers. The air may also be exhausted by means of a fan located in the main flue. In case any of these means for producing drafts by exhausting or rarefying the air in the discharge-ducts is employed, every precaution that has been mentioned in regard to chimney-tops (page 162) should be observed, otherwise a considerable portion of the force may be required to overcome adverse wind currents.

The general remarks regarding hot-blast heating-systems and also the tables of dimensions apply equally well to this case. The tables on page 52 will be useful in proportioning areas of flues and registers for the discharge of a given amount of air; as an allowance for friction add one inch to each lineal dimension.

The blower system of ventilation has been fully described in connection with the hot-blast system of heating, and tables of capacities of various fans given which are applicable to this case. In this system as well as in the hot-blast system of heating especial care should be taken that the resistances in pipes and flues are as small as can be made, that bends are made with a long radius, and that the reduction in size in passing from one pipe to another is as gradual as possible.

157. Heating with Refrigerating Machines.—The refrigerating machine is virtually a pump which removes heat from a body at one temperature and discharges it at a higher temperature. Reckoned on the basis of heat transmitted, it is a very efficient machine, as it may move from a lower to a higher temperature 10 to 20 times as much heat as the mechanical equivalent of the work performed; in all respects this machine is the converse of the steam-engine. By utilizing the heat which is discharged from a machine of this character in warming a building, and also that in the exhaust steam from the engine working the compressor pump, there is a possible efficiency many times greater than that which can be obtained by burning the coal directly.

The practical arrangement of such a machine, if using air as the working fluid, would be such as to draw in air from the outside, compress it to such a point that its temperature would be very high, pass it through circulating pipes and radiating surfaces when still under pressure, and discharge into a chamber from which the pressure has been removed, or in the outside air after being cooled. If the exhaust steam could be used for heating, such a system would be very economical, although it would be costly and take up considerable room. An ammonia refrigerating machine might be used, in which case the heat in the compressed ammonia could be removed by water, which would thus become heated and could be circulated for the purpose of warming. The scheme of using the reversed heat-engine or refrigerating machine as a warming machine was pointed out first by Lord Kelvin in 1852,* and although it presents great advantages economically, the writer has no data showing that it has ever been put to practical use.

158. Cooling of Rooms.—The converse operation of cooling rooms, although at the present not undertaken except in the case of cold-storage plants and warehouses, bids fair to be at some time an industry of considerable importance. Rooms may be artificially cooled by a system constructed similar to that described for hot-blast heating. The coils or radiating surface, however, would need to be replaced by ice or constructed in such a manner that ammonia or some liquid at a very low temperature could be circulated. Over these the air could be driven, its heat would be absorbed, and it could be reduced in temperature to any point desired. In lowering the temperature of the air, a considerable amount of moisture might be precipitated, and some means should be provided for artificially removing it without heating, otherwise the rooms would be made damp. It may be remarked that ordinary pipe-fittings cannot be used with safety for ammonia circulation, and that special fittings are manufactured for this purpose.

---

* Proc. of the Phil. Soc. of Glasgow, Vol. III, p. 269.

CHAPTER XIV.

HEATING WITH ELECTRICITY.

159. Equivalents of Electrical and Heat Energy.—Electrical energy can all be transformed into heat, and as there are certain advantages pertaining to its ready distribution, it is likely to come into more and more extended use for heating, especially where the cost is not of prime importance. The value of mechanical and electrical units has been given on page 5, from which it will be seen that one watt for one hour, which is the ordinary commercial unit for electricity, is equal to 3.41 heat-units; for one minute it is 1/60 and for one second it is 1/3600 this amount. Electricity is usually sold on the basis of 1000* watt-hours as a unit of measurement, the watts being the product obtained by multiplying the amount of current estimated in ampères by the pressure or intensity estimated in volts; on this basis 1000 watt-hours is the equivalent of 3410 heat-units. We have considered in Chapter III the amount of heat required per hour for the purpose of warming. This amount divided by 3410 will give the equivalent value in kilowatt-hours which would need to be supplied for the required amount of heat.

160. Expense of Heating by Electricity.—The expense of electric heating must in every case be very great, unless the electricity can be supplied at an exceedingly low price. Much data exists regarding the cost of electrical energy when it is obtained from steam-power. Estimated † on the basis of

---

* One thousand watts is called a *kilowatt*.

† The mechanical energy in one horse-power is equivalent to 0.707 B. T. U. per second or 2545 per hour. One pound of pure carbon will give off 14,500 heat-units by combustion, which if all utilized would produce 5.7 horse-power

present practice, the average transformation into electricity does not account for more than 4 per cent of the energy in the fuel which is burned in the furnace; although under best conditions 15 per cent has been realized, it would not be safe to assume that in commercial enterprises more than 5 per cent could be transformed into electrical energy. In transmitting this to a point where it could be applied losses will take place amounting to from 10 to 20 per cent, so that the amount of electrical energy which can be usefully applied for heating would probably not average over 4 per cent of that in the fuel. In heating with steam or hot water or hot air the average amount utilized will probably be about 60 per cent, so that the expense of electrical heating is approximately as much greater than that of heating with coal as 60 is greater than 4, or about 15 times. If the electrical current can be furnished by water-power which otherwise would not be usefully applied, these figures can be very much reduced. The above figures are made on the basis of fuel cost of the electrical current, and do not provide for operating, profit, interest, etc., which aggregate many times that of the fuel. With coal at $3.30 per ton this cost on above basis is about .97 cent per thousand watt-hours. The lowest commercial price quoted, known to the writer, for the electric current was 3 cents; per thousand watt-hours the ordinary price for lighting current varies from 10 to 20 cents. It may be said that for lighting purposes 10 cents per thousand watt-hours is considered approximately the equivalent of gas at $1.25 per thousand cubic feet.

It may be a matter of some interest to consider the method of computation employed for some of these quantities. The ordinary steam-engine requires about 4 pounds of coal for each horse-power developed; on account of friction and other losses about 1.5 horse-power are required per kilowatt, or in other

---

for one hour, in which case one horse power could be produced by the combustion of 0.175 lb. of carbon. The best authenticated actual performance is one horse-power for 1.2 lb., corresponding to 14.6 per cent efficiency. The usual consumption is not less than 4 to 6 pounds per indicated horse-power, or from 3 to 5 times the above. A *kilowatt* is very nearly 1⅓ horse-power, but because of friction and other losses requires an engine of 1.5 indicated horse-power.

words 6 pounds of coal are required for each thousand watts of electrical energy. In the very best plants where the output is large and steady this amount is frequently reduced 20 to 30 per cent from the above figures in cost. The cost of 6 pounds of coal at $3.33 per ton is one cent. To this we must add transmission loss about 10 per cent, attendance and interest 20 per cent, making the actual cost per kilowatt 1.3 cents per hour. As one pound of coal represents from 13,000 to 15,000 heat-units, depending upon its quality, and one kilowatt-hour is equivalent to 3415 heat-units, if there were no loss whatever in connection with transformation of heat into electricity, one pound of coal should produce 4 to 5 kilowatts per hour of electrical energy. This discussion is sufficient to show that at cost prices electrical heating obtained from coal will amount under ordinary conditions to 15 to 20 times that of heating with steam or hot water, and at commercial prices which are likely to be charged for current its cost will be from 2 to 10 times this amount.

The following table gives the cost of a given amount of heat,

COST OF HEAT OBTAINED FROM ELECTRICITY.

Heat-units, B. T. U.	Cost per kilowatt hour, cents.									
	1	2	3	4	5	6	7	8	9	10
	Cost of heat obtained, cents.									
10,000	2.93	5.86	8.78	11.71	14.64	17.57	20.50	23.42	26.35	29.28
20,000	5.85	11.68	17.57	23.42	29.28	35.13	40.99	46.84	52.70	58.56
30,000	8.78	17.57	26.35	35.14	43.92	52.70	61.49	70.28	79.06	87.84
40,000	11.71	22.42	35.14	46.84	58.56	70.28	81.98	93.68	105.40	117.12
50,000	14.64	29.28	43.92	58.56	73.20	87.84	102.48	117.12	131.86	146.40
60,000	17.57	35.14	52.70	70.28	87.84	105.40	122.98	140.56	158.12	175.68
70,000	20.50	40.99	61.49	81.98	102.48	122.98	143.47	163.96	184.46	204.96
80,000	23.42	46.84	70.28	93.68	117.12	140.56	163.97	187.36	210.80	234.24
90,000	26.35	52.70	79.06	105.42	131.76	158.10	184.46	210.84	237.17	263.52
100,000	29.28	58.56	87.84	117.12	146.40	175.68	204.96	234.24	263.52	292.80

NOTE.—10,000 heat-units is equal to two thirds the heat contained in one pound of the best coal, and is very near the average amount that can be realized per pound in steam or hot-water heating, hence the table can also be considered as showing the relative price of electricity and coal for the same amount of heating. For instance, if 5 cents per kilowatt hour is charged for electric current, the expense would be the same as that of good coal at 14.64 cents per pound, which is at rate of $392.80 per ton.

if obtained from the electric current, furnished at different prices. Thus 30,000 heat-units if obtained from electric current furnished at 8 cents per *kilowatt* hour would cost 70.28 cents per hour. The amount of heat needed for various buildings can be determined by methods stated in Chap. III.

There are some conditions where the cost is not of moment and where other advantages are such as to make its use desirable. In such cases electricity will be extensively used for heating.

For the purposes of cooking it will be found in many cases that electrical heat, despite its great first cost, is more economical than that obtained directly from coal. This is due to the fact that of the total amount of heat, which is given off from the fuel burned in a cook stove very little, perhaps less than one per cent, is applied usefully in cooking: the principal part is radiated into the room and diffused, being of no use whatever for cooking, while the heat from the electric current can be utilized with scarcely any loss.

161. Formulæ and General Considerations.—The following formulæ express the fundamental conditions relating to the transformation of the electric current into heat:

$$C = \frac{E}{R} \quad \ldots \quad (1) \qquad W = CE = C^2 R \quad \ldots \quad (2)$$

$$R = \frac{kl}{w} \quad \ldots \quad (3) \qquad H = 0.24 C^2 R \quad \ldots \quad (4)$$

$$h_1 = .000000095 C^2 R \quad (5) \qquad h_2 = 3.415 W = 3.415 C^2 R = 3.415 CE \quad (6)$$

In which the symbols represent the following quantities: E, electromotive force in *volts*; C, intensity of current in *amperes*; R, resistance of conductor in *ohms*; l, the length in metres; w, the area of cross-section in square centimetres; k, coefficient of specific resistance; W, kilowatts; H, the heat in minor calories, and h_1 in B. T. U. per second, h_2 the heat in B. T. U. per hour.

The amount of heat given off per hour is given in equation (6), and is seen to be dependent upon both the resistance and the current, and apparently would be increased by increase in either of these quantities. The effect, however, of increasing the resistance as seen by equation (1) will be to reduce the amount of current flowing, so that the total heat supplied would be reduced by this change. On the other hand, if there were no resistance no heat would be given off, for to make $R = 0$ in equation (6) would result in making $h_s = 0$. From these considerations it is seen that in order to obtain the maximum amount of heat, the resistance must have a certain mean value dependent upon the character of material used for the conductor in the heater, its length and diameter.

For purposes of heating, a constant electromotive force or voltage is maintained in the main wire leading to the heater. A very much less voltage is maintained on the return wire, and the current in passing through the heater from the main to the return drops in voltage or pressure. This drop provides the energy which is transformed into heat.

The principle of electric heating is much the same as that involved in the non-gravity return system of steam-heating. In that system the pressure on the main steam-pipes is essentially that at the boiler, that on the return is much less, the reduction of pressure occurring in the passage of the steam through the radiators; the water of condensation is received into a tank and returned to the boiler by a steam-pump. In a system of electric heating the main wires must be sufficiently large, to prevent a sensible reduction in *voltage* or pressure between the dynamo and the heater, so that the pressure in them shall be substantially that in the dynamo. The pressure or voltage in the main return wire is also constant but very low, and the dynamo has an office similar to that of the steam-pump in the system described, viz., that of raising the pressure of the return current up to that in the main. The power which drives the dynamo can be considered synonymous with the boiler in the other case. All the current which passes from the main to the return current must flow through the heater, and in so doing its pressure or voltage falls from that of the main to that of the return.

Thus in Fig. 215 a dynamo is located at D, from which main and return wires are run, much as in the two-pipe system of heating, and these are so proportioned as to carry the required current without sensible drop or loss of pressure. Between these wires are placed the various heaters; these are arranged so that when electric connection is made, they draw current from the main and discharge into the return wire. Connections which are made and broken by switches take the

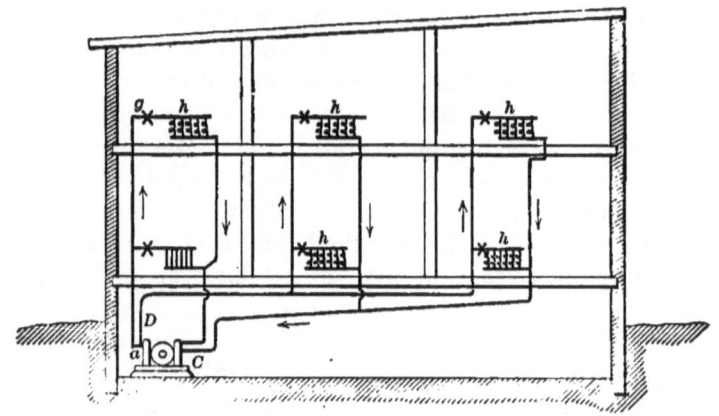

Fig. 215.—Diagram of Electric Heating.

place of valves in steam-heating, no current flowing when the switches are open.

The heating effect is proportional to the current flowing, and this in turn is affected by the length, cross-section, and relative resistance of the material in the heater. The resistance is generally proportioned such as to maintain a constant temperature with the electromotive force available, and the amount of heat is regulated by increasing the number of conductors in the heater.

162. Construction of Electrical Heaters.—Various forms of heaters have been employed. Some of the simplest consist merely of coils or loops of iron wire arranged in parallel rows so that the current can be passed through as many wires as are needed to provide the heat required. In other forms of

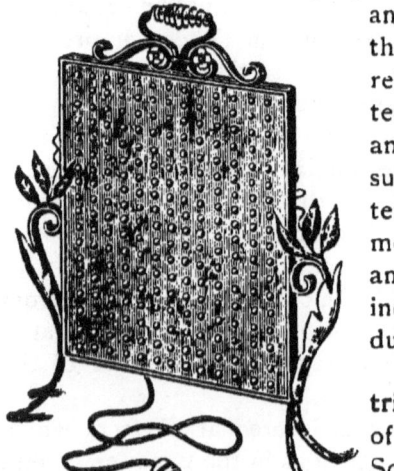

Fig. 216.—Electric Heater at the Vaudeville Theatre, London.

these heaters the heating material has been surrounded with fire-clay, enamel, or some relatively poor conductor, and in other cases the material itself has been such as to give considerable resistance to the current. It is generally conceded that

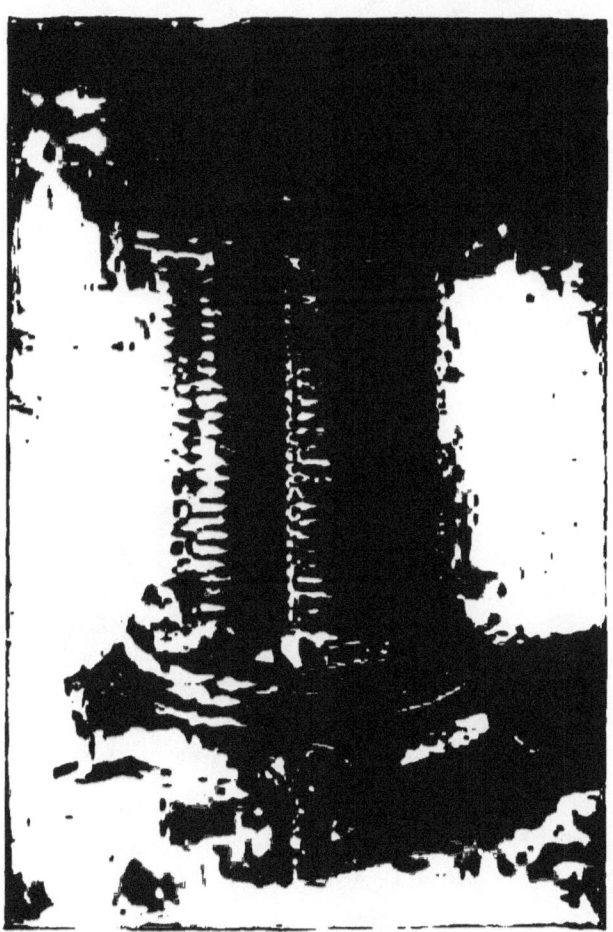

FIG. 217.—OFFICE OR HOUSE HEATER.

the most satisfactory results are obtained with electrical as with other heaters by regulating the resistance, by change of length and cross-section of the conductor, to such an extent as to keep the heating coils at a moderately low temperature. Some of the various forms which have been used are shown in the cuts. Fig. 216 represents a portable form of electrical heater used in the Vaudeville Theatre, London. Fig. 217 shows the interior of an office or house heater made by the Consolidated Car Heating Co., of Albany. The electrical heating surface is made in the latter by a coil of wire wound spirally about an incombustible clay core. The casing is like that for an ordinary

FIG. 218.—CAR HEATER OF CONSOLIDATED CO.

FIG. 219.—AMERICAN CAR HEATER.

stove, and is built so that air will draw in at the bottom and pass out at the top.

The electrical heaters at the present time are used almost exclusively in heating electrical cars, where current is available and room is of considerable value. These heaters are generally located in an inconspicuous place beneath the seats, their general form being shown in Figs. 218 and 219.

163. Connections for Electrical Heaters.—The method of wiring for electrical heaters must be essentially the same as for lights which require the same amount of current. The details of this work pertain rather to the province of the electrician than to that of the steam-fitter or mechanic usually employed for installing heating apparatus. These wires must be run in accordance with the underwriters' specifications, so as not, under any conditions, to endanger the safety of the building from fire.

CHAPTER XV.

TEMPERATURE REGULATORS.

164. General Remarks.—A temperature regulator is an automatic device which will open or close, as required to produce a uniform temperature, the valves which control the supply of heat to the various rooms. Although these regulators are often constructed so as to operate the dampers of the heater, they differ from damper-regulators for steam-boilers, as described in Article 91, by the fact that the latter are unaffected by the temperature of the surrounding air, although acting to maintain a uniform pressure and temperature within the boiler, while the former are put in operation by changes of temperature in the rooms heated.

The temperature regulator, in general, consists of three parts, as follows: First, a *thermostat* which is so constructed that some of its parts will move because of change of temperature in the surrounding air, the motion so produced being used either directly or indirectly to open dampers or valves, and thus to control the supply of heat. Second, means of transmitting and often of multiplying the slight motion of the parts of the thermostat produced by change of temperature in the room, to the valves or dampers controlling the supply of heat. Third, a motor or mechanism for opening the valves or dampers, which may or may not be independent from the thermostat.

In some systems the thermostat is directly connected to the valves or dampers, and no independent motor or mechanism is employed; in this case the power which is used to open or close the valves regulating the heat-supply is generated within the thermostat, and is obtained either from the expansion or contraction of metallic bodies, or by the change in pressure caused by the vaporizing of some liquid which boils at a low temperature. The force generated by slight changes in temperature is comparatively feeble, and the motion produced is generally very slight, so that when no auxiliary motor is employed it is necessary to have the regulating valves constructed so as to move very easily and not be liable to stick or get out of order. In most systems, however, a motor operated by clockwork, water, or compressed air is employed, and the thermostat is required simply to furnish power sufficient to start or stop this motor. The limits of this work do not permit an extended historical sketch of many of the forms which have been tried. The reader is referred to Knight's Mechanical Dictionary, article "Thermostats," and to Péclet's "Traité du la Chaleur," Vol. II, for a description of many of the early forms used. Those which are in use may be classified either according to the general character of the thermostat or the construction of the motor employed to operate the heat-regulating valves as follows:

$$\text{Thermostats.} \begin{cases} \text{Moved by expansion or contraction.} \\ \text{Moved by change of pressure.} \end{cases} \quad \text{Temperature Regulators.} \begin{cases} \text{No auxiliary motor.} \begin{cases} \text{Expansion or contraction.} \\ \text{Pressure.} \end{cases} \\ \text{Motor.} \begin{cases} \text{Clockwork.} \\ \text{Water.} \\ \text{Compressed air.} \end{cases} \end{cases}$$

165. Regulators Acting by Change of Pressure.—A change of temperature acting on any liquid or gaseous body causes a change in volume, which in some instances has been utilized to move the heat-regulating valves so as to maintain a constant temperature. Fig. 220 represents a regulator in which the expansion or contraction of a body of confined air is utilized to control the motion of the dampers to a hot-water heater.

It consists of a vessel containing in its lower portion a jacketed chamber connected to the hot-water heater at points of different elevation so as to secure a circulation from the heater through the lower portion or jacket of the vessel from 2 to 3. Above this is a second chamber which is covered on top with a rubber diaphragm, and which contains a funnel-shaped corrugated brass cup. The opening to the cup is in the lower portion of the chamber, the top and larger surface resting against the rubber diaphragm. Enough water at atmospheric pressure or alcohol is poured into the upper chamber through the opening marked 1 to seal the orifice in

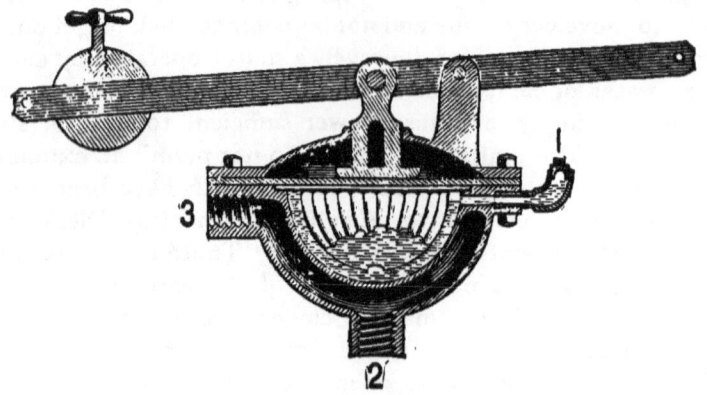

FIG. 220.—LAWLER HOT-WATER DAMPER-REGULATOR.

the inverted cup and confine the air it contains. The regulator acts as follows: The warm water from the heater moving through the lower chamber communicates heat to the water or alcohol in the upper chamber, which in turn warms the air in the inverted cup, causing it to expand. This moves the rubber diaphragm and connected levers leading to the dampers substantially as in the damper-regulator for steam-heaters, already described.

The Powers regulator for hot-water heaters (see Fig. 221) is somewhat similar in construction to the one described, but acts on a different principle. A liquid which will vaporize at a lower temperature than that of the water in the heater is placed in the vessel communicating with the diaphragm, in which case considerable pressure is generated before the water in the heater reaches the boiling-point. As the water in the heater is usually under a pressure of 5 to 10 pounds per square inch, its boiling temperature is from 225 to 240 degrees, water of atmospheric pressure which boils at 212° can be used in the closed vessel, and will generate considerable pressure before that in the heater boils.

The method of construction is shown at the right, in Fig. 221, as applied to a hot-water heater. The diaphragm employed consists of two layers of elastic material with compartments between and beneath; the lower one is connected to the chamber

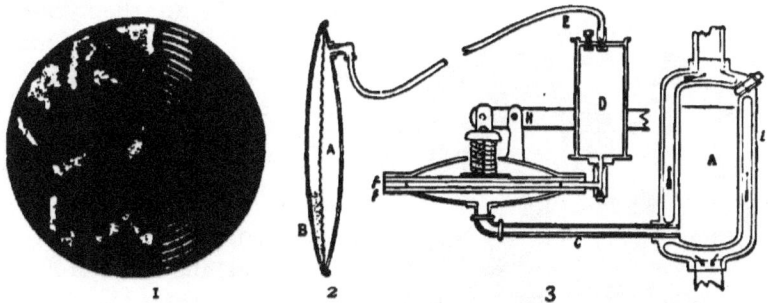

FIG. 221.—THE POWERS THERMOSTAT FOR HOT-WATER HEATERS.

A, which is filled with water at atmospheric pressure and is surrounded by the hot water flowing from the heater. The water in chamber A, being under less pressure, will boil before that in the heater, and will produce sufficient pressure to move the diaphragm and levers so as to close the dampers, before the water in the heater reaches the boiling-point. The compartment between the two diaphragms f, f is in communication with a vessel D, which in turn is connected by a closed pipe E with a thermostat, which may be placed at any point in the house and so arranged that if the temperature becomes too high in that room, the dampers of the heater will be closed. With this apparatus the dampers are closed either by excessive temperature of water at the heater or too great a heat in any room. The intermediate compartment is only required when the dampers are to be operated by change of temperature in the rooms.

The thermostat employed in this apparatus consists of a vessel 2, Fig. 221, separated into two chambers by a diaphragm; one of these chambers, B, is filled with a liquid which will boil at a temperature below that at which the room is to be maintained; the other chamber, A, is filled with a liquid which does not boil, and is connected by a tube to a diaphragm damper-regulator which moves the dampers through the medium of a series of levers.

Fig. 221, 2, shows a transverse section and 1 an elevation with parts broken away of a thermostat, and Fig. 222 an elevation with attached thermometer. The vapor of the liquid in the chamber B produces considerable pressure at the normal temperature of the room, and a slight increase of heat crowds the diaphragm over and forces the liquid in the chamber A outward through a connecting tube which leads to the damper-regulator, one form of which has been described.

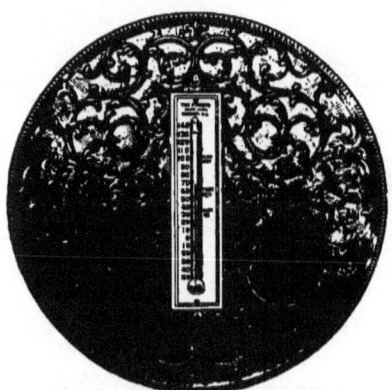

FIG. 222.—ELEVATION OF THERMOSTAT.

The damper-regulator as applied to a steam-heater is provided with a single rubber diaphragm with the parts arranged as shown in the sectional view Fig. 223. In this case the liquid pressure is applied above the diaphragm, its weight being counterbalanced by springs and weights, attached to the levers.

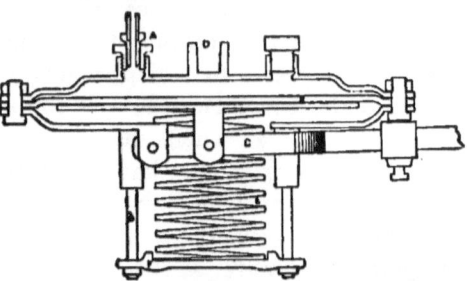

FIG. 223.—DIAPHRAGM DAMPER-REGULATOR.

The liquid used in the thermostat may be any which has a boiling temperature somewhat below that at which the room is to be kept. Many liquids are known which fulfil this condition, of which we may mention etheline, bromine, various petroleum distillates, anhydrous ammonia, and liquid carbonic acid. The liquids employed in the Powers thermostat are said to give pressures as follows at the given temperatures:

At 60°............ 1 pound to the square inch.
" 65°............ 2½ "   "   "   "
" 70°............ 4   "   "   "   "
" 75°............ 5½ "   "   "   "
" 80°............ 7   "   "   "   "
" 90°............ 10 "   "   "   "
" 100°........... 13 "   "   "   "

166. Regulators Operated by Direct Expansion.—
Metals of various kinds expand when heated and contract when cooled, and this fact has often been utilized in the construction of temperature regulators.

A single bar of metal expands so small an amount that it is of little value for this purpose unless very long, or unless its expansion is multiplied by a series of levers. Several forms have been used, of which may be mentioned: a bent rod with its ends confined so that expansion tends to change its curvature; a series of bent rods of oval form resting on each other with the ends confined between two fixed bars; two metallic bars, having different rates of expansion arranged parallel and the variation in length multiplied by a series of connecting levers an amount sufficient to be available in moving dampers; two strips of metal of different kinds bent into the form of an arc and fastened together so as to form a curved bar, with the metal which expands at the greater rate on the inside, so that expansion tends to straighten it when heated; the difference in expansion between an iron rod which is not heated and the flow-pipe of a hot-water heater multiplied by means of a series of levers. The constructions described above have all been tried for the purpose of moving the dampers of heaters or for opening and closing valves. In general, however, they have not proved satisfactory, because of the slight motion caused by expansion, and the uncertainty of operation obtained with multiplying devices.

Certain organic materials have the property of bending or curling when heated, and this has been utilized in the construction of the Howard regulator. This regulator consists of a thermostat in the form of a plaque of triangular form 11 inches long and 9 inches wide (Fig. 224), which is located in any living-room. As the temperature of the room increases the plaque bends. It is connected by means of cords running over pulleys to a very light and easily moved cylinder damper arranged so as to regulate both fire and check drafts. The damper used in connection with this thermostat consists of a slotted cylinder rotating on the inside of a tube which leads in one direction to the ash-pit and in the other to the smoke-pipe. A partition separates the two parts of the tube, and the slots in the cylindric damper are so arranged that when the connection for air to the furnace is open the other is closed, and *vice versa*, a very slight motion serving to completely open or close the damper. The cylinder damper is connected to the plaque by a cord, and is so arranged that the drafts are opened by the motion of the thermostat and closed by gravity.

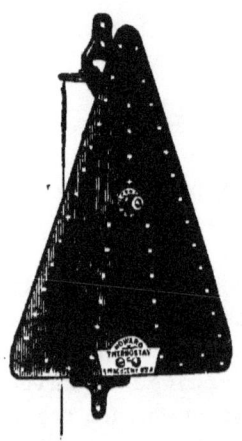

FIG. 224.—HOWARD THERMOSTATIC PLAQUE.

The direct expansion of a liquid or of a gas in a confined vessel has also been utilized to move a diaphragm or piston which is connected by levers to the dampers of heaters, in a manner similar to that described in the preceding article. The writer at one time constructed a regulator for a hot-water system in which the expansion of water in a closed vessel surrounding the return-pipe was employed to operate a damper-regulator similar to those used in steam-heating, page 156. Péclet describes, regulators in which the expansion of air was employed to move a piston connected by cords and pulleys to the dampers.

167. Regulators Operated with Motor—General Types. —The regulators which have been described in the preceding articles operate the regulating valves with a feeble force acting through a considerable range, or with a considerable force acting through a short distance. They are consequently liable to be rendered inoperative by any accident to the levers or connecting tubes, or by any cause which renders the valves difficult to operate. To overcome such difficulties several systems have been devised in which the power for operating the dampers should be obtained from an independent source, and in which the work required of the thermostat would be simply that of starting and stopping an auxiliary motor. In the first systems of this kind the motor employed was a system of clockwork which had to be wound at stated intervals in order to supply the force required for moving the dampers. In recent systems electricity, water, or compressed air is employed to generate the power required, and in some instances regulators are arranged to operate not only the valves which supply heat to the rooms, but also the various dampers for supplying hot or cold air in the ventilating system.

In all of the early forms of this kind of regulator the thermostat consisted of a tube of mercury or a curved strip, made of two metals of different kinds soldered together and arranged so that a given change of temperature would produce sufficient motion to make or break electric contact. A current was obtained from a battery, and connecting wires led to the motor and to the various terminals. When electric contact was made at a position corresponding to the highest temperature, the current would flow in a certain direction and cause a magnet to release a pawl which would start a motor revolving in the proper direction for closing the valves. When the temperature fell below a certain point, the thermostat would make electric connections so that the current would flow in the opposite direction and cause the motor to reverse its motion, thus opening the valve. If the motor was operated by water, the electric current would open and close a valve in the supply-pipe; if the motor was operated by electricity, the current from the battery would move a switch on the wires leading to the motor.

The valves for regulating the heat-supply are made in a great variety of ways. Dampers for regulating the flow in chimneys or flues are generally plain disks, balanced and mounted on a pivot, so that they may be turned very easily; globe- or gate-valves are usually employed in steam-pipes and must, to give satisfactory service, either be closed tight or opened wide. A system in which steam-valves are operated requires much more power than one in which dampers only are moved.

Many systems of heat-regulation employing motors are in use and are doubtless worthy an extended notice, but space will only permit a short description of the one in most extensive use in the larger buildings of this country, namely, the Johnson system of temperature regulation.

168. Pneumatic Motor System.—In the Johnson system of heat-regulation the motive force for opening or closing the valves which regulate the heat-supply is obtained from compressed air which is stored in a reservoir by the action of an automatic motor. The thermostat acts with change of temperature to turn off or on the supply of compressed air. When the air-pressure is on, the valves supplying heat are closed; when off, they are opened by strong springs. The detailed construction of the parts are as follows:

The compressed air is supplied by an automatic air-compressor which is operated in small plants by water-pressure and acts only when the supply of compressed air has fallen below the limit of pressure. The external form of the air-compressor is shown in Fig. 225. It consists of a vessel divided into two chambers by a diaphragm; one chamber is connected to the water-supply, the other to the atmosphere. The water entering on one side crowds the diaphragm over until a certain position is reached when the supply-valve is closed and a discharge-valve is opened, after which the diaphragm returns to its original place. The motion of the diaphragm backward and forward serves to draw in and discharge air from the other chamber in a manner similar to the operation of a piston-pump, valves being provided on both inlet- and discharge-pipes.

Fig. 225.—External View of Small Air-compressor.

When the air-pressure reaches a certain amount, the pump ceases its operation.

An air-pipe leads from the air-compressor to the thermostat, and another from the thermostat to the diaphragms in connection with valves or dampers. The action of the thermostat, as already explained, is simply to operate a minute valve for supplying or wasting, as necessary, compressed air in the pipe leading from the thermostat to the diaphragm-valves.

Fig. 226 is a sectional view of the diaphragm-valve, the

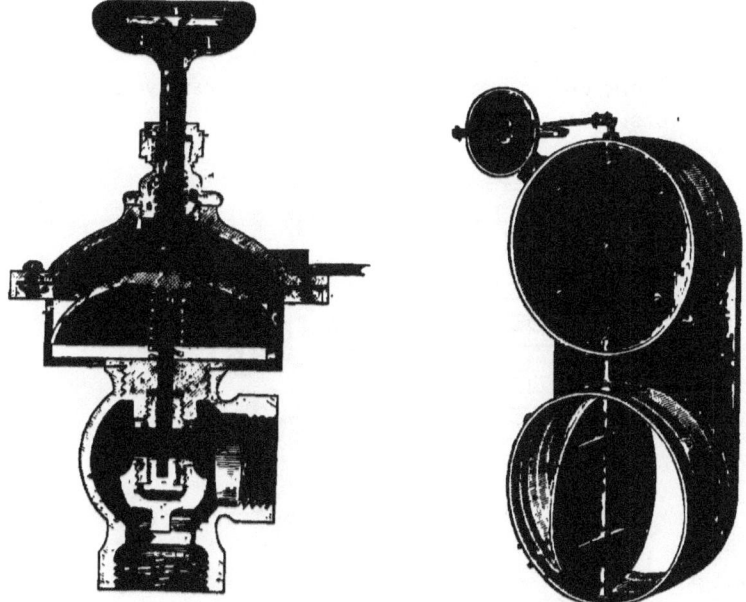

Fig. 226.—Sectional View of Diaphram-Valve.

Fig. 227.—Damper for Hot- and Cold-air Flue.

compressed air being admitted above the valve and acting merely to close it. It can also be closed if necessary by hand. The compressed air can also be made to operate dampers of which various styles are used, and these may be placed in ventilating flues, hot-air pipes, or smoke-flues, and so arranged as to admit either warm or cold air alternately to a room, as may be required to maintain a uniform temperature. Fig. 227 shows a damper for two round flues, one for cold air, the other for hot, connected to a diaphragm and arranged so that when one is open the other will be closed.

This system of heat-regulation has been brought to a very high degree of perfection, and if sufficient heat is supplied the temperature of a room is maintained with certainty within one degree of any required point. Farther than that, the system is so arranged that after all the rooms of the house reach the desired temperature the heat-regulator then acts to close the furnace-dampers. The apparatus is in extensive use for regulating temperature in the hot-blast system of heating. Fig. 228 shows the method adopted of applying a damper-regulator to a stack for indirect heating which is so arranged as to admit either warm or cool air as necessary to maintain a uniform temperature.

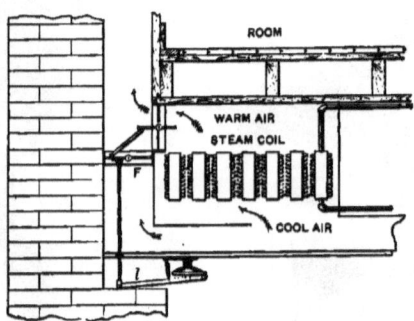

FIG. 228.—DOUBLE DAMPER IN BRICK DUCT.

169. Saving Due to Temperature Regulation.—The expense of constructing a perfect system of heat-regulation is met in a short time by the saving in fuel bills. The writer recently examined the records of the fuel consumed in a building when heated for a series of years without, and afterwards with, the heat-regulating system. He also examined the records showing the coal consumed in two buildings of exactly the same size and class, in the same city, and as nearly as possible with the same exposure. In both these cases the saving was somewhat over 35 per cent annually of the cost of the regulating apparatus.

The saving in any given case must, of course, depend upon

conditions and how carefully the drafts are regulated under ordinary systems of operation. Usually, when the temperature is regulated by hand, the rooms are allowed to become alternately hot and cool, but a greater portion of the time they are much warmer than is necessary, and frequently windows are opened for the escape of the extra heat. The maintenance of a uniform temperature for such cases means a saving of fuel by utilizing the heat better, and usually, also, by a more perfect combustion of fuel. It would seem from these considerations that a reasonable estimate of the saving obtained by the use of a perfect temperature regulator, as compared with ordinary regulation, would run from 15 to 35 per cent of the fuel bills per year.

Construction of Pneumatic Thermostat.—The following diagram and explanation will render the principle of action of

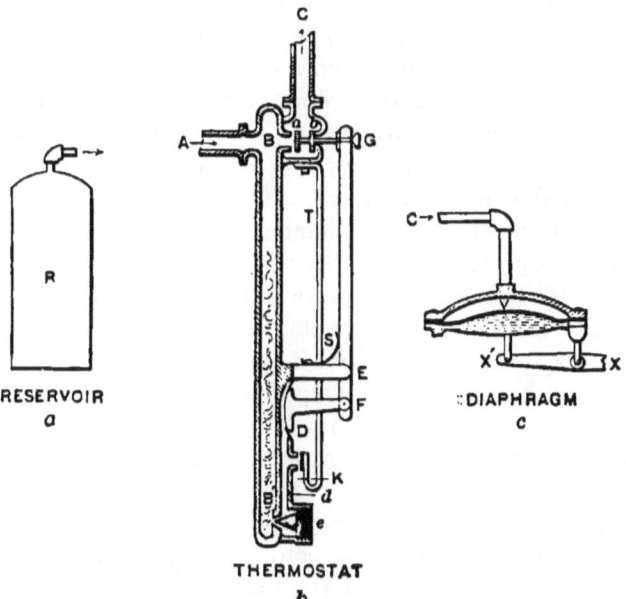

FIG. 229.—DIAGRAM ILLUSTRATING THE PNEUMATIC THERMOSTAT.

the pneumatic thermostat as employed in the Johnson system of heat regulation intelligible.

Fig. 229 shows to different scales the reservoir for compressed air, a diagram of the thermostat and of a diaphragm for operating dampers. The thermostat is drawn relatively to a very large scale. The temperature regulator as a whole consists first of an air compressor, as shown in Fig. 225, or one of similar construction, and arranged so as to maintain a constant pressure in air reservoir R or in the pipes of the building.

The principle of operation of the thermostat is illustrated by the diagram, although the details of construction of the actual instrument are quite different. Compressed air from the reservoir or air-pump passes through the pipe A to the chamber B, thence, if the double valve ab is open, it will pass out through the pipe C to the chamber V above the diaphragm. Its pressure then causes the end X' of the lever $X'X$ to move downward. This lever is connected to the damper in such a manner as to close off the supply of heat when in the lowest position. If the room becomes too cold, mechanism to be hereafter described moves the valve ab into such a position as to close the communication to the compressed air in the chamber B and open communication with the atmosphere at b. This permits the air to escape from the chamber V, through the pipe C and opening b, into the air, the diaphragm in the lower part of the chamber V being moved upward by a spring or weight not shown in the sketch. Thus it is seen that by moving the double valve ab the chamber V is put in communication with the compressed air and the damper moved to close off the heat, or with the outside air, in which case the pressure in the chamber V is lessened and the damper is moved by action of a weight or a spring so as to admit the warm air.

The mechanism for moving the valve ab consists of a thermostat T, which may be made of any two materials having a different rate of expansion, as rubber and brass, zinc and brass, etc. Connected to the thermostatic strip is a small valve K, so adjusted that when the room is too warm the valve will be opened and when too cold it will be closed by the expansion and contraction of the thermostatic strip. Suppose the room too warm and the valve K open, air then flows through the chamber B, through the filtering cotton in the lower part of B', thence through the small tube d and the valve K to the air. The small tube d connects with an expansible chamber D and opens back of a small diaphragm. When the valve K is open the spring S forces the diaphragm into the contracted or collapsed position, causing the lever GF to move the valve ab so as to put the chamber B in communication with chamber V and permit the air-pressure to close the damper connected to the lever $X'X$. If, however, the room becomes too cold, the thermostat T moves so as to close the valve K; this stops the escape of air from the pipe d and causes sufficient pressure to accumulate under the diaphragm at D to move the lever FG, so as to move ab to the left, thus cutting off the supply of compressed air from the chamber V and permitting the air to escape at b. It will be noted that air is continually escaping at K during the time the room is too hot, but this is a very short interval as compared with the entire time, and moreover the orifice at K is exceedingly small, so that the loss of air is quite insignificant. It will also be noted that with this apparatus the damper is quickly moved from a position fully open to shut, or *vice versa*, and that it will not stand in an intermediate position fully open or fully shut.

The manufacturers of the Johnson thermostat have quite recently designed an instrument which will move the adjusting damper connected to the line XX' slowly and will hold it in any intermediate position as desired. This is considered an advantage for systems of ventilation in which it is always desired to admit the same volume of air, but in which the relative amounts of hot and cold air are varied to maintain the desired temperature.

CHAPTER XVI.

SPECIFICATION PROPOSALS AND BUSINESS SUGGESTIONS.

170. General Business Methods. — Nearly all heating-plants are constructed by contractors, who agree for a specified sum to install a heating-plant in accordance with certain specifications, or, in absence of specifiations, one which is guaranteed to fulfil certain stipulations as to warming and ventilating in any stress of weather. Specifications are prepared either by a disinterested third party who is thoroughly familiar with the subject, or by the party submitting the proposal. The first method, although not common except in the case of large buildings, is, when the specifications are properly drawn, satisfactory both to the owner and the contractor. With proper specifications estimates can be obtained from different bidders on work of the same class and quantity, and this is likely to result in a better quality of work, and often in lower prices. Where each contractor bids on his own specifications and arranges for apparatus in accordance with his own judgment, there will be a very great difference in the quality and method of construction proposed, which is likely to result to the advantage of an unscrupulous bidder, who would, if possible, use cheap material and the least possible quantity of heating and radiating surface. It is for these reasons to the advantage of all concerned that full and complete specifications should be provided which will show, accurately, the character, amount and quality of the required work.

The specifications may be written as a part of the tender for the work, or as an independent document to which reference is made in the proposals.

The specifications are often accompanied with drawings which show the location of all the principal parts of the heating apparatus and frequently many details of construction; the drawings are considered in every case a portion of the specifications and are equally binding on the contractor.

After the bid has been accepted a contract is drawn which should contain a full statement of the agreement between contractor and owner, and of all conditions relating to the method of payment, penalties, time of completion of work, etc.

J. J. Blackmore and J. G. Dudley, New York, acting as a committee appointed by the National Association of Manufacturers of Heating Apparatus, have given the matter relating to uniform specifications much study, and we are indebted to them for the following discussion, and also for the copy of the uniform proposals here submitted.

171. General Requirements.[*]—" It is not within the scope of a work such as this, nor have the trade conditions in the heating business advanced to such a point, that all the details of any or every system can be provided for. The following proposed form for uniform standard specifications, however, covers the ground as fully as can be done at this time, as is shown by the recommendation by the National Association of Manufacturers of Heating Apparatus, and if generally accepted by heating contractors, manufacturers, architects, investors, and the laymen installing steam or hot-water heating apparatus, would result in a higher standard of excellence. Much trouble now exists in securing best results, due to ignorance on part of owner, architect, or contractor, as well as to unfair competition or unauthorized substitutions of 'cheap' materials.

" Any specification should set forth unequivocally and in detail (as far as feasible) all that the contractor is to furnish and exactly what is to be accomplished by his guarantee, which should embody a standard of economy as well as one of efficiency. The function of the owner or architect is to stipulate what results must be accomplished according to standards in accepted use, and to give the consulting engineer (when character of heating-plant demands one) or the contractor proper latitude as to *methods* to be pursued. Further than this, it is the office of owner or architect, in justice to himself and to competing bidders, as well as to the successful contractor, to

---

[*] Written for this work by J. J. Blackmore and J. G. Dudley.

see that the provisions of the specifications are carried out, and that the quantity and character of material agreed upon are actually furnished and used. Certificates to that end should be demanded and given, if it is deemed necessary, since much injury is done to a legitimate and beneficial calling by what is termed 'skinning the job,' that is, agreeing to furnish certain things and then by taking advantage of 'lay' ignorance substituting inferior goods or omitting them outright.

"As already shown, the attainment of certain results follows from, and is accomplished by, scientific and mathematical processes, whether actually figured and reasoned out, or arrived at by 'rule of thumb,' as many really excellent contractors are known to do.

"In illustration, imagine a country residence in course of erection after plans by, and under supervision of, a competent architect, and note how a proper heating-plant is installed. To begin with, the owner should learn from his architect or from any other properly informed person that the desired efficiency, sufficiency, and results to be procured by the heating system depends more on amount of investment than on anything else. For instance, the same results can be achieved by employing either steam or water. The first cost, however, is less with steam, while, it is contended by many, the running and ultimate cost is less with water. The reason for this is that with the hot-water system as usually installed, with an open tank for expansion of water, the temperature of the heating medium ranges from 150° to 200° F., while with steam it ranges from 212° to 240° F.; as a consequence more radiating surface is needed for the former than for the latter.

"To continue the illustration, let the owner select steam, and also suppose that he elects to have indirect heating on ground-floor, to obtain extra ventilation (for be it understood that some ventilation, accidental or otherwise, is absolutely necessary to obtain right heating results), while on the upper floors he chooses direct heating. This done, it then devolves on the engineer, contractor, or architect to determine the respective amounts of heating surfaces required to warm the several rooms to the indicated temperature according to an accepted standard. Much harm at present results from demanding and permitting the several bidders to estimate on different amounts of heating-surface for exactly the same work. The minimum amount should be determined by some one individual, who should be recompensed for this service, and he alone held responsible for this estimate. The owner or architect should indicate on the building plans where surfaces shall be placed, bearing in mind always the room required in the allotted spaces and also the requirements of the system. This is necessary for the contractor to know, since on it depend the number of his riser-lines and the amount of piping in his boiler-room.

"When feasible, the owner or architect should indicate all the 'specialties' desired in the apparatus, and each bidder should be compelled to figure as nearly as possible on exactly the same set of specifications. This method is just to those who estimate in good faith, and usually closer and lower figures will be obtained by the owner. The contractor, with these data before him, takes dimensions either from the architect's plans or from the measurements of the building itself; he then computes the quantity and cost of all materials which will be used in the completed apparatus; the method of computation varying from that of pure guesswork or shrewd 'estimating' to that of painstaking measurement and actual figuring out of the exact amount of stock required, together with its purchasable cost from the trade catalogues and price-lists.

"To the net cost for material, including boiler, radiators, pipe, fittings, valves, vents, floor- and ceiling-plates, registers, ducts, covering, painting, bronzing, smoke-pipe, freight and cartage, board, car-fares, labor, and incidentals, is added such a margin of profit as the contractor considers his experience, reputation, and workmanship are entitled to.

"In justice to the bidders the conditions of the award should be clearly set forth beforehand, and it should be stated whether this work will go to the lowest bidder, or whether a 'preference' (often justified) is to be given a certain contractor. When it is known that the preparation of a set of specifications and of an estimate of cost is an expense, and often not a small one, to each and every bidder, the injustice of requiring all to bear this instead of having it done once and for all is too evident for argument. It is for this reason that a uniform standard specification is recommended by the National Association of Manufacturers of Heating Apparatus.

"Suppose now the award be made to the lowest bidder, bids having been made on the same set of specifications which embody full statements in regard to requirements of the completed plant. The owner (or architect) and the contractor are then to execute a proper contract for the performance of the work and for the payments therefor. Then each should be required to fulfil the conditions of said contract. The National Association of Master Steam and Hot-water Fitters has adopted a uniform standard contract which seems to meet the requirements and is quite generally accepted in such cases. The form is given below and may be obtained of the secretary of that association.

172. Form Proposed by the National Association of Manufacturers of Heating Apparatus.—For a *steam-heating plant.*

UNIFORM STANDARD SPECIFICATION FOR A COMPLETE LOW-PRESSURE STEAM OR HOT-WATER HEATING APPARATUS.

NOTE.—All clauses and terms in **this type** and enclosed in brackets [ ] apply only to hot water. All clauses and terms in **this type** and enclosed in parenthesis ( ) apply only to steam. Words in italics are to be supplied in each contract.

TO BE INSTALLED AND ERECTED COMPLETE IN
the three-story stone and frame residence
OWNED BY
I. N. Vestor, No. 75 Broadway, New York City,
LOCATED AT
N. W. Corner of State and Hudson Streets, Yonkers, N. Y.

THE HEATING SYSTEM

shall be erected according to the *single* pipe method of **(steam)** **[water]** heating, the **(steam)** **[water]** to circulate **(under a pressure)** **[at a temperature]** never exceeding (*three* (*3*) **pounds to the square inch at**) [—— **degrees F. in the flow-pipes of**] the boiler, conveyed to heating surfaces by a system of piping so erected that all water (**of condensation**) in the system shall be freely eturned to boiler by gravity alone.

(STEAM GENERATOR.) [WATER HEATER.]

The (**steam**) [**water**] shall be (**generated**) [**heated**] by *one No. 2 Vertical Tubular Sectional* Boiler, manufactured by *C. Iron & Co., N. Y. City*, and by them guaranteed free from all flaws and defects. Said boiler to have a grate area of *700* square inches, capable of burning *all kinds of coal* as fuel, and guaranteed by makers to be capable of supplying (**steam**) [water] to *750* net square feet of direct radiation without " forcing "; boilers to be certified by manufacturer to be able to stand a cold-water pressure of *80* pounds to the square inch.

An opening not less than *two* (*2*) feet by *five* (*5*) feet into the building and boiler-room shall be provided by *owner*.

BOILER SETTING.

Boiler to be placed as near smoke-flue as possible, upon a level concrete or other equally solid foundation provided by *owner*. The top to be not less than *six*—feet from ceiling of boiler-room. All necessary excavating to be done at expense of *contractor*.

When boiler is set in brickwork same shall be not less than *eight* (*8*) inches thick, erected concentric or parallel with external boiler walls as shown by plans of *manufacturer*. Brick to be hard burned, and laid in courses which break joints, with cement mortar not more than one-fourth inch in thickness. Bond courses of headers to be laid once in every five courses. Setting when complete to be air-tight, and guaranteed to stand all strains of expansion and contraction; or, when plastic covering is used for setting, same shall be evenly distributed over external boiler surfaces not less than *two* (*2*) inches thick. The ash-pit shall not be less than *twelve* inches deep, and shall be sloped to edge of clean-out door. Boiler shall be provided with fire, clean-out, and ash-pit doors, of such form, size, structure, and set in such position as shall make accessible all portions of boiler requiring attention. When setting of boiler demands it, same shall be provided with cast-iron front designed by manufacturers for boiler specified. Same to be protected, when necessary, from direct heat of flame by brickwork or other means equally good.

FIXTURES, FIRE TOOLS, AND TRIMMINGS.

Boiler shall be provided with *rocking and dumping* grates designed by manufacturers for boiler specified, together with shaking-lever and all fire tools necessary to care for same, which shall consist of *one* (*1*) poker, *one* (*1*) slice-bar, *one* (*1*) fine brush and handle. Boiler shall be provided with (*one 5" brass*-bound low-pressure *Bourdon* steam-gauge, with stop-cock and siphon), (*one* (*1*) low-pressure safety-valve with *ten* (*10*) pound weight), (*one* (*1*) water-column fitted with *two* (*2*) brass try-cocks) [expansion thermometer registering from 80 degrees F. to 250 degrees F.] (*one Scotch* gauge-glass and *four* (*4*) brass guard-rods), and *one* automatic [   ] damper regulator with connections for operating draft-door and cold-air check; *one* $1\frac{1}{4}$-inch brass steam (blow-off) cock with key; and there shall be provided in addition to above all pipe, fittings, and valves necessary to render connection of all of above to boiler complete.

WATER CONNECTIONS AND BLOW-OFF.

Feed-water with its supply-pipe shall be brought within *six* feet of boiler by *owner*, and left with *one* $1\frac{1}{4}$-inch cast-iron fitting for boiler connection to be made by *contractor*. (Water supply to be controlled by *no* automatic water-feed.) Blow-off cock to be located at lowest point of system, with piping so pitching toward same as to allow of draining boiler and of system, same to be fitted for a hose-nipple connection. The discharge to waste opening (provided by *owner*) shall be always visible.

SMOKE-PIPE AND SMOKE-FLUE.

Contractor shall connect boiler to smoke-flue opening (provided by *owner*) by means of gas-tight pipe *twelve* inches in diameter, built of *No. 14 galvanized iron*, in which shall be placed *one* (*1*) shut-off damper with *wheel* handle attached, together with proper clean-out door. Smoke-flue throughout to be not less than *113* square inches internal area, and *46* feet in height, straight, and presenting no unusual obstructions to gases. Responsibility for proper working to rest on owner.

FLOW, BRANCH, AND RETURN MAINS.

Flow-pipes and branches shall be run on a grade to or from boiler of not less than one inch fall in each ten feet run; size of pipes to be of such area as to quickly, adequately, and noiselessly carry (**steam**) [**water**] by means of branches and risers to heating surfaces, and also to permit an unimpeded flow of all (**water of condensation**) [**return water**] to or from boiler by means of mains, branches, or reliefs. The size of pipes shall be gauged by, and shall in no case be reduced below, standards laid down in Carpenter's "Heating and Ventilating Buildings." All mains are to be so run in straight lines, and junctions so made, as to avoid all traps or pockets which may hold air or (**water of condensation**). (**When pitch of pipes brings level of flow-mains within eighteen inches of water-line of boiler, establish higher level for steam-flow, make connection with proper relief, so as to drip all condensation.**) All expansion and contraction of pipes throughout system must be provided for in joints thereof so as to prevent buckling or bending of same, and all joints made steam *and* water tight. [Note.—**No bushings shall be used on hot-water flow-pipes, whether mains, risers, or radiator connections.**]

This system of piping contemplates *three* (*3*) flow-mains, $1^1/_2$, 2, *and* $2^1/_2$ inches diameter, respectively, pitching *from* the boilers. There shall also be *two* (*2*) return-mains pitching toward the boiler on a grade not less than one inch in (**twenty**) [**fifteen**] feet run; same to be carried to boiler on the *overhead* plan, and to be so connected that there shall be *two* (*2*) return-mains entering boiler of not less than *one and one quarter* inches in diameter. (**Said main to be provided with inch swinging check-valve outside the boiler.**)

RISERS (RELIEFS) AND CONNECTIONS.

All risers shall be erected plumb and straight, and all connections thereto shall be made below or in the floors by means of double joints to allow for expansion. When "offsetting" or coupling parallel lines of pipe, care shall be taken, when possible, to locate centres of like pairs of fittings at same vertical or horizontal level, as case may be. Whenever pitch or size of pipes does not allow full vent of contained air (or **discharge of water of condensation**) proper *automatic* air-vents and reliefs or drip-pipes of sufficient size (according to Carpenter's tables) shall be used. (**When air-vents are fitted with drip-pipes same shall be run plumb and straight and parallel with risers, or return-riser lines, down to boiler-room, where same shall be joined together in one common main, which shall there vent the contained air.**) *No drip-pipes will be attached.*

The within system contemplates *five* (*5*) flow-risers and *no* return-risers.

[EXPANSION-TANK AND CONNECTIONS.]

To be omitted in steam-heating.

[There shall be furnished and connected, at a point not less than twelve inches above highest radiating surface, in best location which conditions of building permit, one ⎯⎯ gallon galvanized-steel expansion-tank, fitted with ⎯⎯ gauge-glass and ⎯⎯ brass guard-rods. Flow connections to tank shall be so made as to maintain a circulation of contained water at all times when apparatus is in use. Tank shall be provided with a ⎯⎯ inch pipe for venting overflow, same to be arranged to waste on roof or other outlet.]

FLANGES AND UNIONS.

At proper points on mains, branches, and return-mains shall be located right and left couplings or flange-unions, so that pipes may be disconnected without injury to balance of apparatus or system. Couplings may be used on all pipes up to two inches in size, all larger pipes to be united with flange-unions made tight with *copper* gaskets.

HANGERS.

All flow and return pipes shall be supported by *chain* adjustable pipe-hangers, securely fastened to building at intervals of not more than ten feet, and so constructed as to permit free expansion and contraction of piping.

FLOOR AND CEILING PLATES, AND PROTECTION.

Wherever pipes pass through floors, floor-plates shall be used. Wherever pipes pass through ceilings, ceiling-plates shall be used, unless other provision for finish be provided for. All protection of woodwork, etc., from heat of pipes shall be done in accordance with rules and regulations of National Board of Fire Underwriters. The return and branch mains and connections in boiler-room shall be covered with *sectional asbestos* covering *one* inch thick [or **with asbestos paper, 1″ Hair Felt,** rosin-sized **paper, and canvas neatly sewed on, to prevent waste of heat**].

RADIATOR-VALVES AND AIR-VENTS.

Each direct or direct-indirect radiator or coil shall be controlled by *one quick-opening* standard *radiator*-valve provided with union and of proper size, made of best (**steam metal, extra heavy, with composition disk or**) [quick-opening] to be *rough*-body *plated all over* and provided with wood handle. All radiators shall be fitted with *automatic* air-vents. [**Radiator return-connections shall be made with** body ell **unions.**]

When indirect radiators are to be controlled same shall be fitted with *iron wheel-gate* valves.

SYSTEM OF WARMING AND DISTRIBUTION OF RADIATION.

The building specified will be heated by means of *ornamental* direct, *prime surface* indirect, *and no* direct-indirect radiation, located in the several rooms to the best advantage, according as conditions of building and will of owner permit, and as shown in following schedule. Indicated temperatures to be maintained during *zero* weather.

Schedule.

Floor.	Room.	Square Feet Direct Radiation.	Square Feet Indirect Radiation.	Sq. Ft. Direct-Indirect Radiation.	Height Radiation.	Temperature F.
First.	*N.W.*	....	165	....	....	70°
	S.W.	....	150	....	....	70°
	S.E.	....	100	....	....	65°
	W.	50	....	....	20″	70°
	S.	24	....	....	38′	70°
Second.	*N.W.*	52	....	....	38″	68°
	S.W.	48	....	....	38″	68°
	S.E.	48	....	....	38″	68°
	W.	36	....	....	24″	68°
	S.	20	....	....	38″	65°
Third.	*W.*	24	....	....	38″	65°
	S.	16	....	....	38″	65°

INDIRECT RADIATION.

(NOTE.—Omitted on specifications when heating system is all direct.)

The indirect radiators shall consist of stacks or clusters of *prime* surfaces connected together with *tight* joints, and firmly suspended from ceiling by *suitable* wrought-iron hangers, as directed by radiator makers, or by other methods equally good. (**There shall be a difference of level of not less than eighteen inches between lowest point of all indirect radiation and the water-line of boiler.**)

All stacks shall be so piped and hung as to permit a quick, noiseless, and constant flow throughout of (**steam and all water of condensation**) [**the heated water**].

COLD-AIR DUCTS, CASINGS, ETC.

(NOTE.—Omitted on specifications when heating system is all direct.)

The area of internal cross-section of fresh-air inlet and duct, as well as registers and warm-air outlet, shall never be less than standards of measurements laid down in Carpenter's "Heating and Ventilating Buildings." Fresh-air inlet shall be of **600** square inches area, and shall be provided with substantial iron wire-gauze screen. Connecting cold-air duct and casing of indirects shall be made of galvanized iron (No. 20 or heavier), provided with door for clean-out and inspection —all joints being made permanently air-tight. Cross-section of duct throughout its length to be as nearly uniform, circular or square, as conditions of building permit. Casing of indirects shall be so erected that all entering air must pass through each stack, and be warmed, before passing to its respective outlet register. Stacks shall be so hung and encased that the full area of inlet and outlet ducts shall be maintained above and below the stack, which space shall in no case be less than ten inches in height, by the length and breadth of stack, and casing shall be so arranged that all inflowing fresh air shall be heated and conveyed to destination without loss through *tin* warm-air ducts, of areas as above provided for, same to be furnished by *owner*, and set in walls or floors by *owner*, as directed by *architects*. Each cold-air inlet shall be provided with one *controlled* damper, fitted with *iron* handle.

REGISTERS AND REGISTER-BOXES.

(NOTE.—Omitted on specifications when heating system is all direct.)

All registers shall be of *Jones* design. The sum of areas of openings in same never to be less than area of warm-air outlet. Registers to be set flush with, and firmly fastened in, openings in floor or wall provided by *owner*, and to be located to best advantage according as conditions of building permit. Proper register-boxes made of *I. X. tin* shall be provided by *contractor* for reception of registers.

CUTTING, PAINTING, BRONZING, ETC.

All cutting and carpenter work shall be done by *owner* as directed by *contractor*. All uncovered exposed piping in boiler-room shall receive *two* coats of best or drying *Japan* paint. All exposed piping and radiation above boiler-room to receive *one* coat of *priming* and *one* coat of *pale-gold bronze*.

EXTRAS.

It is understood and agreed, upon the acceptance of the Proposal accompanying this Specification, that any and all verbal or other agreements, statements, or representations made by any person or persons, for or on behalf of the contractor, shall be considered as absolutely merged in the Proposal and Specification, and that the contract then existing shall be taken and held to be fully set forth and expressed therein.

If any deviation in system, material, or mode of installation is to be made, such change shall be considered an "extra," and must be provided for by a special agreement.

COMPLETION AND TESTING.

If this Specification with accompanying Proposal be accepted notice of date when work may begin shall be given contractor, and same shall be prosecuted with due despatch, and shall be completed on or before   , whereupon notice to that effect shall be served on *architect*. Should any unforeseen or unavoidable delay occur, same shall not constitute a breach of contract on the part of contractor. Upon notification that work as herein provided for has been completed, same shall be promptly inspected, and "accepted" or "rejected," and notice thereof served on the contractor. Acceptance shall in no event waive the guarantee herein below given. Failure to promptly inspect and accept or reject work shall be considered as acceptance, and shall entitle undersigned to payments as provided for. "Testing" shall consist of firing boiler all fuel for which shall be delivered in boiler-room, and furnished by *owner*, and the developing of a (**steam-pressure not exceeding** *fifteen* **pounds to the square inch**) [**flow-temperature not less than   degrees Fahrenheit without boiling over**] and the making tight of all joints in system.

Determination of fulfilment of guarantee shall be gauged by standards set down in Carpenter's "Heating and Ventilating Buildings," page 86. If the condition of building is such that work cannot be completed without delay, and that delay requires running of all or part of apparatus for use or convenience of any one other than the contractor, it will only be so run at the risk and expense of *owner*, and apparatus must be delivered again in as good condition as when taken. A payment of *five (5) dollars* shall be due for each radiator disconnected and reconnected.

IN GENERAL.

Estimates for capacity of within apparatus, as well as this Specification and accompanying Proposal, are all based on dimensions, information, etc., concerning construction of building furnished by *architects;* and if such dimensions, information, etc., are erroneous, or if changes shall be made in construction of building, then in so far as such deviations detract from efficiency of apparatus, the guarantee as to the efficiency thereof which is herein given shall be deemed cancelled. Instructions as to conduct of work must be made to the contractor and not to employees, and all instructions from *architects* shall be considered as final, unless otherwise advised by *owner*.

GUARANTEE.

When the apparatus as herein proposed to be furnished shall be completed, the same is guaranteed to be capable of warming the rooms entered on schedule, to the temperatures specified therein, when apparatus is run as directed, and under the conditions which would maintain in the finished building. Any failure to fulfill this guarantee by reason of any defect of workmanship, material, or efficiency within a period of *one* year will be made good by contractor within a reasonable time after receiving notice of such defect.

N. B. The term "defect" as above used, shall not be construed to cover such imperfections as result from accident, design, or the natural wear and tear of use. The contractor shall have and retain, until the final payment in full shall have been made, a first and valid lien upon all materials (including pipe, fittings, valves, covering, radiators, registers, ducts, boilers, etc.) furnished by *contractor* under terms of this specification and accompanying proposal, and shall have the right at all times prior to such final payment, upon failure on part of *owner*, to make all payments as provided for, to take possession of and remove the said materials, and to retain the possession of same and every part thereof, and also to retain all payments that have been made on account thereof as liquidated damages for non-fulfilment of contract.

SPECIAL NOTE.

This Specification with accompanying Proposal shall be accepted or rejected on or before *inst.*, and notice thereof be served on contractor.

Respectfully submitted,

JOHN G. DOE CO.

October 1, 1895.

173. Form of Uniform Contract.—

UNIFORM CONTRACT FOR THE CONSTRUCTION OF HEATING APPARATUS (TO BE) ADOPTED FOR USE BY THE MASTER STEAM AND HOT-WATER FITTERS' ASSOCIATION OF THE UNITED STATES.*

(Copyright, 1895, by the Master Steam and Hot-Water Fitters' Association of the United States.)

THIS AGREEMENT, made and concluded at *Kalamazoo*, State of *Michigan*, the *first* day of *January*, in the year one thousand eight hundred and ninety-*five*, by and between *Jones & Brown*, of *Chicago*, State of *Illinois*, for *themselves* and *their* legal representatives, parties of the first part (hereinafter designated the Contractor), and *R. I. Peters*, of *Kalamazoo*, State of *Michigan*, for *himself* and *his* legal representatives, party of the second part (hereinafter designated the Owner).

WITNESSETH, That the Contractor, in consideration of the fulfilment of the agreements herein made by the Owner, agrees with the said Owner, as follows:

ARTICLE I. The Contractor, for the consideration hereinafter provided, covenants and agrees, with the Owner, that the Contractor shall and will, within the space of *three* months next, after the date hereof, in a good and workmanlike manner, and at his own proper charge and expense, well and substantially build, furnish, and erect a certain *Steam* Heating Apparatus, at *444 4th Avenue, City of Kalamazoo*, according to the Specifications, Drawings, and Plans designed by *Thomas Robinson, Architect*, which Specifications, Drawings, and Plans are made a part of this Contract and are identified by the signatures of the parties hereto.

---

* Printed words in italics to be supplied in each contract.

ARTICLE II. No alterations shall be made in the work shown or described by the drawings and specifications, except upon a written order of the *Architects*, and when so made, the value of the work added or omitted shall be computed by the *Architects*, and the amount so ascertained shall be added to or deducted from the contract price. In the case of dissent from such award by either party hereto, the valuation of the work added or omitted shall be referred to three (3) disinterested arbitrators, one to be appointed by each of the parties to this Contract, and the third by the two thus chosen; the decision of any two of whom shall be final and binding, and each of the parties hereto shall pay one-half of the expenses of such reference.

ARTICLE III. Should any difference arise in interpreting the Plans or Specifications, involving or assuming additional compensation, the Contractor shall, upon written notice from the Owner, immediately execute such interpretation, the question of compensation to be determined on completion by arbitrators, as provided in Article II.

ARTICLE IV. All of the materials and workmanship of the apparatus to be of the quality as expressed in said Specifications, Drawings, and Plans; said Owner to reserve the right to reject, through himself or his authorized agent, all material or workmanship of an inferior quality, which said Contractor may attempt to use in the erection of said Heating Apparatus, and if the said Contractor, after being notified, neglects or refuses to do the work, or furnish the materials as called for in the Specifications, Drawings, and Plans, then, and in that case, said Owner shall give notice in writing to the Contractor, which notice is to set forth in full the cause or causes of complaint. If the Contractor demurs and refuses to do the work or furnish the materials as directed in the notice of complaint, within three days from the date of said notice, resort to arbitration shall be had as provided in Article II.

ARTICLE V. The Owner shall not, in any manner, be answerable or accountable for any loss or damage that shall or may happen to the said works, or any parts thereof respectively, or for any of the materials or other things used and employed in finishing and completing the same, loss or damage by fire excepted. The Contractor shall be responsible for all damage to the building and adjoining premises, and to individuals, caused by himself or his employees in the course of their employment.

Article VI. It is hereby mutually agreed between the parties hereto, that the sum to be paid by the Owner to the Contractor for said work and materials shall be *Seven Thousand Dollars ($7,000)*, subject to additions and deductions as hereinbefore provided, and that such sum shall be paid in current funds by the Owner to the Contractor, in monthly payments, to the amount of *90* per cent of the value of materials delivered to and labor performed in the said building during the preceding month; and the remaining *10* per cent shall be paid as a final payment within *30* days after this contract is fulfilled.

All payments shall be made upon written certificates of the *Architects* to the effect that such payments have become due.

Article VII. It is mutually agreed that payments for all additional work shall be made at the same time and in the same manner as contract payments, Article VI.

Article VIII. It is mutually agreed that should default be made in any of the payments as herein provided, the Contractor shall have the right to stop work and withdraw all unused materials until such payment is properly made, or may at his option cancel the contract.

Article IX. It is further mutually agreed that the essence of this Agreement is that the Owner purchasing this apparatus and paying therefor will receive full value to the extent that it will warm the subdivisions of the building indicated on the plans to *70* degrees Fahrenheit in the coldest weather; but nothing herein contained, or in the Specification accompanying the same, shall prevent the Contractor from receiving from the Owner a final payment for the work herein and at the time stipulated.

Article X. The Contractor guarantees his workmanship and materials, the capacity of the boiler, the circulation of the system and the efficiency of the heating surfaces, all as called for in the Specifications hereto attached, and should any defects or deficiencies occur, other than from neglect on the part of the Owner or his employees, within the term of one year from the above date, the Contractor agrees to make good the same upon a written notice from the Owner at the Contractor's expense.

ARTICLE XI. If at any time there shall be evidence of any lien or claim for which, if established, the Owner of the said premises might become liable, and which is chargeable to the Contractor, the Owner shall have the right to retain out of any payment then due, or thereafter to become due, an amount sufficient to completely indemnify himself against such lien or claim. Should there prove to be any such claim after all payments are made, the Contractor shall refund to the Owner all moneys that the latter may be compelled to pay in discharging any lien on said premises made obligatory in consequence of the Contractor's default.

ARTICLE XII. It is further mutually agreed, between the parties hereto, that no certificate given or payment made under this Contract, except the final certificate or final payment, shall be conclusive evidence of the performance of this Contract, either wholly or in part, and that no partial payment shall be construed to be an acceptance of defective work or improper materials.

ARTICLE XIII. The said parties for themselves, their heirs, executors, administrators, and assigns, do hereby agree to the full performance of the covenants herein contained.

IN WITNESS WHEREOF, the parties to these presents have hereunto set their hands and seals, the day and year first above written.

In presence of
J. B. Saxe

    Jones & Brown   (SEAL)
    R. J. Peters   (SEAL)
              (SEAL)
              (SEAL)

ALTERNATE FOR ARTICLE VI.

It is hereby mutually agreed, between the parties hereto, that the sum to be paid by the Owner to the Contractor for said work and materials shall be *Seven Thousand Dollars ($7.000)* subject to additions and deductions as hereinbefore provided, and that such sum shall be paid in current funds by the Owner to the Contractor in instalments, as follows:

When	*The Boilers are delivered and set,*	*$1,500*
When	*Steam Mains and Risers are in place,*	*$1,500*
When	*The Radiators are delivered,*	*$1,500*
When	*The Radiators are connected,*	*$1,500*

And the balance of *$1,000* as a final payment to be made within *30* days after this contract is fulfilled.

All payments shall be made upon written certificates of the *Architects* to the effect that such payments have become due.

174. Specifications for Plain Tubular and Water-tube Boilers.—This boiler is employed extensively for heating large buildings. The boiler is described on page 130, and several methods of setting are shown on page 145. The following specifications represent the best practice of to-day in the construction of plain tubular boilers employed for heating. They are in each case to be set in brickwork, substantially as described on page 143.

STANDARD SPECIFICATION FOR HORIZONTAL TUBULAR BOILERS FOR WORKING STEAM-PRESSURES OF 100 LBS. AND 125 LBS. PER SQUARE INCH.*

Dimensions and Capacity of Boilers. (See also table, page 131.)

Nominal horse-power			25	30	35	40	45	50	60	70	80	100	125	150
Diameter shell		inches	42	44	44	44	48	54	54	60	60	66	72	72
Diameter dome		"	22	22	22	22	26	30	30	32	32	36	36	36
Height dome		"	24	24	24	24	28	34	34	36	36	40	40	40
Thickness shell	100 lbs. working pressure	No.	5/16	5/16	5/16	5/16	5/16	5/16	5/16	11/32	11/32	3/8	7/16	7/16
"	125 "	"	3	3	3	3	11/32	11/32	11/32	11/32	11/32	13/32	1/2	1/2
" dome	100 "	"	3/8	3/8	3/8	3/8	3/8	3/8	3/8	3/8	3/8	7/16	7/16	7/16
"	125 "	"	7/16	7/16	7/16	7/16	7/16	7/16	7/16	7/16	7/16	9/16	9/16	1/2
" heads	100 "	"	7/16	7/16	7/16	7/16	7/16	7/16	7/16	7/16	7/16	7/16	9/16	9/16
"	125 "	"	1/2	1/2	1/2	1/2	1/2	1/2	1/2	1/2	1/2	1/2	1/2	1/2
" dome hds.	100 "	"	7/16	7/16	7/16	7/16	7/16	7/16	7/16	7/16	7/16	7/16	9/16	9/16
"	125 "	"												
Number of tubes			38	46	46	46	52	64	60	82	82	98	120	84
Length of tubes		feet	10	12	12	14	14	12	16	14	16	16	16	18
Diameter of tubes		inches	3	3	3	3	3	3	3	3	3	3	3	4
Gauge of iron in tubes			12	12	12	12	12	12	12	12	12	12	12	10
Diameter rivets, 100 lbs. working pressure		inches	5/8	5/8	5/8	5/8	5/8	5/8	5/8	11/16	11/16	11/16	3/4	3/4
"	125 "	"	11/16	11/16	11/16	11/16	11/16	3/4	3/4	3/4	3/4	13/16	7/8	7/8

* Prepared especially for this work by Messrs. F. J. Frary and J. G. Dudley.

Materials.—Best open-hearth "flange" steel, having a tensile strength of 55,000 to 60,000 lbs. per square inch, a reduction of area of 45% to 50%, an elongation of 20% to 25%, and may be bent and closed down upon itself when cold without fracture, and must not blister.

All plates plainly stamped with maker's name and brand, and tensile strength.

Rivets made of best refined iron.

Tubes of charcoal iron or steel, lap-welded.

Braces of best refined iron.

CONSTRUCTION.

Shell.—Shells preferably made with but one plate in the bottom and one or more plates in the top, and without circular seams on the bottom of boilers except at the heads; or shells may be made in not to exceed three courses and with but one plate in each course, with but one horizontal seam, and this above reach of fire.

Riveting for a Working Pressure of 100 *lbs.*—Horizontal seams double-staggered riveted, lap-joint; pitch of rivets 3" longitudinally and 2¼" diagonally. Circular seams single-riveted, lap-joint; pitch of rivets 2¼". Flange seam and vertical seam of dome double-staggered riveted; pitch of rivets 3" longitudinally and 2¼" diagonally. Circular seam at dome head single-riveted; pitch of rivets 2¼".

For a Working Pressure of 125 *lbs.*—Horizontal seams triple-riveted; lap-joints required except for boilers exceeding 66" diameter, when horizontal seams shall be made with butt-joint, with inside and outside lap strips covering the joint, these strips same thickness as plate in shell of boilers; three rows of rivets each side of joint; pitch of rivets on triple lap-joints 3¼" longitudinally, 2" diagonally, 2⅜" transversely. Pitch of rivets on butt-strapped joints 3¼" and 6½" longitudinally, 2" diagonally, and 2⅜" transversely. Circular seams single-riveting, lap-joint; pitch of rivets 2¼". Flange seam, *od*, dome triple-riveting staggered; pitch of rivets 3" longitudinally and 2" diagonally; vertical seam of dome double-staggered riveting; pitch of rivets 3" longitudinally and 2¼" diagonally. Circular seam at dome head single-riveting; pitch of rivets 2¼".

Bracing.—All braces to have a sectional area of 1¼ square inches and to be of the solid crowfoot style, and riveted to heads and shell with two rivets in each end; pitch of rivets 4". On heads of boiler these braces to be set radially and spaced about 7" centres, and to lead from head to shell and to be at least 3 ft. in length and preferably longer. Braces in dome to lead from shell of dome to shell of boiler, spaced about 18" centres, two rivets in each end spaced 4" centres; braces as long as height of dome will permit. Head of dome may be convex and without braces.

Tube Setting.—Tubes to be set in straight horizontal and vertical rows, one inch apart each way, and no tube nearer shell than three inches. Distance from top of upper row of tubes to shell not less than one third the diameter of boiler. Tubes to extend through heads, and be carefully expanded and beaded to the heads.

Calking.—Calking edges of each seam to be bevelled by machine before plates are put together, and calking tool driven straight.

Manholes.—A suitable manhole in top of shell, having an internal opening 11" × 15", reinforced with strong internal frame of forged iron. Manhole to be provided with suitable plate, bolt, guard, and gasket. For large boilers a manhole shall be left in front head beneath the tubes.

Hand-holes.—A suitable hand-hole, 4¼" × 6", in each head under tubes, provided with suitable plate, bolt, guard, and gasket.

Outlets.—Outlet for steam should be on top of the dome, the opening into dome to be reinforced with wrought-iron flange properly threaded and riveted to the head; the safety-valve to be attached to this opening.

The opening for blow-off should be in the back head at the side of hand-hole. The opening for surface blow shall be in the top of the shell, and provided with pipe having a trumpet shaped mouth ending at water-line.

The opening for feed connection should be in the top of shell and reinforced. The feed-pipe is to be extended downward below the water-line, and at least four feet horizontally.

The upper connection for water-column should be in front head near top. The lower connection for water-column should be in front head, about on the centre line of the boiler.

Wall-brackets.—There should be two heavy cast-iron wall-brackets riveted to each side of shell for supporting boiler on masonry. These brackets should be at least 9 inches wide with foot 12 inches long, and 14 inches on the boiler and $1\frac{1}{4}$ inches thick, with heavy rib through the centre. These, and all other castings riveted to the shell, to conform to the shape of same and fit accurately without linings of any kind.

Testing.—For a working pressure of 100 lbs. the boiler should be tested to a hydrostatic pressure of 150 lbs. per square inch, and for a working pressure of 125 lbs. it should be tested to a hydrostatic pressure of 200 lbs. per square inch, and should be perfectly tight under each test.

Castings.—The boiler should be provided with a cast-iron front at least $\frac{1}{2}$" thick, with double flue, fire and ash-pit doors swinging right and left. Fire-doors should be provided with perforated liners and air-registers. Provide heavy cast-iron dead-plate, arch-plate over fire-door, and cast-iron plates at each side of fire-door opening, to protect the fire-brick.

The grates* should equal in width the full diameter of the boiler, and should be in two lengths, with necessary bearing bars; the entire length of the grate surface should equal about one third the length of the tubes.

The air-space in the grates for soft coal should be from $\frac{1}{4}$" to $\frac{3}{8}$", and for hard coal from $\frac{1}{8}$' 'to $\frac{1}{4}$".

Two heavy cast-iron arch bars for supporting brick at rear of boiler.

One back door and frame of cast iron, to provide access to rear of the setting.

All necessary anchor-bolts for holding front and back doors in position, and at least four long tie-bolts extending full length of the setting, with cast-iron washers for rear end.

Four heavy cast-iron buck-stays with rods, extending crosswise of the setting, for supporting side walls.

Four cast-iron wall-plates with rollers for supporting brackets to rest upon.

Fittings.—One steam-gauge   inches diameter. One lever safety-valve. One water-gauge fitted to cast-iron water-column, with three gauge-cocks. One steam-cock for blow-off. One globe-valve, and one check-valve for feed-pipe connections. One set of fire tools, slice-bar, and rake. One damper with suitable handles and with automatic regulator.

---

* If rocking gates are desired, name of manufacturer should be specified.

It is generally desirable to provide two independent methods of feeding, so that an accident will not affect the supply of feed-water; but specifications for the feed-pumps are not often included with those for the boiler.

175. Protection from Fire—Hot Air and Steam Heating.—Where hot-air stacks or steam-pipes pass up through partitions near woodwork there is considerable danger of fire, and for this reason certain requirements have been made both as to the position of hot-air pipes in furnace-heating and steam pipes in steam-heating. The following digest, compiled by H. A. Phillips, of the municipal laws relating to hot pipes in buildings, in force in some of the principal cities of the United States, appeared in the *American Architect and Building News*, Feb. 1893, and is useful in preparing specifications. They are as follows:

Boston.—1. Hot-air pipes shall be at least 1 inch from woodwork. (This may be modified by inspector in first-class buildings.)

2. Any metal pipe conveying heated air or steam shall be kept 1 inch from any woodwork, unless pipe is protected by soapstone or earthen tube or ring, or metal casing.

Baltimore.—1. Metal flue for hot air may be of one thickness of metal, if built into stone or brick wall.

2. Otherwise it must be double, the two pipes separated by 1 inch air-space.

3. No woodwork shall be placed against any flue or metal pipe used for conveying hot air.

Chicago.—1. Hot-air conductors placed within 10 inches of woodwork shall be made double, one within the other, with at least $\frac{1}{2}$ inch air-space between the two.

2. All hot-air flues and appendages shall be made of IC or IX bright tin.

3. Steam-pipes shall be kept at least 2 inches from woodwork, unless protected by soapstone, earthen ring or tube, or rest on iron supports.

Cincinnati.—No pipes conveying heated air or steam shall be placed nearer than 6 inches to any unprotected combustible material. All subject to approval of inspector.

Cleveland.—1. Hot-air conductors placed within 10 inches of woodwork shall be made double, one within the other, with at least $\frac{1}{2}$ inch air-space between the two.

2. No pipes conveying heated air or steam shall be placed nearer than 6 inches to any unprotected combustible material.

Denver.—Metal flue for hot air may be of one thickness of metal, if built into stone or brick wall; otherwise it shall be made double or wrapped in incombustible material.

Detroit.—No metal pipe for conveying hot air shall be placed nearer than 3 inches to any woodwork. Such pipes over 15 feet long shall be safely stayed by wire or metal rods.

District of Columbia.—1. Hot-air pipes shall be at least 1 inch from woodwork.

2. Pipes passing through stud or wooden partitions shall be guarded by double collar of metal, "giving at least 2 inches air-space, having holes for ventilation, or other device equally secure, to be approved by inspector."

3. Metal pipe double, with the space filled with 1 inch of non-combustible, non-conducting material, or a single pipe surrounded by 1 inch of plaster of Paris or other non-conducting material between pipe and timber.

Kansas City.—1. Any metal pipe conveying heated air or steam shall be kept 1 inch from any woodwork, unless pipe is protected by soapstone or earthen tube or ring, or metal casing, or otherwise protected to satisfaction of superintendent.

2. No wooden flue or air-duct for heating or ventilation shall be placed in any building.

Memphis.—1. All stone or brick hot-air flues and shafts shall be lined with tin pipes.

2. No wooden casing, furring, or lath shall be placed against or over any smoke-flue or metal pipe used to convey hot air or steam.

3. No metal flues or pipes to convey heated air shall be allowed unless inclosed with 4 inches thickness of hard, incombustible material, except horizontal pipes in stud partitions, which shall be built in the following manner: The pipes shall be double, one inside the other, and $\frac{1}{4}$ inch apart, and with 3 inches space between pipe and stud on each side; the inside faces of said stud well lined with tin plate, and the outside face with iron lath or slate. Where hot-air pipe passes through partition shall be at least 8 feet from furnace.

4. Horizontal hot-air pipes shall be kept 6 inches below floor-beams or ceiling. If floor-beams or ceiling are plastered or protected by metal shield, then distance shall not be less than 3 inches.

5. Where hot-air pipes pass through wooden or stud partition, they shall be guarded by double collar of metal with 2-inch air-space and holes for ventilation, or by 4 inches of brickwork.

6. No hot-air flues or pipes shall be allowed between any combustible floor or ceiling.

7. Steam-pipe shall not be placed less than 2 inches from woodwork unless wood is protected by metal shield, and then distance shall not be less than 1 inch.

8. Steam-pipes passing through floors and ceilings or lath-and-plaster

partitions shall be protected by metal tube 2 inches larger in diameter than pipe.

9. Wooden boxes or casings inclosing steam-pipes and all covers to recesses shall be lined with iron or tin plate.

Milwaukee.—1. Hot-air conductors placed within 10 inches of woodwork shall be made double, one within the other, with at least ¼ inch air-space between them.

2. All hot-air flues and appendages shall be made of IC or IX bright tin.

Nashville.—1. Sheet-iron flue running through floor or roof shall have a sheet-iron or terra-cotta guard at least 2 inches larger than flue.

2. Steam-pipes shall be kept at least 2 inches from woodwork.

3. All steam and hot-air flues and pipes must be suspended by iron brackets.

Newark.—1. Hot-air pipes shall be set at least 2 inches from woodwork and the woodwork protected with tin.

2. Such pipes placed in lath-and-plaster partitions must be covered with iron, tin, or other fire-proof material.

New York.—(Same regulations as noted under heading of "Memphis.")

No hot-air flue or pipe allowed between combustible floor or ceiling.

Omaha.—1. Steam-pipe shall not be placed less than 2 inches from woodwork unless wood is protected by metal shield; and then distance shall not be less than 1 inch.

2. Steam-pipes passing through floors and ceilings, or lath-and-plaster partitions, shall be protected by metal tube 2 inches larger in diameter than pipe.

3. Wooden boxes or casings inclosing steam-pipes and all covers to recesses shall be lined with iron or tin plate.

4. Stud partitions in which hot-air pipes are placed to be at least 5 inches wide, and the space between studs at least 14 inches.

5. Hot-air pipes shall not be placed between floor-joists unless same are doubled and the joists 14 inches apart.

6. Bright tin shall be used in construction of all hot-air flues and appendages.

Providence.—1. Hot-air pipes shall be at least 1 inch from woodwork, unless protected by soapstone or earthen ring, or metal casing permitting circulation of air around pipe.

2. Steam-pipes must be kept at least 1 inch from woodwork, or supported by incombustible tubes or rest on iron supports.

St. Louis.—1. Hot-air pipes shall be at least 1 inch from woodwork, unless protected by soapstone or earthen ring or metal casing permitting circulation of air around pipe.

2. Steam or hot-water pipes carried through wooden partition or between joists, or in other close proximity to woodwork, shall be

inclosed in clay pipe or covered with felting or other non-conducting material.

San Francisco.—1. Metal flue for hot air may be of one thickness of metal, if built into stone or brick wall; otherwise double, one pipe within the other, ¼ inch apart, and space filled with fire-proof material.

2. No woodwork shall be placed against any flue or metal pipe used for conveying hot air.

3. Steam-pipes shall be placed at least 3 inches from woodwork, or protected by ring of soapstone or earthenware.

Wilmington.—Metal pipes to carry hot air shall be double, one inside the other, ¼ inch apart; or, if single, have a thickness of 2 inches of plaster of Paris between pipe and woodwork adjoining same.

176. Duty of the Architect.—The heating system is an essential part of the building in this latitude, and it should be the duty of the architect to provide building designs of such character that it can be readily and economically installed. The architect's specifications for the buildings hould provide for the construction of ventilating, heating, and smoke flues, and his plans should show the location, including pipe-lines, of every essential part of the heating apparatus. All responsibility regarding flues and the general adaptability of the heating system to the building should be assumed by the architect, and not shifted to the contractor. If the heating system is designed at the same time as the building, slight changes can be made in arrangements of details, partitions, doors, etc., that will tend to cheapen construction, and will add to the efficiency of operation and the general appearance of the heating apparatus. If steam or water pipes are required to be erected out of sight, conduits should be provided, so that they will be readily accessible for inspection and repairs.

177. Methods of Estimating Cost of Construction.—In estimating the cost of construction of any system of heating apparatus the contractor must depend largely upon his own experience and knowledge. No general directions can be given, but a few suggestions are offered which may aid in adopting a systematic method of proceeding. Determine first the amount and character of radiation to be placed in each room by the methods which have already been given fully in Chapter X. Second, determine the position and sizes of pipes leading from the heater to the various radiating surfaces by methods given in Chapter XI.

To facilitate the above work, a set of floor drawings of each story should be obtained, and on these there should be carefully laid out the position of all radiators, flues, pipe-lines, etc. After determining the amount required, a schedule of material should be made and the cost should be computed.

The manufacturers have adopted a price, which is changed very rarely, for all standard fittings, pipes, etc., and from which a discount is given which varies with the condition of the market, cost of material, labor, etc. The discount is usually large upon cast-iron fittings and brass goods, being seldom less than 70 per cent, and sometimes 80 per cent and even greater. The discount on piping, especially the smaller sizes, is much less, ordinarily ranging from 40 to 70 per cent.

The cost of labor will vary greatly in different localities, so that no general method of estimating can be given. It must be determined largely by experience in each locality and with a given set of men. The cost of heaters of any given type, with fittings, etc., can only be determined accurately by correspondence with manufacturers.

Table XXII may frequently be useful, as it gives the list-price of the principal standard fittings, pipes, and valves (see appendix to book).

178. Suggestions for Pipe-fitting.—Certain suggestions are here made relating to the actual work of pipe-construction which may be useful to those not having an extended experience.

In the actual construction of steam-heating or hot-water heating systems it is usually customary to send a supply of pipe and fittings to the building somewhat greater than is required, and the workman, after receiving plans of construction which show the location and sizes of the various pipes to be erected, makes his own measurements, cuts the pipes to the proper length in the building, threads them, and proceeds to screw them into place. In some rare instances all lengths of pipe are purchased the proper length, and the workman has merely to put them in the proper position. The skill required for pipe-fitting may seem to the novice to be easily acquired:

this is not true, as it is a trade requiring as much training and experience as any with which the writer is familiar.

The tools belonging to this trade consist of tongs or wrenches for screwing the pipe together, cutters for cutting, taps and dies for threading the pipe, and vises for holding it in position while cutting or threading. A very great variety of tongs and wrenches is to be found on the market, some of which are adjustable to various sizes of pipe, and others are suited for only one size. For rapid work no tool is perhaps superior to the plain tongs, and one or more sets especially for the smaller sizes of pipes should always be available. For large pipes, chain tongs of some pattern will be found strong and convenient, and can be used with little danger of crushing the pipe. A form of adjustable wrench known from the inventor as the Stilson wrench has proved a very excellent and durable tool, and is well worthy a place in the chest of any fitter. Other wrenches of value are also on the market, one with a triangular head and projecting teeth being especially valuable for small pipes. The wrenches or tongs which are used for turning the pipe in most cases exert more or less lateral pressure, and if too great strength is applied at the handles there is a tendency to split the pipe. It is an advantage to have the tongs or wrenches catch on the outer circumference of the pipe with as little lateral pressure as posible, and to this end the projecting edges should be kept sharp and clean.

The cutter ordinarily employed for small pipe consists of one or more sharp-edged steel wheels, which are held in an adjustable frame, the cutting being performed by applying pressure and revolving it around the pipe. With this instrument the cutting is accomplished by simply crowding the metal to one side, and hence burrs of considerable magnitude will be formed both on the outside and inside of the pipe. The outside burr must usually be removed by filing before the pipe can be threaded. The inside burr forms a great obstruction to the flow of steam or water, and should in every case be removed by the use of a reamer. Workmen quite often neglect to remove the inside burr. A cutter consisting of a cape chisel set in a frame is more difficult to use and keep in order, although it makes cleaner cuts; it can be had in connection with some of the adjustable die-stocks, but is rarely used. . Pipes, especially the larger sizes, are sometimes cut by expert workmen with *diamond-pointed* or *cape chisels*, but this process requires too much time to be applicable to small pipes.

The *hack-saw* is coming into use to some extent for cutting pipes, and is an excellent instrument for this purpose, as it does not tend to burr or crush the pipe, and is quite as rapid as the wheel-cutter.

The dies for threading the pipes are of a solid form, each die fitting into a stock or holder with handles, or of an adjustable form, the dies being made of chasers, which are held where wanted and can be set in various positions by a cam. The adjustable dies can be run over the pipes several times, and hence work easier than solid ones; but in their use great care should be taken that the exterior diameter of the pipe is not made less than the standard size. The cutting edges of the dies should be kept very sharp and clean, otherwise perfect threads cannot be cut. In the use of the dies some lubricant, as oil or grease, kept on the iron will be found to add materially to the ease with which the work can be done, and will tend to prevent heating and crumbling of the pipe and injury to the threads.

Taps are required for cutting threads in openings or couplings into which pipes must be screwed—an operation which the pipe-fitter seldom has to perform, unless a thread has been injured. The vises for holding the pipe should be such as will prevent it from turning without crushing it under any circumstances. Adjustable vises with triangular-shaped jaws on which teeth are cut are usually employed.

In the erection of pipe great care should be taken to preserve the proper pitch and alignment, and the pipes should, to appear well, be screwed together until no threads are in sight. Every joint should be screwed six to eight complete turns for the smaller sizes, 2" and under, and eight to twelve turns for the larger sizes, otherwise there will be danger of leakage. It is a good plan to test the threads on all pipes before erection by unscrewing the coupling and screwing it back with the ends reversed. It is also advisable to look through each length of pipe and see if it is clear before erecting in place; serious trouble has been caused by dirt or waste in pipes, which would have been removed had this precaution been taken.

In screwing pipes together, red or white lead is often used; the writer believes this practice to be generally objectionable, and to be of no especial benefit in preventing leaks. The lead acts as a lubricant, and consequently aids by reducing the force required to turn the pipe. It will generally be found, however, that linseed or some good lubricating oil will be equally valuable in that respect, and will have the advantage of not discoloring the work.

If possible, arrange the work so that it can " be made up " with right and left elbows, or right and left couplings. Packed joints, especially *unions*, are objectionable, and likely to leak after use. Flange-unions, packed with copper gaskets, should be used on heavy work.

Good workmanship in pipe-fitting is shown by the perfection with which small details are executed, and it should be remembered that bad workmanship in any of the particulars mentioned may defeat the perfect operation of the best-designed plant.

APPENDIX

CONTAINING

REFERENCES AND TABLES.

LITERATURE AND REFERENCES.

The literature devoted to the subject of warming and ventilation is quite extensive, dating back to a treatise on the economy of fuel and management of heat by Buchanan in 1815. A most excellent compilation of this literature was made by Hugh J. Barron of New York, in a paper presented to the American Society of Heating and Ventilating Engineers at its first meeting in January, 1895, from which the following list of books has been copied:

A Treatise on the Economy of Fuel and Management of Heat. Robertson Buchanan, C.E. Glasgow, 1815.

Conducting of Air by Forced Ventilation. Marquis de Chabannes. London, 1818.

The Principles of Warming and Ventilating Public Buildings, Dwelling-houses, etc. Thos. Tredgold, C.E. London, 1824.

Warming, Ventilation, and Sound. W. S. Inman. London, 1836.

The Principles of Warming and Ventilating, by Thos. Tredgold, with an appendix. T. Bramah, C.E. London, 1836.

Heating by the Perkins System. C. J. Richardson. London, 1840.

Illustrations of the Theory and Practice of Ventilation, with Remarks on Warming. David Boswell Reid, M.D. London, 1844.

A Practical Treatise on Warming by Hot Water. Chas. Hood, F.R.S. London, 1844.

History and Art of Warming and Ventilating. Walter Bernan, C.E. London, 1845.

Warming and Ventilation. Chas. Tomlinson. London, 1844.

Walker's Hints on Ventilation. London, 1845.

Practical Treatise on Ventilation. Morrill Wyman. Boston, 1846.

Traité de la Chaleur. E. Péclet. Paris. First edition, 1848; second edition, 3 vols, 1859.

Practical Method of Ventilating Buildings, with an appendix on Heating by Steam and Water. Dr. Luther V. Bell. Boston, 1848.

Warming and Ventilation. Chas. Tomlinson. London, 1850.

Practical Ventilation. Robert Scott Burns. Edinburgh, 1850.

Ventilation and Warming. Henry Ruttan. New York, 1862.

A Treatise on Ventilation. Robert Richey. London, 1862.

American edition of Dr. Reid's Ventilation as Applied to American Houses, edited by Dr. Harris. New York, 1864.

A Treatise on Ventilation. Lewis W. Leeds. Philadelphia, 1868; New York, 1871.

Observations on the Construction of Healthy Dwellings. Capt. Douglas Galton. Oxford, 1875.

Practical Ventilating and Warming. Jos. Constantine. London, 1875.

Warming and Ventilation. Chas. Tomlinson. London, 1876. Sixth edition.

Mechanics of Ventilating. Geo. W. Rafter, C.E. New York, 1878.

Ventilation. H. A. Gouge. New York, 1881.

Ventilation. R. S. Burns. Edinburgh, 1882.

American Practice in Warming Buildings by Steam. Robert Briggs. Edited by A. R. Wolf, with additions. New York, 1882.

Steam-heating for Buildings. W. J. Baldwin. New York, 1883. Thirteenth edition published in 1893.

The Principles of Ventilation and Heating. John S. Billings, M.D. New York, 1884.

Heating by Hot Water. Walter Jones. London, 1884.

A Manual of Heating and Ventilation. F. Schuman. New York, 1886.

Ventilation. W. Butler. Edited by Greenleaf. New York, 1888.

Steam-heating Problems from the Sanitary Engineer. New York, 1888.

Metal Worker Essays on House Heating. New York, 1890.

Heat—Its Application to the Warming and Ventilation of Buildings. John H. Mills. Boston, 1890.

Ventilation and Heating. T. Edwards. London, 1890.

Ventilation—A Text-book to the Art of Ventilating Buildings. Wm. Paton Buchan. London, 1891.

The Ventilating and Warming of School Buildings. Gilbert B. Morrison. New York, 1892.

Hot-water Heating. Wm. J. Baldwin. New York, 1893.

Ventilation and Heating. John S. Billings, M.D. New York, 1893.

Warming by Hot Water. Chas. Hood, C.E. Edited by F. Dye. London, 1894.

In addition to this list of books a large number of pamphlets have been printed containing valuable articles on special subjects. The scope of this work does not permit any

APPENDIX CONTAINING REFERENCES AND TABLES.

historical review of the literature or of progress and improvements in the art of heating.

CURRENT LITERATURE OF THE DAY.

The current literature of the day relating to this subject is very extensive and is mainly found in magazines or papers published either weekly or monthly and devoted to the whole or special portions of this industry. In these are to be found the best available descriptions of plants, of new and improved methods and appliances, and in general all that relates to the best systems of construction. The journals devoted to this industry provide an invaluable literature to those engaged in the art of constructing heating and ventilating apparatus.

REFERENCES.

Information which has been obtained from other works has generally been credited in the body of the book. The writer wishes, however, to express special thanks for substantial assistance to the publishers of the various papers, and to J. J. Blackmore and J. G. Dudley, members of the Committee of the Boiler Manufacturers' Association, as well as to other engineers who have given cordial help in the preparation of the work. It may be stated that Messrs. Blackmore and Dudley read and revised all proofs and contributed considerable matter of practical and general interest.

LIST OF TABLES IN BODY OF BOOK.

	PAGE		PAGE
Air delivered in pipes of different diameters...................	286	Electrical heat, expense of.......	303
Air discharged at different heights and temperatures.............	45	Equalization of pipe areas for air	287
		Exhaust-steam heating..........	251
		Flue for indirect heating, area of.	233
Air discharged under pressure....	42	Flues, area of...................	53
Air-flues, area of, residence heating	234	Forced-blast heating surface, heat emitted	84
Air-pipes, various diameters, capacity of......................	286	Forced-blast test................	80
Air required per person for various standards of purity............	32	Greenhouse heating.............	241
		Heat emitted, Peclet's table.....	64–66
Blowers or fans, capacity of......	296	Heat emitted, Tredgold's experiments......................	76
Boiler explosions................	174	Heat transmitted, different media.	69
Boilers, steam, proportion of parts	125	Hot-air heating.	275
Boiling-point, different pressures.	159	Hot-water heaters, proportion of parts...	125
Boiling temperature of water.....	22	
Building loss....................	56	
Chimney, diameter of............	162	Hot-water heating, data.........	229
Conduction of heat, absolute.....	18	Hot-water heating, main-pipe diameter.	231
Conduction of heat, relative......	18	
Drip-pipe, diameter.............	228	Hot-water heating, proportions..	237

356 APPENDIX CONTAINING REFERENCES AND TABLES.

	PAGE		PAGE
Hot-water pipes, velocity in feet per second	221	Relation between velocity and pressure of air	45
Indirect radiators, air heated	213	Relation between temperature and color	12
Indirect radiators, cubic feet heated	214	Return-pipe, diameter	227
Indirect radiator tests	81, 82	Size of room influence on ventilation	34
Indirect radiators, heat emitted	84	Stacks, area of, hot-air heating	278
Moisture in air	30	Steam-boiler, energy in	173
Pipe-coverings, tests of	199	Steam-heating, proportions of	237
Pipe diameter for great lengths	226	Steam-heating boilers, proportions in use	136
Power-transmission, loss in	264	Steam-heating boilers, proportions of parts	125
Radiant heat, amount transmitted	17	Steam-pipe, area and diameter	223
Radiant heat, diffusion of	17	Steam-pipe, diameter for different lengths	226
Radiant heat, relative emissive powers	16	Temperature produced by radiation in warm weather	86
Radiant heat, relative reflecting powers	16	Thermometric scales	8
Radiator tests	77–79	Tubular boiler, dimensions of	131
Radiators, cubic feet of space heated	208, 209	Ventilation-flues, indirect heating	238
Radiators, diameter of openings	119	Windows	54
Radiators, direct proportioning of	205	Wrought-iron pipe	91
Radiators, indirect, factors for	211		
Registers, areas of	53		
Registers, commercial sizes	280		

LIST OF TABLES IN APPENDIX.

Table No. I. United States standard weights and measures.
   II. The equivalent value of units in British and metric system, and (IIA) of properties of gases.
   III. Table of circles, squares, and cubes.
   IV. Circumferences and areas of circles.
   V. Logarithms of numbers.
   VI. Important properties of familiar substances.
   VII. Coefficients, strength of materials.
   VIII. Properties of air.
   IX. Moisture absorbed by air.
   X. Relative humidity of the air.
   XI. Properties of saturated steam.
   XII. Composition and value of various fuels of the United States
   XIII. Reducing barometric observations to the freezing-point.
   XIV. Thermal conductivities.
   XV. Dimensions of wrought-iron, steam, gas, and water pipe.
   XVI. Weight of water per cubic foot.
   XVII. Pressure of water per square inch per different heights in feet.
   XVIII. Contents of pipes in cubic feet and gallons.
   XIX. Equalization of pipe areas.
   XX. Temperatures of various localities.
   XXI. Price of pipe and fittings.

EXPLANATION OF TABLES.

Of the tables which have been given a few only need special explanation in order to fully understand their use. These are as follows: Table No. V, Logarithms of numbers. This table will be found of very great convenience in facilitating any operation involving multiplication and division. Thus it will be found in every case that the sum of two logarithms is the logarithm of a number equal to the product of the two numbers whose sum was taken, and the difference of two logarithms is the logarithm of the quotient obtained by dividing one by the other. Every logarithm consists of two parts: a decimal part, which is given in the table, and an index or characteristic, which must be prefixed. The index or characteristic is a whole number and is one less than the number of integral places; for a decimal number it is negative and one more than the number of ciphers between the decimal point and the first significant figure. Thus, to find the logarithm of 254, a number containing 3 integral places, the index is 2, the decimal part of this logarithm found opposite 25 and under 4 in the table is 4048, making the full logarithm 2.4048. If the number had been 25.4 the index would have been 1, the decimal part as before. If the number had been 0.0254, the index would have been minus 2, the decimal part the same as before.

As an illustration showing how to multiply by logarithms, multiply 254 by 2.48. We have:

$$\begin{aligned} \text{The logarithm of } 254 &= 2.4048 \\ \text{``  `` `` } 2.48 &= 0.3945 \\ \hline \text{Log. of product} &= 2.7993 \end{aligned}$$

The sum of these two logarithms, which is the logarithm of the product, is equal to 2.7993. The index, or number 2, is of use in showing that there are three figures or integral places in the result. To find the logarithm, look in the table for the number next smaller than 7993; in this case the result is exact and is found opposite 63 in the column marked zero, indicating that the product is 630; the actual product of these numbers is slightly less than this, the difference, however, being scarcely ever of any practical importance. Had our number been 7994, it would have been one greater than 7993 and 6 less than the logarithm of the next number. In that case our number would have been 630¼, which, reduced to a decimal, would have been the number to consider as the product. The logarithm of a power can be found by multiplying the logarithm by the number which represents the power and the logarithm of a root by dividing by the index of the root.

Thus, to raise 368 to the fifth power, we have :
$$\text{Log. } 368 = 2.5658$$
$$\text{Multiply by } 5$$
$$\text{Log. 5th power} = \overline{12.8290}$$
No. = 674½ expanded to 13 places = 6745000000000.
To extract 5th root: 368 :—
$$\text{Log. } 368 = 2.5658$$
$$\text{Divide by } 5 = 0.51316 = \text{log. of root}$$
$$\text{Root} = 3.259$$

In general the table will be found to afford an easy method of dividing or multiplying, and it will be well worth while to become master of its use.

The table which is printed in the book is correct for 4 places of figures only, but tables of 7 and even 13 places have been printed.

The four-place table can be used with confidence for all operations not requiring extreme accuracy. It will in almost every case be found sufficiently accurate for all practical problems of designing.

The method of using Tables Nos. IX and X to determine the amount of moisture in the air has been quite fully explained on page 30. The method of using Table No. XI. (properties of saturated steam) has been fully explained on page 120. The reader should note that the steam-pressure tabulated is that above a vacuum, and not the reading of a pressure-gauge. The pressure-gauge reads from the atmosphere, which is generally 14.7 pounds above zero; hence, in order to use the table, add 14.7 pounds to the steam-gauge reading for the pressure above zero. The other quantities will be quite readily understood.

The table for equalization of pipe areas has been quite fully explained on page 287. The number of pipes of the size, as shown in the side column, required to give an equivalent area to the one in the top column is given by the numbers. Thus 14.7 pipes 1 inch in diameter have a carrying capacity equivalent to that of one pipe 3 inches in diameter.

TABLE No. I.
UNITED STATES STANDARD WEIGHTS AND MEASURES.
By T. C. MENDENHALL, Superintendent, U. S. Coast Survey.

LINEAR.

	Inches to millimetres.	Feet to Metres.	Yards to Metres.	Miles to Kilometres.
1=	25.4001	0.304801	0.914402	1.60935
2=	50.8001	0.609601	1.828804	3.21869
3=	76.2001	0.914402	2.743205	4.82804
4=	101.6002	1.219202	3.657607	6.43739
5=	127.0002	1.524003	4.572009	8.04674
6=	152.4003	1.828804	5.486411	9.65608
7=	177.8003	2.133604	6.400813	11.26543
8=	203.2004	2.438405	7.315215	12.87478
9=	228.6004	2.743205	8.229616	14.48412

SQUARE.

	Sq. Ins. to Sq. Centimetres.	Square Ft. to Square Decimetres.	Square Yards to Square Metres.	Acres to Hectares.
1=	6.452	9.290	0.836	0.4047
2=	12.903	18.581	1.672	0.8094
3=	19.355	27.871	2.508	1.2141
4=	25.807	37.161	3.344	1.6187
5=	32.258	46.452	4.181	2.0234
6=	38.710	55.742	5.017	2.4281
7=	45.161	65.032	5.853	2.8328
8=	51.613	74.323	6.689	3.2375
9=	58.065	83.613	7.525	3.6422

CUBIC

	Cu. Ins. to Cubic Centimetres.	Cubic Feet to Cubic Metres.	Cubic Yards to Cubic Metres.	Bushels to Hectolitres.
1=	16.387	0.02832	0.765	0.35242
2=	32.774	0.05663	1.529	0.70485
3=	49.161	0.08495	2.294	1.05727
4=	65.549	0.11327	3.058	1.40969
5=	81.936	0.14158	3.823	1.76211
6=	98.323	0.16990	4.587	2.11454
7=	114.710	0.19822	5.352	2.46696
8=	131.097	0.22654	6.116	2.81938
9=	147.484	0.25485	6.881	3.17181

CAPACITY.

	Flu.l Drams to Millilitres or Cu.Centimetres.	Fluid Ounces to Millilitres.	Quarts to Litres.	Gallons to Litres.
1=	3.70	29.57	0.94636	3.78544
2=	7.39	59.15	1.89272	7.57088
3=	11.09	88.72	2.83908	11.35632
4=	14.79	118.30	3.78544	15.14176
5=	18.48	147.87	4.73180	18.92720
6=	22.18	177.44	5.67816	22.71264
7=	25.88	207.02	6.62452	26.49806
8=	29.57	236.59	7.57088	30.28352
9=	33.28	266.16	8.51724	34.06896

WEIGHT.

	Grains to Milligrammes.	Avoirdupois Ounces to Grammes.	Avoirdupois Pounds to Kilogrammes.	Troy Ounces to Grammes.
1=	64.7989	28.3495	0.45359	31.10348
2=	129.5978	56.6991	0.90719	62.20696
3=	194.3968	85.0486	1.36078	93.31044
4=	259.1957	113.3981	1.81437	124.41392
5=	323.9940	141.7476	2.26796	155.51740
6=	388.7935	170.0972	2.72156	186.62089
7=	453.5924	198.4467	3.17515	217.72437
8=	518.3914	226.7962	3.62874	248.82785
9=	583.1903	255.1457	4.08233	279.93133

1 chain = 20.1169 metres
1 square mile = 259 hectares
1 fathom = 1.829 metres
1 nautical mile = 1853.27 metres
1 foot = 0.304801 metre, 9.484o158 log
1 avoir. pound = 453.5924277 gram.
5432.35699 grains = 1 kilogramme

The only authorized material standard of customary length is the Troughton scale belonging to the Coast Survey office, whose length at 59°.62 Fahr. conforms to the British standard. The yard in use in the United States is therefore equal to the British yard.
The only authorized material standard of customary weight is the Troy pound of the Mint. It is of brass of unknown density, and therefore not suitable for a standard of mass. It was derived from the British standard Troy pound of 1758 by direct comparison. The British Avoirdupois pound was also derived from the latter, and contains 7000 grains Troy.
The grain Troy is therefore the same as the grain Avoirdupois, and the pound Avoirdupois in use in the United States is equal to the British pound Avoirdupois. The British gallon = 4.54346 litres. The British bushel = 36.3477 litres.

TABLE No. I.—Continued.

LINEAR.

	Metres to Inches.	Metres to Feet.	Metres to Yards.	Kilometres to Miles.
1 =	39.3700	3.28083	1.093611	0.62137
2 =	78.7400	6.56167	2.187222	1.24274
3 =	118.1100	9.84250	3.280833	1.86411
4 =	157.4800	13.12333	4.374444	2.48548
5 =	196.8500	16.40417	5.468056	3.10685
6 =	236.2200	19.68500	6.561667	3.72822
7 =	275.5900	22.96583	7.655278	4.34959
8 =	314.9600	26.24667	8.748889	4.97096
9 =	354.3300	29.52750	9.842500	5.59233

SQUARE.

	Sq. Centimetres to Sq. Inches.	Square Metres to Square Feet.	Square Metres to Square Yards.	Hectares to Acres.
1 =	0.1550	10.764	1.196	2.471
2 =	0.3100	21.528	2.392	4.942
3 =	0.4650	32.292	3.588	7.413
4 =	0.6200	43.055	4.784	9.884
5 =	0.7750	53.819	5.980	12.355
6 =	0.9300	64.583	7.176	14.826
7 =	1.0850	75.347	8.372	17.297
8 =	1.2400	86.111	9.568	19.768
9 =	1.3950	96.874	10.764	22.239

CUBIC.

	Cu. Centimetres to Cu. Inches.	Cubic Decimetres to Cubic Inches.	Cubic Metres to Cubic Feet.	Cubic Metres to Cubic Yards.
1 =	0.0610	61.023	35.314	1.308
2 =	0.1220	122.047	70.629	2.616
3 =	0.1831	183.070	105.943	3.924
4 =	0.2441	244.093	141.258	5.232
5 =	0.3051	305.117	176.572	6.540
6 =	0.3661	366.140	211.887	7.848
7 =	0.4272	427.163	247.201	9.156
8 =	0.4882	488.187	282.516	10.464
9 =	0.5492	549.210	317.830	11.771

CAPACITY.

	Millilitres or Cubic Centilitres to Fluid Drams.	Centilitres to Fluid Ounces.	Litres to Quarts.	Decalitres to Gallons.	Hectolitres to Bushels.
1 =	0.27	0.338	1.0567	2.6417	2.8375
2 =	0.54	0.676	2.1134	5.2834	5.6750
3 =	0.81	1.014	3.1700	7.9251	8.5125
4 =	1.08	1.352	4.2267	10.5668	11.3500
5 =	1.35	1.691	5.2834	13.2085	14.1875
6 =	1.62	2.029	6.3401	15.8502	17.0250
7 =	1.89	2.368	7.3968	18.4919	19.8625
8 =	2.16	2.706	8.4534	21.1336	22.7000
9 =	2.43	3.043	9.5101	23.7753	25.5375

WEIGHT.

	Milligrammes to Grains.	Kilogrammes to Grains.	Hectogrammes (100 grammes) to Ounces Av.	Kilogrammes to Pounds Av.	Quintals to Pounds Av.	Milliers or Tonnes to Pounds	Grammes to Ounces Troy.
1 =	0.01543	15432.36	3.5274	2.20462	220.46	2204.6	0.03215
2 =	0.03086	30864.71	7.0548	4.40924	440.92	4409.2	0.06430
3 =	0.04630	46297.07	10.5822	6.61386	661.38	6613.8	0.09645
4 =	0.06173	61729-43	14.1096	8.81849	881.84	8818.4	0.12860
5 =	0.07716	77161.78	17.6370	11.02311	1102.30	11023.0	0.16075
6 =	0.09259	92594.14	21.1644	13.22773	1322.76	13227.6	0.19290
7 =	0.10803	108026.49	24.6918	15.43235	1543.22	15432.2	0.22505
8 =	0.12346	123458.85	28.2192	17.63697	1763.68	17636.8	0.25721
9 =	0.13889	138891.21	31.7466	19.84159	1984.14	19841.4	0.28936

By the concurrent action of the principal governments of the world, an International Bureau of Weights and Measures has been established near Paris. Under the direction of the International Committee, two ingots were cast of pure platinum-iridium in the proportion of 9 parts of the former to 1 of the latter metal. From one of these a certain number of kilogrammes were prepared, from the other a definite number of metre bars. These standards of weight and length were intercompared, without preference, and certain ones were selected as International prototype standards. The others were distributed by lot to the different governments, and are called national prototype standards. Those apportioned to the United States are in the keeping of this office.

The metric system was legalized in the United States in 1866.

The International Standard Metre is derived from the Metre des Archives, and its length is defined by the distance between two lines at 0° Centigrade, on a platinum-iridium bar deposited at the International Bureau of Weights and Measures.

The International Standard Kilogramme is a mass of platinum-iridium deposited at the same place, and its weight in vacuo is the same as that of the Kilogramme des Archives.

The litre is equal to a cubic decimetre of water, and it is measured by the quantity of distilled water which, at its maximum density, will counterpoise the standard kilogramme in a vacuum, the volume of such a quantity of water being, as nearly as has been ascertained, equal to a

Table No. II.

EQUIVALENT VALUE OF UNITS IN BRITISH AND METRIC SYSTEMS.

One foot = 12 inches = 30.48 centimetres = 0.3048 metre.
One metre = 100 centimetres = 3.2808 ft. = 1.936 yd.
One mile = 5280 ft. = 1750 yd. = 1609.3 metre.
One foot = 144 sq. in. = 1/9 sq. yd. = 929 sq. centimetres = .0929 sq. metre.
One sq. metre = 10000 sq. centimetres = 1.1960 sq. yds. = 10.764 sq. ft.
One cubic foot = 1728 sq. in. = 2832 cu. centimetres = 0.02832 cu. metres.
One cubic metre = 35.314 cu. ft. = 1.3079 cu. yds.
One pound adv. = 7000 grains = 16 oz. = 453.59 grains = 0.45359 kilograms.
One kilogram = 1000 grams = 2.2046 lbs. = 15432 grains = 35.27 oz. adv.

COMPOUND UNITS.

One foot-pound = 0.13826 kg.-mt. = 1,3826 gr.-c. = 1/778 B. T. U.
One horse-power = 33000 ft.-pound per minute = 746 Watts.
One kilogram-metre = 7.233 ft.-lb = 723.300 gr.-c. = 1/426 calorie.
One gram-centimetre = 1/100000 kg.-mt. = .00007233 ft.-lb.
One calorie = 426.10 kg.-mt. = 3.9672 B. T. U. = 42000 million ergs per second = 42 Watts.
One B. T. U. = 778 ft.-lbs. = 0.2521 cal. = 10820 mil. ergs. = 107.37 kg.-m.
One calorie per sq. metre = 0.3686 B. T. U. per sq. ft.

C. G. S. SYSTEM.

One dyne = one gram /981 = 0.00215 lb.
One erg. = 1 dyne × 1 cent. = 0.0000707 ft.-lb.
One Watt = 10 mil. ergs. per sec. = 0738 ft.-lbs. per sec. = h. p. /746.
One h. p. = 746 Watts.

Table No. IIA.

TABLE OF PROPERTIES OF GASES.

Element or Compound.	Symbol by Volume.	Atomic Weights.	Cubic feet per lb. at 62°.	Weight per. cu. ft. at 62°. Lbs.	Specific Gravity at 62°. Water = 1	Relative Density.
Oxygen	O	16	11.88	0.0814	0.001350	1.10563
Nitrogen	N	14	13.54	0.0738	0.001185	0.97137
Hydrogen	H	1	189.7	0.00527	0.0000846	0.06926
Argon		19			0.001607	1.3118
Carbon	C	12	15.84	0.63131	0.001013	0.82323
Phosphorus	P	31	6.119	0.16337	0.0026221	2.1877
Sulphur	S	32	5.932	0.16861	0.002705	2.2150
Silicon	Si	14	13.55*	0.07378	0.001184	1.01032
Air	$79N + 21O$		13.14	0.0761	0.001221	1.0000
Water vapor	H_2O	18	21.07	0.04745	0.0007613	0.6253
Ammonia	NH_3	17	22.3	0.0448	0.00118	0.5892
Carbon monoxide (Carbonic oxide)	CO	28	13.6	0.07364	0.002369	0.9674
Carbon dioxide (Carbonic acid)	CO_2	44	8.64	0.11631	0.00187	1.52901
Olefiant gas	CH_2	14	13.587	0.0736	0.001181	0.967104
Marsh gas	CH_4	16	23.757	0.04209	0.000675	0.55306
Sulphurous acid	SO_2	64	6.463	0.15536	0.002493	1.54143
Sulphuretted hydrogen	SH_2	34	5.582	0.17918	0.002877	2.3943
Bisulphuret of carbon	S_2C	76	2.487	0.40052	0.00643	5.3007
Ozone	$\bar{O}_3$	24	7.97	0.12648	0.00203	1.64656

* By this table there would be 12.75 cubic feet of air at 32° per pound.

TABLE No. III.

TABLE OF CIRCLES, SQUARES, AND CUBES.

n Diam.	$n\pi$ Circumf.	$n^2 \dfrac{\pi}{4}$ Area.	n^2 Square.	n^3 Cube.	$\sqrt{n}$ Sq. Root.	$\sqrt[3]{n}$ Cub. Rt.
1.0	3.142	0.7854	1.000	1.000	1.0000	1.0000
1.1	3.456	0.9503	1.210	1.331	1.0488	1.0323
1.2	3.770	1.1310	1.440	1.728	1.0955	1.0627
1.3	4.084	1.3273	1.690	2.197	1.1402	1.0914
1.4	4.398	1.5394	1.960	2.744	1.1832	1.1187
1.5	4.712	1.7672	2.250	3.375	1.2247	1.1447
1.6	5.027	2.0106	2.560	4.096	1.2649	1.1696
1.7	5.341	2.2698	2.890	4.913	1.3038	1.1935
1.8	5.655	2.5447	3.240	5.832	1.3416	1.2164
1.9	5.969	2.8353	3.610	6.859	1.3784	1.2386
2.0	6.283	3.1416	4.000	8.000	1.4142	1.2599
2.1	6.597	3.4636	4.410	9.261	1.4491	1.2806
2.2	6.912	3.8013	4.840	10.648	1.4832	1.3006
2.3	7.226	4.1546	5.290	12.167	1.5166	1.3200
2.4	7.540	4.5239	5.760	13.824	1.5492	1.3389
2.5	7.854	4.9087	6.250	15.625	1.5811	1.3572
2.6	8.168	5.3093	6.760	17.576	1.6125	1.3751
2.7	8.482	5.7256	7.290	19.683	1.6432	1.3925
2.8	8.797	6.1575	7.840	21.952	1.6733	1.4095
2.9	9.111	6.6052	8.410	24.389	1.7029	1.4260
3.0	9.425	7.0686	9.00	27.000	1.7321	1.4422
3.1	9.739	7.5477	9.61	29.791	1.7607	1.4581
3.2	10.053	8.0425	10.24	32.768	1.7889	1.4736
3.3	10.367	8.5530	10.89	35.937	1.8166	1.4888
3.4	10.681	9.0792	11.56	39.304	1.8439	1.5037
3.5	10.996	9.6211	12.25	42.875	1.8708	1.5183
3.6	11.310	10.179	12.96	46.656	1.8974	1.5326
3.7	11.624	10.752	13.69	50.653	1.9235	1.5467
3.8	11.938	11.341	14.44	54.872	1.9494	1.5605
3.9	12.252	11.946	15.21	59.319	1.9748	1.5741
4.0	12.566	12.566	16.00	64.000	2.0000	1.5874
4.1	12.881	13.203	16.81	68.921	2.0249	1.6005
4.2	13.195	13.854	17.64	74.088	2.0494	1.6134
4.3	13.509	14.522	18.49	79.507	2.0736	1.6261
4.4	13.823	15.205	19.36	85.184	2.0976	1.6386
4.5	14.137	15.904	20.25	91.125	2.1213	1.6510
4.6	14.451	16.619	21.16	97.336	2.1448	1.6631
4.7	14.765	17.349	22.09	103.823	2.1680	1 6751

CIRCLES, SQUARES, AND CUBES—*Continued*.

n Diam.	$n\pi$ Circumf.	$n^2\frac{\pi}{4}$ Area.	n^2 Square.	n^3 Cube.	$\sqrt{n}$ Sq. Root.	$\sqrt[3]{n}$ Cub. Rt.
4.8	15.080	18.096	23.04	110.592	2.1909	1.6869
4.9	15.394	18.857	24.01	117.649	2.2136	1.6985
5.0	15.708	19.635	25.00	125.000	2.2361	1.7100
5.1	16.022	20.428	26.01	132.651	2.2583	1.7213
5.2	16.336	21.237	27.04	140.608	2.2804	1.7325
5.3	16.650	22.062	28.09	148.877	2.3022	1.7435
5.4	16.965	22.902	29.16	157.464	2.3238	1.7544
5.5	17.279	23.758	30.25	166.375	2.3452	1.7652
5.6	17.593	24.630	31.36	175.616	2.3664	1.7758
5.7	17.907	25.518	32.49	185.193	2.3875	1.7863
5.8	18.221	26.421	33.64	195.112	2.4083	1.7967
5.9	18.535	27.340	34.81	205.379	2.4290	1.8070
6.0	18.850	28.274	36.00	216.000	2.4495	1.8171
6.1	19.164	29.225	37.21	226.981	2.4698	1.8272
6.2	19.478	30.191	38.44	238.328	2.4900	1.8371
6.3	19.792	31.173	39.69	250.047	2.5100	1.8469
6.4	20.106	32.170	40.96	262.144	2.5298	1.8566
6.5	20.420	33.183	42.25	274.625	2.5495	1.8663
6.6	20.735	34.212	43.56	287.496	2.5691	1.8758
6.7	21.049	35.257	44.89	300.763	2.5884	1.8852
6.8	21.363	36.317	46.24	314.432	2.6077	1.8945
6.9	21.677	37.393	47.61	328.509	2.6268	1.9038
7.0	21.991	38.485	49.00	343.000	2.6458	1.9129
7.1	22.305	39.592	50.41	357.911	2.6646	1.9220
7.2	22.619	40.715	51.84	373.248	2.6833	1.9310
7.3	22.934	41.854	53.29	389.017	2.7019	1.9399
7.4	23.248	43.008	54.76	405.224	2.7203	1.9487
7.5	23.562	44.179	56.25	421.875	2.7386	1.9574
7.6	23.876	45.365	57.76	438.976	2.7568	1.9661
7.7	24.190	46.566	59.29	456.533	2.7749	1.9747
7.8	24.504	47.784	60.84	474.552	2.7929	1.9832
7.9	24.819	49.017	62.41	493.039	2.8107	1.9916
8.0	25.133	50.266	64.00	512.000	2.8284	2.0000
8.1	25.447	51.530	65.61	531.441	2.8461	2.0083
8.2	25.761	52.810	67.24	551.468	2.8636	2.0165
8.3	26.075	54.106	68.89	571.787	2.8810	2.0247
8.4	26.389	55.418	70.56	592.704	2.8983	2.0328
8.5	26.704	56.745	72.25	614.125	2.9155	2.0408
8.6	27.018	58.088	73.96	636.056	2.9326	2.0488
8.7	27.332	59.447	75.69	658.503	2.9496	2.0567
8.8	27.646	60.821	77.44	681.473	2.9665	2.0646
8.9	27.960	62.211	79.21	704.969	2.9833	2.0724

CIRCLES, SQUARES, AND CUBES—*Continued.*

n Diam.	$n\pi$ Circumf.	$n^2\frac{\pi}{4}$ Area.	n^2 Square.	n^3 Cube.	$\sqrt{n}$ Sq. Root.	$\sqrt[3]{n}$ Cub. Rt.
9.0	28.274	63.617	81.00	729.000	3.0000	2.0801
9.1	28.588	65.039	82.81	753.571	3.0166	2.0878
9.2	28.903	66.476	84.64	778.688	3.0332	2.0954
9.3	29.217	67.929	86.49	804.357	3.0496	2.1029
9.4	29.531	69.398	88.36	830.584	3.0659	2.1105
9.5	29.845	70.882	90.25	857.375	3.0822	2.1179
9.6	30.159	72.382	92.16	884.736	3.0984	2.1253
9.7	30.473	73.898	94.09	912.673	3.1145	2.1327
9.8	30.788	75.430	96.04	941.192	3.1305	2.1400
9.9	31.102	76.977	98.01	970.299	3.1464	2.1472
10.0	31.416	78.540	100.00	1000.000	3.1623	2.1544
10.1	31.730	80.119	102.01	1030.301	3.1780	2.1616
10.2	32.044	81.713	104.04	1061.208	3.1937	2.1687
10.3	32.358	83.323	106.09	1092.727	3.2094	2.1757
10.4	32.673	84.949	108.16	1124.863	3.2249	2.1828
10.5	32.987	86.590	110.25	1157.625	3.2404	2.1897
10.6	33.301	88.247	112.36	1191.016	3.2558	2.1967
10.7	33.615	89.920	114.49	1225.043	3.2711	2.2036
10.8	33.929	91.609	116.64	1259.712	3.2863	2.2104
10.9	34.243	93.313	118.81	1295.029	3.3015	2.2172
11.0	34.558	95.033	121.00	1331.000	3.3166	2.2239
11.1	34.872	96.769	123.21	1367.631	3.3317	2.2307
11.2	35.186	98.520	125.44	1404.928	3.3466	2.2374
11.3	35.500	100.29	127.69	1442.897	3.3615	2.2441
11.4	35.814	102.07	129.96	1481.544	3.3764	2.2506
11.5	36.128	103.87	132.25	1520.875	3.3912	2.2572
11.6	36.442	105.68	134.56	1560.896	3.4059	2.2637
11.7	36.757	107.51	136.89	1601.613	3.4205	2.2702
11.8	37.071	109.36	139.24	1643.032	3.4351	2.2766
11.9	37.385	111.22	141.61	1685.159	3.4496	2.2831
12.0	37.699	113.10	144.00	1728.000	3.4641	2.2894
12.1	38.013	114.99	146.41	1771.561	3.4785	2.2957
12.2	38.327	116.90	148.84	1815.848	3.4928	2.3021
12.3	38.642	118.82	151.29	1860.867	3.5071	2.3084
12.4	38.956	120.76	153.76	1906.624	3.5214	2.3146
12.5	39.270	122.72	156.25	1953.125	3.5355	2.3208
12.6	39.584	124.69	158.76	2000.376	3.5496	2.3270
12.7	39.898	126.68	161.29	2048.383	3.5637	2.3331
12.8	40.212	128.68	163.84	2097.152	3.5777	2.3392
12.9	40.527	130.70	166.41	2146.689	3.5917	2.3453
13.0	40.841	132.73	169.00	2197.000	3.6056	2.3513
13.1	41.155	134.78	171.61	2248.091	3.6194	2.3573
13.2	41.469	136.85	174.24	2299.968	3.6332	2.3633

CIRCLES, SQUARES, AND CUBES—*Continued.*

n Diam.	$n\pi$ Circumf.	$n^2\dfrac{\pi}{4}$ Area.	n^2 Square.	n^3 Cube.	$\sqrt{n}$ Sq. Root.	$\sqrt[3]{n}$ Cub. Rt.
13.3	41.783	138.93	176.89	2352.637	3.6469	2.3693
13.4	42.097	141.03	179.56	2406.104	3.6606	2.3752
13.5	42.412	143.14	182.25	2460.375	3.6742	2.3811
13.6	42.726	145.27	184.96	2515.456	3.6878	2.3870
13.7	43.040	147.41	187.69	2571.353	3.7013	2.3928
13.8	43.354	149.57	190.44	2628.072	3.7148	2.3986
13.9	43.668	151.75	193.21	2685.619	3.7283	2.4044
14.0	43.982	153.94	196.00	2744.000	3.7417	2.4101
14.1	44.296	156.15	198.81	2803.221	3.7550	2.4159
14.2	44.611	158.37	201.64	2863.288	3.7683	2.4216
14.3	44.925	160.61	204.49	2924.207	3.7815	2.4272
14.4	45.239	162.86	207.36	2985.984	3.7947	2.4329
14.5	45.553	165.13	210.25	3048.625	3.8079	2.4385
14.6	45.867	167.42	213.16	3112.136	3.8210	2.4441
14.7	46.181	169.72	216.09	3176.523	3.8341	2.4497
14.8	46.496	172.03	219.04	3241.792	3.8471	2.4552
14.9	46.810	174.37	222.01	3307.949	3.8600	2.4607
15.0	47.124	176.72	225.00	3375.000	3.8730	2.4662
15.1	47.438	179.08	228.01	3442.951	3.8859	2.4717
15.2	47.752	181.46	231.04	3511.808	3.8987	2.4772
15.3	48.066	183.85	234.09	3581.577	3.9115	2.4825
15.4	48.381	186.27	237.16	3652.264	3.9243	2.4879
15.5	48.695	188.69	240.25	3723.875	3.9370	2.4933
15.6	49.009	191.13	243.36	3796.416	3.9497	2.4986
15.7	49.323	193.59	246.49	3869.893	3.9623	2.5039
15.8	49.637	196.07	249.64	3944.312	3.9749	2.5092
15.9	49.951	198.56	252.81	4019.679	3.9875	2.5146
16.0	50.265	201.06	256.00	4096.000	4.0000	2.5198
16.1	50.580	203.58	259.21	4173.281	4.0125	2.5251
16.2	50.894	206.12	262.44	4251.528	4.0249	2.5303
16.3	51.208	208.67	265.69	4330.747	4.0373	2.5355
16.4	51.522	211.24	268.96	4410.944	4.0497	2.5406
16.5	51.836	213.83	272.25	4492.125	4.0620	2.5458
16.6	52.150	216.42	275.56	4574.296	4.0743	2.5509
16.7	52.465	219.04	278.89	4657.463	4.0866	2.5561
16.8	52.779	221.67	282.24	4741.632	4.0988	2.5612
16.9	53.093	224.32	285.61	4826.809	4.1110	2.5663
17.0	53.407	226.98	289.00	4913.000	4.1231	2.5713
17.1	53.721	229.66	292.41	5000.211	4.1352	2.5763
17.2	54.035	132.35	295.84	5088.448	4.1473	2.5813
17.3	54.350	235.06	299.29	5177.717	4.1593	2.5863
17.4	54.664	237.79	302.76	5268.024	4.1713	2.5913

APPENDIX CONTAINING REFERENCES AND TABLES.

CIRCLES, SQUARES, AND CUBES—*Continued*.

n Diam.	$n\pi$ Circumf.	$n^2\frac{\pi}{4}$ Area.	n^2 Square.	n^3 Cube.	$\sqrt{n}$ Sq. Root.	$\sqrt[3]{n}$ Cub. Rt.
17.5	54.978	240.53	306.25	5359.375	4.1833	2.5963
17.6	55.292	243.29	309.76	5451.776	4.1952	2.6012
17.7	55.606	246.06	313.29	5545.233	4.2071	2.6061
17.8	55.920	248.85	316.84	5639.752	4.2190	2.6109
17.9	56.235	251.65	320.41	5735.339	4.2308	2.6158
18.0	56.549	254.47	324.00	5832.000	4.2426	2.6207
18.1	56.863	257.30	327.61	5929.741	4.2544	2.6256
18.2	57.177	260.16	331.24	6028.568	4.2661	2.6304
18.3	57.491	263.02	334.89	6128.487	4.2778	2.6352
18.4	57.805	265.90	338.56	6229.504	4.2895	2.6401
18.5	58.119	268.80	342.25	6331.625	4.3012	2.6448
18.6	58.434	271.72	345.96	6434.856	4.3128	2.6495
18.7	58.748	274.65	349.69	6539.203	4.3243	2.6543
18.8	59.062	277.59	353.44	6644.672	4.3359	2.6590
18.9	59.376	280.55	357.21	6751.269	4.3474	2.6637
19.0	59.690	283.53	361.00	6859.000	4.3589	2.6684
19.1	60.004	286.52	364.81	6967.871	4.3703	2.6731
19.2	60.319	289.53	368.64	7077.888	4.3818	2.6777
19.3	60.633	292.55	372.49	7189.057	4.3932	2.6824
19.4	60.947	295.59	376.36	7301.384	4.4045	2.6869
19.5	61.261	298.65	380.25	7414.875	4.4159	2.6916
19.6	61.575	301.72	384.16	7529.536	4.4272	2.6962
19.7	61.889	304.81	388.09	7645.373	4.4385	2.7008
19.8	62.204	307.91	392.04	7762.392	4.4497	2.7053
19.9	62.518	311.03	396.01	7880.599	4.4609	2.7098
20.0	62.832	314.16	400.00	8000.000	4.4721	2.7144
20.1	63.146	317.31	404.01	8120.601	4.4833	2.7189
20.2	63.460	320.47	408.04	8242.408	4.4944	2.7234
20.3	63.774	323.66	412.09	8365.427	4.5055	2.7279
20.4	64.088	326.85	416.16	8489.664	4.5166	2.7324
20.5	64.403	330.06	420.25	8615.125	4.5277	2.7368
20.6	64.717	333.29	424.36	8741.816	4.5387	2.7413
20.7	65.031	336.54	428.49	8869.743	4.5497	2.7457
20.8	65.345	339.80	432.64	8998.912	4.5607	2.7502
20.9	65.659	343.07	436.81	9129.329	4.5716	2.7545
21.0	65.973	346.36	441.00	9261.000	4.5826	2.7589
21.1	66.288	349.67	445.21	9393.931	4.5935	2.7633
21.2	66.602	352.99	449.44	9528.128	4.6043	2.7676
21.3	66.916	356.33	453.69	9663.597	4.6152	2.7720
21.4	67.230	359.68	457.96	9800.344	4.6260	2.7763
21.5	67.544	363.05	462.25	9938.375	4.6368	2.7806
21.6	67.858	366.44	466.56	10077.696	4.6476	2.7849
21.7	68.173	369.84	470.89	10218.313	4.6585	2.7893

CIRCLES, SQUARES, AND CUBES—*Continued.*

n Diam.	$n\pi$ Circumf.	$n^2 \dfrac{\pi}{4}$ Area.	n^2 Square.	n^3 Cube.	$\sqrt{n}$ Sq. Root.	$\sqrt[3]{n}$ Cub. Rt.
21.8	68.487	373.25	475.24	10360.232	4.6690	2.7935
21.9	68.801	376.69	479.61	10503.459	4.6797	2.7978
22.0	69.115	380.13	484.00	10648.000	4.6904	2.8021
22.1	69.429	383.60	488.41	10793.861	4.7011	2.8063
22.2	69.743	387.08	492.84	10941.048	4.7117	2.8105
22.3	70.058	390.57	497.29	11089.567	4.7223	2.8147
22.4	70.372	394.08	501.76	11239.424	4.7329	2.8189
22.5	70.686	397.61	506.25	11390.625	4.7434	2.8231
22.6	71.000	401.15	510.76	11543.176	4.7539	2.8273
22.7	71.314	404.71	515.29	11697.083	4.7644	2.8314
22.8	71.628	408.28	519.84	11852.352	4.7749	2.8356
22.9	71.942	411.87	524.41	12008.989	4.7854	2.8397
23.0	72.257	415.48	529.00	12167.000	4.7958	2.8438
23.1	72.571	419.10	533.61	12326.391	4.8062	2.8479
23.2	72.885	422.73	538.24	12487.168	4.8166	2.8521
23.3	73.199	426.39	542.89	12649.337	4.8270	2.8562
23.4	73.513	430.05	547.56	12812.904	4.8373	2.8603
23.5	73.827	433.74	552.25	12977.875	4.8477	2.8643
23.6	74.142	437.44	556.96	13144.256	4.8580	2.8684
23.7	74.456	441.15	561.69	13312.053	4.8683	2.8724
23.8	74.770	444.88	566.44	13481.272	4.8785	2.8765
23.9	75.084	448.63	571.21	13651.919	4.8888	2.8805
24.0	75.398	452.39	576.00	13824.000	4.8990	2.8845
24.1	75.712	456.17	580.81	13997.521	4.9092	2.8885
24.2	76.027	459.96	585.64	14172.488	4.9193	2.8925
24.3	76.341	463.77	590.49	14348.907	4.9295	2.8965
24.4	76.655	467.60	595.36	14526.784	4.9396	2.9004
24.5	76.969	471.44	600.25	14706.125	4.9497	2.9044
24.6	77.283	475.29	605.16	14886.936	4.9598	2.9083
24.7	77.597	479.16	610.09	15069.223	4.9699	2.9123
24.8	77.911	483.05	615.04	15252.992	4.9799	2.9162
24.9	78.226	486.96	620.01	15438.249	4.9899	2.9201
25.0	78.540	490.87	625.00	15625.000	5.0000	2.9241
25.1	78.854	494.81	630.01	15813.251	5.0099	2.9279
25.2	79.168	498.76	635.04	16003.008	5.0199	2.9318
25.3	79.482	502.73	640.09	16194.277	5.0299	2.9356
25.4	79.796	506.71	645.16	16387.064	5.0398	2.9395
25.5	80.111	510.71	650.25	16581.375	5.0497	2.9434
25.6	80.425	514.72	655.36	16777.216	5.0596	2.9472
25.7	80.739	518.75	660.49	16974.593	5.0695	2.9510
25.8	81.053	522.79	665.64	17173.512	5.0793	2.9549
25.9	81.367	526.85	670.81	17373.979	5.0892	2.9586

CIRCLES, SQUARES, AND CUBES—*Continued*.

n Diam.	$n\pi$ Circumf.	$n^2\frac{\pi}{4}$ Area.	n^2 Square.	n^3 Cube.	$\sqrt{n}$ Sq. Root.	$\sqrt[3]{n}$ Cub. Rt.
26.0	81.681	530.93	676.00	17576.000	5.0990	2.9624
26.1	81.996	535.02	681.21	17779.581	5.1088	2.9662
26.2	82.310	539.13	686.44	17984.728	5.1185	2.9701
26.3	82.624	543.25	691.69	18191.447	5.1283	2.9738
26.4	82.938	547.39	696.96	18399.744	5.1380	2.9776
26.5	83.252	551.55	702.25	18609.625	5.1478	2.9814
26.6	83.566	555.72	707.56	18821.096	5.1575	2.9851
26.7	83.881	559.90	712.89	19034.163	5.1672	2.9888
26.8	84.195	564.10	718.24	19248.832	5.1768	2.9926
26.9	84.509	568.32	723.61	19465.109	5.1865	2.9963
27.0	84.823	572.56	729.00	19683.000	5.1962	3.0000
27.1	85.137	576.80	734.41	19902.511	5.2057	3.0037
27.2	85.451	581.07	739.84	20123.648	5.2153	3.0074
27.3	85.765	585.35	745.29	20346.417	5.2249	3.0111
27.4	86.080	589.65	750.76	20570.824	5.2345	3.0147
27.5	86.394	593.96	756.25	20796.875	5.2440	3.0184
27.6	86.708	598.29	761.76	21024.576	5.2535	3.0221
27.7	87.022	602.63	767.29	21253.933	5.2630	3.0257
27.8	87.336	606.99	772.84	21484.952	5.2725	3.0293
27.9	87.650	611.36	778.41	21717.639	5.2820	3.0330
28.0	87.965	615.75	784.00	21952.000	5.2915	3.0366
28.1	88.279	620.16	789.61	22188.041	5.3009	3.0402
28.2	88.593	624.58	795.24	22425.768	5.3103	3.0438
28.3	88.907	629.02	800.89	22665.187	5.3197	3.0474
28.4	89.221	633.47	806.56	22906.304	5.3291	3.0510
28.5	89.535	637.94	812.25	23149.125	5.3385	3.0546
28.6	89.850	642.42	817.96	23393.656	5.3478	3.0581
28.7	90.164	646.93	823.69	23639.903	5.3572	3.0617
28.8	90.478	651.44	829.44	23887.872	5.3665	3.0652
28.9	90.792	655.97	835.21	24137.569	5.3758	3.0688
29.0	91.106	660.52	841.00	24389.000	5.3852	3.0723
29.1	91.420	665.08	846.81	24642.171	5.3944	3.0758
29.2	91.735	669.66	852.64	24897.088	5.4037	3.0794
29.3	92.049	674.26	858.49	25153.757	5.4129	3.0829
29.4	92.363	678.87	864.36	25412.184	5.4221	3.0864
29.5	92.677	683.49	870.25	25672.375	5.4313	3.0899
29.6	92.991	688.13	876.16	25934.336	5.4405	3.0934
29.7	93.305	692.79	882.09	26198.073	5.4497	3.0968
29.8	93.619	697.47	888.04	26463.592	5.4589	3.1003
29.9	93.934	702.15	894.01	26730.899	5.4680	3.1038
30.0	94.248	706.86	900.00	27000.000	5.4772	3.1072
30.1	94.562	711.58	906.01	27270.901	5.4863	3.1107
30.2	94.876	716.32	912.04	27543.608	5.4954	3.1141

CIRCLES, SQUARES, AND CUBES—*Continued*.

n Diam.	$n\pi$ Circumf.	$n^2\dfrac{\pi}{4}$ Area.	n^2 Square.	n^3 Cube.	$\sqrt{n}$ Sq. Root.	$\sqrt[3]{n}$ Cub. Rt.
30.3	95.190	721.07	918.09	27818.127	5.5045	3.1176
30.4	95.505	725.83	924.16	28094.464	5.5136	3.1210
30.5	95.819	730.62	930.25	28372.625	5.5226	3.1244
30.6	96.133	735.42	936.36	28652.616	5.5317	3.1278
30.7	96.447	740.23	942.49	28934.443	5.5407	3.1312
30.8	96.761	745.06	948.64	29218.112	5.5497	3.1346
30.9	97.075	749.91	954.81	29503.629	5.5587	3.1380
31.0	97.389	754.77	961.00	29791.000	5.5678	3.1414
31.1	97.704	759.65	967.21	30080.231	5.5767	3.1448
31.2	98.018	764.54	973.44	30371.328	5.5857	3.1481
31.3	98.332	769.45	979.69	30664.297	5.5946	3.1515
31.4	98.646	774.37	985.96	30959.144	5.6035	3.1548
31.5	98.960	779.31	992.25	31255.875	5.6124	3.1582
31.6	99.274	784.27	998.56	31554.496	5.6213	3.1615
31.7	99.588	789.24	1004.89	31855.013	5.6302	3.1648
31.8	99.903	794.23	1011.24	32157.432	5.6391	3.1681
31.9	100.22	799.23	1017.61	32461.759	5.6480	3.1715
32.0	100.53	804.25	1024.00	32768.000	5.6569	3.1748
32.1	100.85	809.28	1030.41	33076.161	5.6656	3.1781
32.2	101.16	814.33	1036.84	33386.248	5.6745	3.1814
32.3	101.47	819.40	1043.29	33698.267	5.6833	3.1847
32.4	101.79	824.48	1049.76	34012.224	5.6921	3.1880
32.5	102.10	829.58	1056.25	34328.125	5.7008	3.1913
32.6	102.42	834.69	1062.76	34645.976	5.7096	3.1945
32.7	102.73	839.82	1069.29	34965.783	5.7183	3.1978
32.8	103.04	844.96	1075.84	35287.552	5.7271	3.2010
32.9	103.36	850.12	1082.41	35611.289	5.7358	3.2043
33.0	103.67	855.30	1089.00	35937.000	5.7446	3.2075
33.1	103.99	860.49	1095.61	36264.691	5.7532	3.2108
33.2	104.30	865.70	1102.24	36594.368	5.7619	3.2140
33.3	104.62	870.92	1108.89	36926.037	5.7706	3.2172
33.4	104.93	876.16	1115.56	37259.704	5.7792	3.2204
33.5	105.24	881.41	1122.25	37595.375	5.7879	3.2237
33.6	105.56	886.68	1128.96	37933.056	5.7965	3.2269
33.7	105.87	891.97	1135.69	38272.753	5.8051	3.2301
33.8	106.19	897.27	1142.44	38614.472	5.8137	3.2332
33.9	106.50	902.59	1149.21	38958.219	5.8223	3.2364
34.0	106.81	907.92	1156.00	39304.000	5.8310	3.2396
34.1	107.13	913.27	1162.81	39651.821	5.8395	3.2428
34.2	107.44	918.63	1169.64	40001.688	5.8480	3.2460
34.3	107.76	924.01	1176.49	40353.607	5.8566	3.2491
34.4	108.07	929.41	1183.36	40707.584	5.8651	3.2522

CIRCLES, SQUARES, AND CUBES—Continued.

n Diam.	$n\pi$ Circumf.	$n^2\frac{\pi}{4}$ Area.	n^2 Square.	n^3 Cube.	$\sqrt{n}$ Sq. Root.	$\sqrt[3]{n}$ Cub. Rt.
34.5	108.38	934.82	1190.25	41063.625	5.8730	3.2554
34.6	108.70	940.25	1197.16	41421.736	5.8821	3.2586
34.7	109.01	945.69	1204.09	41781.923	5.8906	3.2617
34.8	109.33	951.15	1211.04	42144.192	5.8991	3.2648
34.9	109.64	956.62	1218.01	42508.549	5.9076	3.2679
35.0	109.96	962.11	1225.00	42875.000	5.9161	3.2710
35.1	110.27	967.62	1232.01	43243.551	5.9245	3.2742
35.2	110.58	973.14	1239.04	43614.208	5.9329	3.2773
35.3	110.90	978.68	1246.09	43986.977	5.9413	3.2804
35.4	111.21	984.23	1253.16	44361.864	5.9497	3.2835
35.5	111.53	989.80	1260.25	44738.875	5.9581	3.2866
35.6	111.84	995.38	1267.36	45118.016	5.9665	3.2897
35.7	112.15	1000.98	1274.49	45499.293	5.9749	3.2927
35.8	112.47	1006.60	1281.64	45882.712	5.9833	3.2958
35.9	112.78	1012.23	1288.81	46268.279	5.9916	3.2989
36.0	113.10	1017.88	1296.00	46656.000	6.0000	3.3019
36.1	113.41	1023.54	1303.21	47045.881	6.0083	3.3050
36.2	113.73	1029.22	1310.44	47437.928	6.0166	3.3080
36.3	114.04	1034.91	1317.69	47832.147	6.0249	3.3111
36.4	114.35	1040.62	1324.96	48228.544	6.0332	3.3141
39.5	114.67	1046.35	1332.25	48627.125	6.0415	3.3171
36.6	114.98	1052.09	1339.56	49027.896	6.0497	3.3202
36.7	115.30	1057.84	1346.89	49430.863	6.0580	3.3232
36.8	115.61	1063.62	1354.24	49836.032	6.0663	3.3262
36.9	115.92	1069.41	1361.61	50243.409	6.0745	3.3292
37.0	116.24	1075.21	1369.00	50653.000	6.0827	3.3322
37.1	116.55	1081.03	1376.41	51064.811	6.0909	3.3352
37.2	116.87	1086.87	1383.84	51478.848	6.0991	3.3382
37.3	117.18	1092.72	1391.29	51895.117	6.1073	3.3412
37.4	117.50	1098.58	1398.76	52313.624	6.1155	3.3442
37.5	117.81	1104.47	1406.25	52734.375	6.1237	3.3472
37.6	118.12	1110.36	1413.76	53157.376	6.1318	3.3501
37.7	118.44	1116.28	1421.29	53582.633	6.1400	3.3531
37.8	118.75	1122.21	1428.84	54010.152	6.1481	3.3561
37.9	119.07	1128.15	1436.41	54439.939	6.1563	3.3590
38.0	119.38	1134.11	1444.00	54872.000	6.1644	3.3620
38.1	119.69	1140.09	1451.61	55306.341	6.1725	3.3649
38.2	120.01	1146.08	1459.24	55742.968	6.1806	3.3679
38.3	120.32	1152.09	1466.89	56181.887	6.1887	3.3708
38.4	120.64	1158.12	1474.56	56623.104	6.1967	3.3737
38.5	120.95	1164.16	1482.25	57066.625	6.2048	3.3767
38.6	121.27	1170.21	1489.96	57512.456	6.2129	3.3796
38.7	121.58	1176.28	1497.69	57960.603	6.2209	3.3825

CIRCLES, SQUARES, AND CUBES—Continued.

n Diam.	$n\pi$ Circumf.	$n^2\dfrac{\pi}{4}$ Area.	n^2 Square.	n^3 Cube.	$\sqrt{n}$ Sq. Root.	$\sqrt[3]{n}$ Cub. Rt.
38.8	121.89	1182.37	1505.44	58411.072	6.2289	3.3854
38.9	122.21	1188.47	1513.21	58863.869	6.2370	3.3883
39.0	122.52	1194.59	1521.00	59319.000	6.2450	3.3912
39.1	122.84	1200.72	1528.81	59776.471	6.2530	3.3941
39.2	123.15	1206.87	1536.64	60236.288	6.2610	3.3970
39.3	123.46	1213.04	1544.49	60698.457	6.2689	3.3999
39.4	123.78	1219.22	1552.36	61162.984	6.2769	3.4028
39.5	124.09	1225.42	1560.25	61629.875	6.2849	3.4056
39.6	124.41	1231.63	1568.16	62099.136	6.2928	3.4085
39.7	124.72	1237.86	1576.09	62570.773	6.3008	3.4114
39.8	125.04	1244.10	1584.04	63044.792	6.3087	3.4142
39.9	125.35	1250.36	1592.01	63521.199	6.3166	3.4171
40.0	125.66	1256.64	1600.00	64000.000	6.3245	3.4200
40.1	125.98	1262.93	1608.01	64481.201	6.3325	3.4228
40.2	126.29	1269.23	1616.04	64964.808	6.3404	3.4256
40.3	126.61	1275.56	1624.09	65450.827	6.3482	3.4285
40.4	126.92	1281.90	1632.16	65939.264	6.3561	3.4313
40.5	127.23	1288.25	1640.25	66430.125	6.3639	3.4341
40.6	127.55	1294.62	1648.36	66923.416	6.3718	3.4370
40.7	127.86	1301.00	1656.49	67419.143	6.3796	3.4398
40.8	128.18	1307.41	1664.64	67911.312	6.3875	3.4426
40.9	128.49	1313.82	1672.81	68417.929	6.3953	3.4454
41.0	128.81	1320.25	1681.00	68921.000	6.4031	3.4482
41.1	129.12	1326.70	1689.21	69426.531	6.4109	3.4510
41.2	129.43	1333.17	1697.44	69934.528	6.4187	3.4538
41.3	129.75	1339.65	1705.69	70444.997	6.4265	3.4566
41.4	130.06	1346.14	1713.96	70957.944	6.4343	3.4594
41.5	130.38	1352.65	1722.25	71473.375	6.4421	3.4622
41.6	130.69	1359.18	1730.56	71991.296	6.4498	3.4650
41.7	131.00	1365.72	1738.89	72511.713	6.4575	3.4677
41.8	131.32	1372.28	1747.24	73034.632	6.4653	3.4705
41.9	131.63	1378.85	1755.61	73560.059	6.4730	3.4733
42.0	131.95	1385.44	1764.00	74088.000	6.4807	3.4760
42.1	132.26	1392.05	1772.41	74618.461	6.4884	3.4788
42.2	132.58	1398.67	1780.84	75151.448	6.4961	3.4815
42.3	132.89	1405.31	1789.29	75686.967	6.5038	3.4843
42.4	133.20	1411.96	1797.76	76225.024	6.5115	3.4870
42.5	133.52	1418.63	1806.25	76765.625	6.5192	3.4898
42.6	133.83	1425.31	1814.76	77308.776	6.5268	3.4925
42.7	134.15	1432.01	1823.29	77854.483	6.5345	3.4952
42.8	134.46	1438.72	1831.84	78402.752	6.5422	3.4980
42.9	134.77	1445.45	1840.41	78953.589	6.5498	3.5007

APPENDIX CONTAINING REFERENCES AND TABLES. 373

CIRCLES, SQUARES, AND CUBES—*Continued.*

n Diam.	$n\pi$ Circumf.	$n^2 \frac{\pi}{4}$ Area.	n^2 Square.	n^3 Cube.	$\sqrt{n}$ Sq. Root.	$\sqrt[3]{n}$ Cub. Rt.
43.0	135.09	1452.20	1849.00	79507.000	6.5574	3.5034
43.1	135.40	1458.96	1857.61	80062.991	6.5651	3.5061
43.2	135.72	1465.74	1866.24	80621.568	6.5727	3.5088
43.3	136.03	1472.54	1874.89	81182.737	6.5803	3.5115
43.4	136.35	1479.34	1883.56	81746.504	6.5879	3.5142
43.5	136.66	1486.17	1892.25	82312.875	6.5954	3.5169
43.6	136.97	1493.01	1900.96	82881.856	6.6030	3.5196
43.7	137.29	1499.87	1909.69	83453.453	6.6106	3.5223
43.8	137.60	1506.74	1918.44	84027.672	6.6182	3.5250
43.9	137.92	1513.63	1927.21	84604.519	6.6257	3.5277
44.0	138.23	1520.53	1936.00	85184.000	6.6333	3.5303
44.1	138.54	1527.45	1944.81	85766.121	6.6408	3.5330
44.2	138.86	1534.39	1953.64	86350.888	6.6483	3.5357
44.3	139.17	1541.34	1962.49	86938.307	6.6558	3.5384
44.4	139.49	1548.30	1971.36	87528.384	6.6633	3.5410
44.5	139.80	1555.28	1980.25	88121.125	6.6708	3.5437
44.6	140.12	1562.28	1989.16	88716.536	6.6783	3.5463
44.7	140.43	1569.30	1998.09	89314.623	6.6858	3.5490
44.8	140.74	1576.33	2007.04	89915.392	6.6933	3.5516
44.9	141.06	1583.37	2016.01	90518.849	6.7007	3.5543
45.0	141.37	1590.43	2025.00	91125.000	6.7082	3.5569
45.1	141.69	1597.51	2034.01	91733.851	6.7156	3.5595
45.2	142.00	1604.60	2043.04	92345.408	6.7231	3.5621
45.3	142.31	1611.71	2052.09	92959.677	6.7305	3.5648
45.4	142.63	1618.83	2061.16	93576.664	6.7379	3.5674
45.5	142.94	1625.97	2070.25	94196.375	6.7454	3.5700
45.6	143.26	1633.13	2079.36	94818.816	6.7528	3.5726
45.7	143.57	1640.30	2088.49	95443.993	6.7602	3.5752
45.8	143.88	1647.48	2097.64	96071.912	6.7676	3.5778
45.9	144.20	1654.68	2106.81	96702.579	6.7749	3.5805
46.0	144.51	1661.90	2116.00	97336.000	6.7823	3.5830
46.1	144.83	1669.14	2125.21	97972.181	6.7897	3.5856
46.2	145.14	1676.39	2134.44	98611.128	6.7971	3.5882
46.3	145.46	1683.65	2143.69	99252.847	6.8044	3.5908
46.4	145.77	1690.93	2152.96	99897.344	6.8117	3.5934
46.5	146.08	1698.23	2162.25	100544.625	6.8191	3.5960
46.6	146.40	1705.54	2171.56	101194.696	6.8264	3.5986
46.7	146.71	1712.87	2180.89	101847.563	6.8337	3.6011
46.8	147.03	1720.21	2190.24	102503.232	6.8410	3.6037
46.9	147.34	1727.57	2199.61	103161.709	6.8484	3.6063
47.0	147.65	1734.94	2209.00	103823.000	6.8556	3.6088
47.1	147.97	1742.34	2218.41	104487.111	6.8629	3.6114
47.2	148.28	1749.74	2227.84	105154.048	6.8702	3.6139

CIRCLES, SQUARES, AND CUBES—Continued.

n Diam.	$n\pi$ Circumf.	$n^2 \frac{\pi}{4}$ Area.	n^2 Square.	n^3 Cube.	$\sqrt{n}$ Sq. Root.	$\sqrt[3]{n}$ Cub. Rt.
47.3	148.60	1757.16	2237.29	105823.817	6.8775	3.6165
47.4	148.91	1764.60	2246.76	106496.424	6.8847	3.6190
47.5	149.23	1772.05	2256.25	107171.875	6.8920	3.6216
47.6	149.54	1779.52	2265.76	107850.176	6.8993	3.6241
47.7	149.85	1787.01	2275.29	108531.333	6.9065	3.6267
47.8	150.17	1794.51	2284.84	109215.352	6.9137	3.6292
47.9	150.48	1802.03	2294.41	109902.239	6.9209	3.6317
48.0	150.80	1809.56	2304.00	110592.000	6.9282	3.6342
48.1	151.11	1817.11	2313.61	111284.641	6.9354	3.6368
48.2	151.42	1824.67	2323.24	111980.168	6.9426	3.6393
48.3	151.74	1832.25	2332.89	112678.587	6.9498	3.6418
48.4	152.05	1839.84	2342.56	113379.904	6.9570	3.6443
48.5	152.37	1847.45	2352.25	114084.125	6.9642	3.6468
48.6	152.68	1855.08	2361.96	114791.256	6.9714	3.6493
48.7	153.00	1862.72	2371.69	115501.303	6.9785	3.6518
48.8	153.31	1870.38	2381.44	116214.272	6.9857	3.6543
48.9	153.62	1878.05	2391.21	116930.169	6.9928	3.6568
49.0	153.94	1885.74	2401.00	117649.000	7.0000	3.6593
49.1	154.25	1893.45	2410.81	118370.771	7.0071	3.6618
49.2	154.57	1901.17	2420.64	119095.488	7.0143	3.6643
49.3	154.88	1908.90	2430.49	119823.157	7.0214	3.6668
49.4	155.19	1916.65	2440.36	120553.784	7.0285	3.6692
49.5	155.51	1924.42	2450.25	121287.375	7.0356	3.6717
49.6	155.82	1932.21	2460.16	122023.936	7.0427	3.6742
49.7	156.14	1940.00	2470.09	122763.473	7.0498	3.6767
49.8	156.45	1947.82	2480.04	123505.992	7.0569	3.6791
49.9	156.77	1955.65	2490.01	124251.499	7.0640	3.6816
50.0	157.08	1963.50	2500.00	125000.000	7.0711	3.6840
51.0	160.22	2042.82	2601.00	132651.000	7.1414	3.7084
52.0	163.36	2123.72	2704.00	140608.000	7.2111	3.7325
53.0	166.50	2206.19	2809.00	148877.000	7.2801	3.7563
54.0	169.64	2290.22	2916.00	157464.000	7.3485	3.7798
55.0	172.78	2375.83	3025.00	166375.000	7.4162	3.8030
56.0	175.93	2463.01	3136.00	175616.000	7.4833	3.8259
57.0	179.07	2551.76	3249.00	185193.000	7.5498	3.8485
58.0	182.21	2642.08	3364.00	195112.000	7.6158	3.8709
59.0	185.35	2733.97	3481.00	205379.000	7.6811	3.8930
60.0	188.49	2827.44	3600.00	216000.000	7.7460	3.9149
61.0	191.63	2922.47	3721.00	226981.000	7.8102	3.9365
62.0	194.77	3019.07	3844.00	238328.000	7.8740	3.9579
63.0	197.92	3117.25	3969.00	250047.000	7.9373	3.9791
64.0	201.06	3216.99	4096.00	262144.000	8.0000	4.0000
65.0	204.20	3318.31	4225.00	274625.000	8.0623	4.0207
66.0	207.34	3421.20	4356.00	287496.000	8.1240	4.0412

CIRCLES, SQUARES, AND CUBES—*Continued.*

n Diam.	$n\pi$ Circumf.	$n^2\frac{\pi}{4}$ Area.	n^2 Square.	n^3 Cube.	$\sqrt{n}$ Sq. Root.	$\sqrt[3]{n}$ Cub. Rt.
67.0	210.48	3525.66	4489.00	300763.000	8.1854	4.0615
68.0	213.63	3631.69	4624.00	314432.000	8.2462	4.0817
69.0	216.77	3739.29	4761.00	328509.000	8.3066	4.1016
70.0	219.91	3848.46	4900.00	343000.000	8.3666	4.1213
71.0	223.05	3959.20	5041.00	357911.000	8.4261	4.1408
72.0	226.19	4071.51	5184.00	373248.000	8.4853	4.1602
73.0	229.33	4185.39	5329.00	389017.000	8.5440	4.1793
74.0	232.47	4300.85	5476.00	405224.000	8.6023	4.1983
75.0	235.62	4417.87	5625.00	421875.000	8.6603	4.2172
76.0	238.76	4536.47	5776.00	438976.000	8.7178	4.2358
77.0	241.90	4656.63	5929.00	456533.000	8.7750	4.2543
78.0	245.04	4778.37	6084.00	474552.000	8.318	4.2727
79.0	248.18	4901.68	6241.00	493039.000	8.8882	4.2908
80.0	251.32	5026.56	6400.00	512000.000	8.9443	4.3089
81.0	254.47	5153.01	6561.00	531441.000	9.0000	4.3267
82.0	257.61	5281.03	6724.00	551368.000	9.0554	4.3445
83.0	260.75	5410.62	6889.00	571787.000	9.1104	4.3621
84.0	263.89	5541.78	7056.00	592704.000	9.1652	4.3795
85.0	267.03	5674.50	7225.00	614125.000	9.2195	4.3968
86.0	270.17	5808.81	7396.00	636056.000	9.2736	4.4140
87.0	273.32	5944.69	7569.00	658503.000	9.3274	4.4310
88.0	276.46	6082.13	7744.00	681472.000	9.3808	4.4480
89.0	279.60	6221.13	7921.00	704969.000	9.4340	4.4647
90.0	282.74	6361.74	8100.00	729000.000	9.4868	4.4814
91.0	285.88	6503.89	8281.00	753571.000	9.5394	4.4979
92.0	289.02	6647.62	8464.00	778688.000	9.5917	4.5144
93.0	292.17	6792.92	8649.00	804357.000	9.6437	4.5307
94.0	295.31	6939.78	8836.00	830584.000	9.6954	4.5468
95.0	298.45	7088.23	9025.00	857375.000	9.7468	4.5629
96.0	301.59	7238.24	9216.00	884736.000	9.7980	4.5789
97.0	304.73	7389.83	9409.00	912673.000	9.8489	4.5947
98.0	307.87	7542.98	9604.00	941192.000	9.8995	4.6104
99.0	311.02	7697.68	9801.00	970299.000	9.9499	4.6261
100.0	314.16	7854.00	10000.00	1000000.000	10.0000	4.6416

TABLE No. IV.

CIRCUMFERENCES AND AREAS OF CIRCLES.*

Diam.	Circum.	Area.	Diam.	Circum.	Area.	Diam.	Circum.	Area.
1	3.1416	0.7854	65	204.20	3318.31	129	405.27	13069.81
2	6.2832	3.1416	66	207.34	3421.19	130	408.41	13273.23
3	9.4248	7.0686	67	210.49	3525.65	131	411.55	13478.22
4	12.5664	12.5664	68	213.63	3631.68	132	414.69	13684.78
5	15.7080	19.635	69	216.77	3739.28	133	417.83	13892.91
6	18.850	28.274	70	219.91	3848.45	134	420.97	14102.61
7	21.991	38.485	71	223.05	3959.19	135	424.12	14313.88
8	25.133	50.266	72	226.19	4071.50	136	427.26	14526.72
9	28.274	63.617	73	229.34	4185.39	137	430.40	14741.14
10	31.416	78.540	74	232.48	4300.84	138	433.54	14957.12
11	34.558	95.033	75	235.62	4417.86	139	436.68	15174.68
12	37.699	113.10	76	238.76	4536.46	140	439.82	15393.80
13	40.841	132.73	77	241.90	4656.63	141	442.96	15614.50
14	43.982	153.94	78	245.04	4778.36	142	446.11	15836.77
15	47.124	176.71	79	248.19	4901.67	143	449.25	16060.61
16	50.265	201.06	80	251.33	5026.55	144	452.39	16286.02
17	53.407	226.98	81	254.47	5153.00	145	455.53	16513.00
18	56.549	254.47	82	257.61	5281.02	146	458.67	16741.55
19	59.690	283.53	83	260.75	5410.61	147	461.81	16971.67
20	62.832	314.16	84	263.89	5541.77	148	464.96	17203.36
21	65.973	346.36	85	267.04	5674.50	149	468.10	17436.62
22	69.115	380.13	86	270.18	5808.80	150	471.24	17671.46
23	72.257	415.48	87	273.32	5944.68	151	474.38	17907.86
24	75.398	452.39	88	276.46	6082.12	152	477.52	18145.84
25	78.540	490.87	89	279.60	6221.14	153	480.66	18385.39
26	81.681	530.93	90	282.74	6361.73	154	483.81	18626.50
27	84.823	572.56	91	285.88	6503.88	155	486.95	18869.19
28	87.965	615.75	92	289.03	6647.61	156	490.09	19113.45
29	91.106	660.52	93	292.17	6792.91	157	493.23	19359.28
30	94.248	706.86	94	295.31	6939.78	158	496.37	19606.68
31	97.389	754.77	95	298.45	7088.22	159	499.51	19855.65
32	100.53	804.25	96	301.59	7238.23	160	502.65	20106.19
33	103.67	855.30	97	304.73	7389.81	161	505.80	20358.31
34	106.81	907.92	98	307.88	7542.96	162	508.94	20611.99
35	109.96	962.11	99	311.02	7697.69	163	512.08	20867.24
36	113.10	1017.88	100	314.16	7853.98	164	515.22	21124.07
37	116.24	1075.21	101	317.30	8011.85	165	518.36	21382.46
38	119.38	1134.11	102	320.44	8171.28	166	521.50	21642.43
39	122.52	1194.59	103	323.58	8332.29	167	524.65	21903.97
40	125.66	1256.64	104	326.73	8494.87	168	527.79	22167.08
41	128.81	1320.25	105	329.87	8659.01	169	530.93	22431.76
42	131.95	1385.44	106	333.01	8824.73	170	534.07	22698.01
43	135.09	1452.20	107	336.15	8992.02	171	537.21	22965.83
44	138.23	1520.53	108	339.29	9160.88	172	540.35	23235.22
45	141.37	1590.43	109	342.43	9331.32	173	543.50	23506.18
46	144.51	1661.90	110	345.58	9503.32	174	546.64	23778.71
47	147.65	1734.94	111	348.72	9676.89	175	549.78	24052.82
48	150.80	1809.56	112	351.86	9852.03	176	552.92	24328.49
49	153.94	1885.74	113	355.00	10028.75	177	556.06	24605.74
50	157.08	1963.50	114	358.14	10207.03	178	559.20	24884.56
51	160.22	2042.82	115	361.28	10386.89	179	562.35	25164.94
52	163.36	2123.72	116	364.42	10568.32	180	565.49	25446.90
53	166.50	2206.18	117	367.57	10751.32	181	568.63	25730.43
54	169.65	2290.22	118	370.71	10935.88	182	571.77	26015.53
55	172.79	2375.83	119	373.85	11122.02	183	574.91	26302.20
56	175.93	2463.01	120	376.99	11309.73	184	578.05	26590.44
57	179.07	2551.76	121	380.13	11499.01	185	581.19	26880.25
58	182.21	2642.08	122	383.27	11689.87	186	584.34	27171.63
59	185.35	2733.97	123	386.42	11882.29	187	587.48	27464.59
60	188.50	2827.43	124	389.56	12076.28	188	590.62	27759.11
61	191.64	2922.47	125	392.70	12271.85	189	593.76	28055.21
62	194.78	3019.07	126	395.84	12468.98	190	596.90	28352.87
63	197.92	3117.25	127	398.98	12667.69	191	600.04	28652.11
64	201.06	3216.99	128	402.12	12867.96	192	603.19	28952.92

* From Kent's Pocket-book for Mechanical Engineers.

TABLE No. V.

LOGARITHMS OF NUMBERS.

No.	0	1	2	3	4	5	6	7	8	9
10	0000	0043	0086	0128	0170	0212	0253	0294	0334	0374
11	0414	0453	0492	0531	0569	0607	0645	0682	0719	0755
12	0792	0828	0864	0899	0934	0969	1004	1038	1072	1106
13	1139	1173	1206	1239	1271	1303	1335	1367	1399	1430
14	1461	1492	1523	1553	1584	1614	1644	1673	1703	1732
15	1761	1790	1818	1847	1875	1903	1931	1959	1987	2014
16	2041	2068	2095	2122	2148	2175	2201	2227	2253	2279
17	2304	2330	2355	2380	2405	2430	2455	2480	2504	2529
18	2553	2577	2601	2625	2648	2672	2695	2718	2742	2765
19	2788	2810	2833	2856	2878	2900	2923	2945	2967	2989
20	3010	3032	3054	3075	3096	3118	3139	3160	3181	3201
21	3222	3243	3263	3284	3304	3324	3345	3365	3385	3404
22	3424	3444	3464	3483	3502	3522	3541	3560	3579	3598
23	3617	3636	3655	3674	3692	3711	3729	3747	3766	3784
24	3802	3820	3838	3856	3874	3892	3909	3927	3945	3962
25	3979	3997	4014	4031	4048	4065	4082	4099	4116	4133
26	4150	4166	4183	4200	4216	4232	4249	4265	4281	4298
27	4314	4330	4346	4362	4378	4393	4409	4425	4440	4456
28	4472	4487	4502	4518	4533	4548	4564	4579	4594	4609
29	4624	4639	4654	4669	4683	4698	4713	4728	4742	4757
30	4771	4786	4800	4814	4829	4843	4857	4871	4886	4900
31	4914	4928	4942	4955	4969	4983	4997	5011	5024	5038
32	5051	5065	5079	5092	5105	5119	5132	5145	5159	5172
33	5185	5198	5211	5224	5237	5250	5263	5276	5289	5302
34	5315	5328	5340	5353	5366	5378	5391	5403	5416	5428
35	5441	5453	5465	5478	5490	5502	5514	5527	5539	5551
36	5563	5575	5587	5599	5611	5623	5635	5647	5658	5670
37	5682	5694	5705	5717	5729	5740	5752	5763	5775	5786
38	5798	5809	5821	5832	5843	5855	5866	5877	5888	5899
39	5911	5922	5933	5944	5955	5966	5977	5988	5999	6010
40	6021	6031	6042	6053	6064	6075	6085	6096	6107	6117
41	6128	6138	6149	6160	6170	6180	6191	6201	6212	6222
42	6232	6243	6253	6263	6274	6284	6294	6304	6314	6325
43	6335	6345	6355	6365	6375	6385	6395	6405	6415	6425
44	6435	6444	6454	6464	6474	6484	6493	6503	6513	6522
45	6532	6542	6551	6561	6571	6580	6590	6599	6609	6618
46	6628	6637	6646	6656	6665	6675	6684	6693	6702	6712
47	6721	6730	6739	6749	6758	6767	6776	6785	6794	6803
48	6812	6821	6830	6839	6848	6857	6866	6875	6884	6893
49	6902	6911	6920	6928	6937	6946	6955	6964	6972	6981
50	6990	6998	7007	7016	7024	7033	7042	7050	7059	7067
51	7076	7084	7093	7101	7110	7118	7126	7135	7143	7152
52	7160	7168	7177	7185	7193	7202	7210	7218	7226	7235
53	7243	7251	7259	7267	7275	7284	7292	7300	7308	7316
54	7324	7332	7340	7348	7356	7364	7372	7380	7388	7396
No.	0	1	2	3	4	5	6	7	8	9

LOGARITHMS OF NUMBERS—Continued.

No.	0	1	2	3	4	5	6	7	8	9
55	7404	7412	7419	7427	7435	7443	7451	7459	7466	7474
56	7482	7490	7497	7505	7513	7520	7528	7536	7543	7551
57	7559	7566	7574	7582	7589	7597	7604	7612	7619	7627
58	7634	7642	7649	7657	7664	7672	7679	7686	7694	7701
59	7709	7716	7723	7731	7738	7745	7752	7760	7767	7774
60	7782	7789	7796	7803	7810	7818	7825	7832	7839	7846
61	7853	7860	7868	7875	7882	7889	7896	7903	7910	7917
62	7924	7931	7938	7945	7952	7959	7966	7973	7980	7987
63	7993	8000	8007	8014	8021	8028	8035	8041	8048	8055
64	8062	8069	8075	8082	8089	8096	8102	8109	8116	8122
65	8129	8136	8142	8149	8156	8162	8169	8176	8182	8189
66	8195	8202	8209	8215	8222	8228	8235	8241	8248	8254
67	8261	8267	8274	8280	8287	8293	8299	8306	8312	8319
68	8325	8331	8338	8344	8351	8357	8363	8370	8376	8382
69	8388	8395	8401	8407	8414	8420	8426	8432	8439	8445
70	8451	8457	8463	8470	8476	8482	8488	8494	8500	8506
71	8513	8519	8525	8531	8537	8543	8549	8555	8561	8567
72	8573	8579	8585	8591	8597	8603	8609	8615	8621	8627
73	8633	8639	8645	8651	8657	8663	8669	8675	8681	8686
74	8692	8698	8704	8710	8716	8722	8727	8733	8739	8745
75	8751	8756	8762	8768	8774	8779	8785	8791	8797	8802
76	8808	8814	8820	8825	8831	8837	8842	8848	8854	8859
77	8865	8871	8876	8882	8887	8893	8899	8904	8910	8915
78	8921	8927	8932	8938	8943	8949	8954	8960	8965	8971
79	8976	8982	8987	8993	8998	9004	9009	9015	9020	9025
80	9031	9036	9042	9047	9053	9058	9063	9069	9074	9079
81	9085	9090	9096	9101	9106	9112	9117	9122	9128	9133
82	9138	9143	9149	9154	9159	9165	9170	9175	9180	9186
83	9191	9196	9201	9206	9212	9217	9222	9227	9232	9238
84	9243	9248	9253	9258	9263	9269	9274	9279	9284	9289
85	9294	9299	9304	9309	9315	9320	9325	9330	9335	9340
86	9345	9350	9355	9360	9365	9370	9375	9380	9385	9390
87	9395	9400	9405	9410	9415	9420	9425	9430	9435	9440
88	9445	9450	9455	9460	9465	9469	9474	9479	9484	9489
89	9494	9499	9504	9509	9513	9518	9523	9528	9533	9538
90	9542	9547	9552	9557	9562	9566	9571	9576	9581	9586
91	9590	9595	9600	9605	9609	9614	9619	9624	9628	9633
92	9638	9643	9647	9652	9657	9661	9666	9671	9675	9680
93	9685	9689	9694	9699	9703	9708	9713	9717	9722	9727
94	9731	9736	9741	9745	9750	9754	9759	9763	9768	9773
95	9777	9782	9786	9791	9795	9800	9805	9809	9814	9818
96	9823	9827	9832	9836	9841	9845	9850	9854	9859	9863
97	9868	9872	9877	9881	9886	9890	9894	9899	9903	9908
98	9912	9917	9921	9926	9930	9934	9939	9943	9948	9952
99	9956	9961	9965	9969	9974	9978	9983	9987	9991	9996
No.	0	1	2	3	4	5	6	7	8	9

APPENDIX CONTAINING REFERENCES AND TABLES.

TABLE No. VI.

IMPORTANT PROPERTIES OF FAMILIAR SUBSTANCES.

	Specific Gravity. Water, 1.	Specific Heat. Water, 1.	Absorbing and Radiating Power of Bodies in Units of Heat per Square Foot for Difference of 1°.	Conducting Power in Units of Heat per Square Foot of Surface with Difference of 1°.	Weight in Pounds	Melting Points. Degrees Fahr.
Metals from 32° to 212°—					Per cu. in.	
Aluminium	2.61 to 2.65	.212			0.0956	
Antimony	6.712	.0508			0.2428	810
Bismuth	9.823	.0308			0.3533	476
Brass	8.1	.0939	.049		0.2930	1692
Copper	8.788	.092	.0327	515.0	0.3179	1996
Iron, cast	7.5	.1298	.648	103.0	0.2707	2250
Iron, wrought	7.744	.1138	.566	103.0	0.2801	3700
Gold	19.258	.0324			0.6965	2590
Lead	11.352	.0314	.1329	50.0	0.4106	608
Mercury at 32°	13.598	.0333			0.4918	−39
Nickel	8.800	.1086			0.3183	...
Platinum	16.000	.0324			0.5787	3700
Silver	10.474	.056	.0265		0.3788	2000
Steel	7.834	.1165			0.2916	4000
Tin	7.291	.0562	.0439		0.2637	446
Zinc	7.191	.0953	.049	102.0	0.26	680
Stones—					Per cu. ft.	
Chalk	2.784	.2149	.6786		174.0	
Limestone	3.156	.2174	.735		197.0	
Masonry	2.240	.2	.735		140.0	
Marble, gray	2.686	.2694	.735	5.6	168.0	
Marble, white	2.650	.2158	.735	4.4	165.0	
Woods—						
Oak	.86	.57	.73	0.4	54.0	
Pine, white	.55	.65	.73	.17	34.6	
Mineral substances—						
Charcoal, pine	.44	.2415			27.5	
Coal, anthracite	1.43	.2411			88.7	
Coke	1.00	.203			62.5	
Glass, white	2.89	.1977	.5948	1.5	180.7	
Sulphur	2.03	.2026			127.0	
Liquids—						
Alcohol, mean	.9	.6588			57.5	
Oil, petroleum	.88	.31	1.480		55.0	
Steam at 212°	.0006	.847			.050	
Turpentine	.87	.416			54.37	
Water at 62°	1.000	1.000	1.0853		62.35	
Solid—						
Ice at 32°	.922	.504			57.5	
Gases—						
Air at 32°	.00122	.238			.0807	
Oxygen	.00127	.2412			.0892	
Hydrogen	.000089	3.2936			.00559	
Carbonic acid	.00198	.2210			.1234	

Table No. VII.

COEFFICIENTS, STRENGTH OF MATERIALS.

	Ultimate Strength. Tons per Square Inch.			Moduli. Tons per Sq. Inch.	
	Tension.	Compression.	Shearing.	Elasticity.	Rig.
	T	C	S	E	E_1
Cast-iron...............	5½–10¼	25–65	9–13	5000 to 6000	1300 to 2500
Average...............	7	42	11		
American ordnance.......	14	36–58			
Repeatedly melted........	15–20	60–75			
Wrought-iron—					
Finest Low-moor plates: with grain..	27–29		18–22	12,000 to 13,000	5000
across " ..	24				
Bridge-iron: with " ..	22				
across " ..	19				
Bars, finest............	27–29				
Bars, ordinary...........	25	20			
Bars, soft Swedish........	19–24				
Wire...................	25–50				
Steel—					
Mild-steel plates..........	26–32		⅘ of Tension.	12,000 to 13,000	5000 to 5200
Axle and rail steel......	30–45				
Crucible tool- "	40–65				
Chrome "	80				
Tungsten "	72				
Steel wire..........	70				
Piano-wire...............	150			13,000	
Copper—					
Cast....................	10–14			7000	
Rolled.................	15–16	35	10–14	8000	2800
Wire, hard drawn..........	28				
Brass.....................	8–13	5		5500	1500
Wire....................	22			6400	2200
Gun-metal................	11–23			4500–6000	1700
Phosphor bronze...........	15–26			6000	2400
Zinc, cast.................	2–3				
Zinc, rolled...............	7–10			5500	
Tin.......................	2				
Lead......................	0.9	3		1000	
Timber—					
Oak....................	3–7	4	1	800	
White pine...............	1½–3½	2½		600	
Pitch-pine................	4			950	
Ash....................	4–7	2–4	⅓	750	
Beech..................	4–6	4			
Mahogany................	4–7	3¼		650	
Stone—					
Granite..................		2½–5			
Sandstone................		1½–2½			
Limestone...............		1½–3			
Brick...................		¾–6			

From Vol. XXII., Encyc. Britannica.

TABLE No. VIII.

PROPERTIES OF AIR.

Of the Weights of Air, Vapor of Water, and Saturated Mixtures of Air and Vapor of Different Temperatures, under the Ordinary Atmospheric Pressure of 29.921 Inches of Mercury.

Temperature Fahr.	Volume of Dry Air at different temperatures, the volume at 32° being 1000.	Weight of a cubic foot of Dry Air at different temperatures in pounds.	Elastic Face of Vapor in inches of Mercury (Regnault).	Elastic Force of the Air in the mixture of Air and Vapor in inches of Mercury.	Mixtures of Air saturated with Vapor.		
					Weight of the Air in pounds.	Weight of the Vapor in pounds.	Total weight of mixture in pounds.
1	2	3	4	5	6	7	8
0°	.935	.0864	0.044	29.877	.0863	.000079	.086379
12	.960	.0842	.074	29.849	.0840	.000130	.084130
22	.980	.0824	.118	29.803	.0821	.000202	.082302
32	1.000	.0807	.181	29.740	.0802	.000304	.080504
42	1.020	.0791	.267	29.654	.0784	.000440	.078840
52	1.041	.0776	.388	29.533	.0766	.000627	.077227
60	1.057	.0764	.522	29.399	.0751	.000830	.075252
62	1.061	.0761	.556	29.365	.0747	.000881	.075581
70	1.078	.0750	.754	29.182	.0731	.001153	.073509
72	1.082	.0747	.785	29.136	.0727	.001221	.073921
82	1.102	.0733	1.092	28.829	.0706	.001667	.072267
92	1.122	.0720	1.501	28.420	.0684	.002250	.070717
100	1.139	.0710	1.929	27.992	.0664	.002848	.069261
102	1.143	.0707	2.036	27.885	.0659	.002997	.068897
112	1.163	.0694	2.731	27.190	.0631	.003946	.067042
122	1.184	.0682	3.621	26.300	.0599	.005142	.065046
132	1.204	.0671	4.752	25.169	.0564	.006639	.063039
142	1.224	.0660	6.165	23.756	.0524	.008473	.060873
152	1.245	.0649	7.930	21.991	.0477	.010716	.058416
162	1.265	.0638	10.099	19.822	.0423	.013415	.055715
172	1.285	.0628	12.758	17.163	.0360	.016682	.052682
182	1.306	.0618	15.960	13.961	.0288	.020536	.049336
192	1.326	.0609	19.828	10.093	.0205	.025142	.045642
202	1.347	.0600	24.450	5.471	.0109	.030545	.041445
212	1.367	.0591	29.921	0.000	.0000	.036820	.036820

TABLE No. VIII.—Continued.

PROPERTIES OF AIR.

Temperature Fahr.	Mixture of Air saturated with Vapor.		Cubic feet of Vapor from one pound of Water at pressure as in column 4.	B. T. U. absorbed by one cubic foot Dry Air per degree F.	B. T. U. absorbed by one cubic foot Saturated Air per degree F.	Cubic feet Dry Air warmed one degree per B. T. U.	Cubic feet Saturated Air warmed one degree per B. T. U.
	Ratio of Water to Dry Air.	Ratio of Dry Air to Water Vapor					
1	9	10	11	12	13	14	15
0°	.00092	1092.4	....	.02056	.02054	48.5	48.7
12	.00115	646.1	....	.02004	.02006	50.1	50.0
22	.00245	406.4	....	.01961	.01963	51.1	51.0
32	.00379	263.81	3289	.01921	.01924	52.0	51.8
42	.00561	178.18	2252	.01882	.01884	53.2	52.8
52	.00819	122.17	1595	.01847	.01848	54.0	53.8
60	.01251	92.27	1227	.01818	.01822	55.0	54.9
62	.01179	84.79	1135	.01811	.01812	56.2	55.7
70	.01780	64.59	882	.01777	.01794	57.3	56.5
72	.01680	59.54	819	.01777	.01790	58.5	56.8
82	.02361	42.35	600	.01744	.01770	57.2	56.5
92	.03289	30.40	444	.01710	.01751	58.5	57.1
100	.04495	23.66	356	.01690	.01735	59.1	57.8
102	.04547	21.98	334	.01682	.01731	59.5	57.8
112	.06253	15.99	253	.01651	.01711	60.6	58.5
122	.08584	11.65	194	.01623	.01691	61.7	59.1
132	.11771	8.49	151	.01596	.01670	62.5	59.9
142	.16170	6.18	118	.01571	.01652	63.7	60.6
152	.22465	4.45	93.3	.01544	.01654	65.0	60.5
162	.31713	3.15	74.5	.01518	.01656	62.2	60.4
172	.46338	2.16	59.2	.01494	.01658	67.1	60.3
182	.71300	1.402	48.6	.01471	.01687	68.0	59.5
192	1.22643	.815	39.8	.01449	....	68.9	....
202	2.80230	.357	32.7	.01466	....	68.5	....
212	Infinite	.000	27.1	.01406	....	71.4	....

TABLE No. IX.

MOISTURE ABSORBED BY AIR.

THE QUANTITY OF WATER WHICH AIR IS CAPABLE OF ABSORBING TO THE POINT OF MAXIMUM SATURATION, IN GRAINS PER CUBIC FOOT FOR VARIOUS TEMPERATURES.

Degrees Fahr.	Grains in a Cubic Foot.	Degrees Fahr.	Grains in a Cubic Foot.
10	1.1	85	12.43
15	1.31	90	14.38
20	1.56	95	16.60
25	1.85	100	19.12
30	2.19	105	22.0
32	2.35	110	25.5
35	2.59	115	30.0
40	3.06	130	42.5
45	3.61	141	58.0
50	4.24	157	85.0
55	4.97	170	112.5
60	5.82	179	138.0
65	6.81	188	166.0
70	7.94	195	194.0
75	9.24	212	265.0
80	10.73		

TABLE No. X.

RELATIVE HUMIDITY OF THE AIR.

Difference of Temperature of the Air and Dew-point.	Temperature of Air. 32° Fahr.	Temperature of Air. 70° Fahr.	Temperature of Air. 95° Fahr.
0	100	100	100
1	96	97	97
2	92	93	94
3	88	90	91
4	85	87	88
5	81	84	86
6	78	81	83
7	74	78	80
8	71	76	78
9	68	73	75
10	65	71	73
12	60	66	68
14	54	61	64
16	50	57	60
18	45	53	56
20	41	49	53
22	38	46	49
24	34	42	46

TABLE No. XI.

PROPERTIES OF SATURATED STEAM.

[From Charles T. Porter's treatise on *The Richards Steam-engine Indicator.*]

Pressure above zero.	Temperature.	Sensible Heat above zero Fahr.	Latent Heat.	Total Heat above zero Fahr.	Weight of One Cubic Foot.
Lbs. per sq. in.	Fahr. Deg.	B.T.U.	B.T.U.	B.T.U.	Lbs.
1	102.00	102.08	1042.96	1145.03	.0030
2	126.26	126.44	1026.01	1152.45	.0058
3	141.62	141.87	1015.26	1157.13	.0085
4	153.07	153.39	1007.22	1160.62	.0112
5	162.33	162.72	1000.72	1163.44	.0137
6	170.12	170.57	995.24	1165.82	.0163
7	176.91	177.42	990.47	1167.89	.0189
8	182.91	183.48	986.24	1169.72	.0214
9	188.81	188.94	982.43	1171.37	.0239
10	193.24	193.92	978.95	1172.87	.0264
11	197.76	198.49	975.76	1174.25	.0289
12	201.96	202.73	972.80	1175.53	.0313
13	205.88	206.70	970.02	1176.73	.0337
14	209.56	210.42	967.42	1177.85	.0362
15	213.02	213.93	964.97	1178.91	.0387
16	216.29	217.25	962.65	1179.90	.0413
17	219.41	220.40	960.45	1180.85	.0437
18	222.37	223.41	958.34	1181.76	.0462
19	225.20	226.28	956.34	1182.62	.0487
20	227.91	229.03	954.41	1183.45	.0511
21	230.51	231.67	952.57	1184.24	.0536
22	233.01	234.21	950.79	1185.00	.0561
23	235.43	236.67	949.07	1185.74	.0585
24	237.75	239.02	947.42	1186.45	.0610
25	240.00	241.31	945.82	1187.13	.0634
26	242.17	243.52	944.27	1187.80	.0658
27	244.28	245.67	942.77	1188.44	.0683
28	246.32	247.74	941.32	1189.06	.0707
29	248.31	249.76	939.90	1189.67	.0731
30	250.24	251.73	938.52	1190.26	.0755
31	252.12	253.64	937.18	1190.83	.0779
32	253.95	255.51	935.88	1191.39	.0803
33	255.73	257.32	934.60	1191.93	.0827
34	257.47	259.10	933.36	1192.46	.0851
35	259.17	260.83	932.15	1192.98	.0875
36	260.83	262.52	930.96	1193.49	.0899
37	262.45	264.18	929.80	1193.98	.0922
38	264.04	265.80	928.67	1194.47	.0946
39	265.59	267.38	927.56	1194.94	.0970
40	267.12	268.93	926.47	1195.41	.0994
41	268.61	270.46	925.40	1195.86	.1017
42	270.07	271.95	924.35	1196.31	.1041
43	271.50	273.41	923.33	1196.74	.1064
44	272.91	274.85	922.32	1197.17	.1088
45	274.29	276.26	921.33	1197.60	.1111
46	275.65	277.65	920.36	1198.01	.1134

Properties of Saturated Steam—*Continued*.

Pressure above zero.	Temperature.	Sensible Heat above zero Fahr.	Latent Heat.	Total Heat above zero Fahr.	Weight of One Cubic Foot.
Lbs. per sq. in.	Fahr. Deg.	B.T.U.	B.T.U.	B.T.U.	Lbs.
47	276.98	279.01	919.40	1198.42	.1158
48	278.29	280.35	918.46	1198.82	.1181
49	279.58	281.67	917.54	1199.21	.1204
50	280.85	282.96	916.63	1199.60	.1227
51	282.09	284.24	915.73	1199.98	.1251
52	283.32	285.49	914.85	1200.35	.1274
53	284.53	286.73	913.98	1200.72	.1297
54	285.72	287.95	913.13	1201.08	.1320
55	286.89	289.15	912.29	1201.44	.1343
56	288.05	290.33	911.46	1201.79	.1366
57	289.11	291.50	910.64	1202.14	.1388
58	290.31	292.65	909.83	1202.48	.1411
59	291.42	293.79	909.03	1202.82	.1434
60	292.52	294.91	908.24	1203.15	.1457
61	293.59	296.01	907.47	1203.48	.1479
62	294.66	297.10	906.70	1203.81	.1502
63	295.71	298.18	905.94	1204.13	.1524
64	296.75	299.24	905.20	1204.44	.1547
65	297.77	300.30	904.46	1204.76	.1569
66	298.78	301.33	903.73	1205.07	.1592
67	299.78	302.36	903.01	1205.37	.1614
68	300.77	303.37	902.29	1205.67	.1637
69	301.75	304.38	901.50	1205.97	.1659
70	302.71	305.37	900.89	1206.20	.1681
71	303.67	306.35	900.21	1206.56	.1703
72	304.61	307.32	899.52	1206.84	.1725
73	305.55	308.27	898.85	1207.13	.1748
74	306.47	309.22	898.18	1207.41	.1770
75	307.38	310.16	897.52	1207.69	.1792
76	308.29	311.09	896.87	1207.96	.1814
77	309.18	312.01	896.23	1208.24	.1836
78	310.06	312.92	895.59	1208.51	.1857
79	310.94	313.82	894.95	1208.77	.1879
80	311.81	314.71	894.33	1209.04	.1901
81	312.67	315.59	893.70	1209.30	.1923
82	313.52	316.46	893.09	1209.56	.1945
83	314.36	317.33	892.48	1209.82	.1967
84	315.19	318.19	891.88	1210.07	.1988
85	316.02	319.04	891.28	1210.32	.2010
86	316.83	319.88	890.69	1210.57	.2032
87	317.65	320.71	890.10	1210.82	.2053
88	318.45	321.54	889.52	1211.06	.2075
89	319.24	322.36	888.94	1211.31	.2097
90	320.03	323.17	888.37	1211.55	.2118
91	320.82	323.98	887.80	1211.79	.2139
92	321.59	324.78	887.24	1212.02	.2160
93	322.36	325.57	886.68	1212.26	.2182
94	323.12	326.35	886.13	1212.49	.2204
95	323.88	327.13	885.59	1212.72	.2224

Properties of Saturated Steam—*Continued.*

Pressure above zero. Lbs. per sq. in.	Temperature. Fahr. Deg.	Sensible Heat above zero Fahr. B.T.U.	Latent Heat. B.T.U.	Total Heat above zero Fahr. B.T.U.	Weight of One Cubic Foot. Lbs
96	324.63	327.90	885.04	1212.95	.2245
97	325.37	328.67	884.50	1213.18	.2266
98	326.11	329.43	883.97	1213.40	.2288
99	326.84	330.18	883.44	1213.62	.2309
100	327.57	330.93	882.91	1213.84	.2330
101	328.29	331.67	882.39	1214.06	.2351
102	329.00	332.41	881.87	1214.28	.2371
103	329.71	333.14	881.35	1214.50	.2392
104	330.41	333.86	880.84	1214.71	.2413
105	331.11	334.58	880.34	1214.92	.2434
106	331.80	335.30	879.84	1215.14	.2454
107	332.49	336.00	879.34	1215.35	.2475
108	333.17	336.71	878.84	1215.55	.2496
109	333.85	337.41	878.35	1215.76	.2516
110	334.52	338.10	877.86	1215.97	.2537
111	335.19	338.79	877.37	1216.17	.2558
112	335.85	339.47	876.89	1216.37	.2578
113	336.51	340.15	876.41	1216.57	.2599
114	337.16	340.83	875.94	1216.77	.2619
115	337.81	341.50	875.47	1216.97	.2640
116	338.45	342.16	875.00	1217.17	.2661
117	339.10	342.83	874.53	1217.36	.2681
118	339.73	343.48	874.07	1217.56	.2702
119	340.36	344.14	873.61	1217.75	.2722
120	340.99	344.78	873.15	1217.94	.2742
121	341.61	345.43	872.70	1218.13	.2762
122	342.23	346.07	872.25	1218.32	.2783
123	342.85	346.70	871.80	1218.51	.2802
124	343.46	347.34	871.35	1218.69	.2822
125	344.07	347.97	870.91	1218.88	.2842
126	344.67	348.50	870.47	1219.06	.2862
127	345.27	349.21	870.03	1219.25	.2882
128	345.87	349.83	869.59	1219.43	.2902
129	346.45	350.44	869.16	1219.61	.2922
130	347.05	351.05	868.73	1219.79	.2942
131	347.64	351.66	868.30	1219.97	.2961
132	348.22	352.26	867.88	1220.15	.2981
133	348.80	352.86	867.46	1220.32	.3001
134	349.38	353.46	867.03	1220.50	.3020
135	349.95	354.05	866.62	1220.67	.3040
136	350.52	354.64	866.20	1220.85	.3060
137	351.08	355.23	866.79	1221.02	.3079
138	351.75	355.81	865.38	1221.19	.3099
139	352.21	356.39	864.97	1221.36	.3118
140	352.76	356.96	864.56	1221.53	.3138
141	353.31	357.54	864.16	1221.70	.3158
142	353.86	358.11	863.76	1221.87	.3178
143	354.41	358.67	863.36	1222.03	.3199
144	354.96	359.24	862.96	1222.20	.3219

Properties of Saturated Steam—*Continued*.

Pressure above zero Lbs. per sq. in.	Temperature Fahr. Deg.	Sensible Heat above zero Fahr. B.T.U.	Latent Heat. B.T.U.	Total Heat above zero Fahr. B.T.U.	Weight of One Cubic Foot. Lbs.
145	355.50	359.80	862.56	1222.36	.3239
146	356.03	360.85	862.17	1222.53	.3259
147	356.57	360.91	861.78	1222.69	.3279
148	357.10	361.46	861.39	1222.85	.3290
149	357.63	362.01	861.00	1223.01	.3319
150	358.16	362.55	860.62	1223.18	.3340
151	358.68	363.10	860.23	1223.33	.3358
152	359.20	363.64	859.85	1223.49	.3376
153	359.72	364.17	859.47	1223.65	.3394
154	360.23	364.71	859.10	1223.81	.3412
155	360.74	365.24	858.72	1223.97	.3430
156	361.26	365.77	858.35	1224.12	.3448
157	361.76	366.30	857.98	1224.28	.3466
158	362.27	366.82	857.61	1224.43	.3484
159	362.77	367.34	857.24	1224.58	.3502
160	363.27	367.86	856.87	1224.74	.3520
161	363.77	368.38	856.50	1224.89	.3539
162	364.27	368.89	856.14	1225.04	.3558
163	364.76	369.41	855.78	1225.19	.3577
164	365.25	369.92	855.42	1225.34	.3596
165	365.74	370.42	855.06	1225.49	.3614
166	366.23	370.93	854.70	1225.64	.3633
167	366.71	371.43	854.35	1225.78	.3652
168	367.19	371.93	853.99	1225.93	.3671
169	367.68	372.43	853.64	1226.08	.3690
170	368.15	372.93	853.29	1226.22	.3709
171	368.63	373.42	852.94	1226.37	.3727
172	369.10	373.91	852.59	1226.51	.3745
173	369.57	374.40	852.25	1226.66	.3763
174	370.04	374.89	851.90	1226.80	.3781
175	370.51	375.38	851.56	1226.94	.3799
176	370.97	375.86	851.22	1227.08	.3817
177	371.44	376.34	850.88	1227.23	.3835
178	371.90	376.82	850.54	1227.37	.3853
179	372.36	377.30	850.20	1227.51	.3871
180	372.82	377.78	849.86	1227.65	.3889
181	373.27	378.25	849.53	1227.78	.3907
182	373.73	378.72	849.20	1227.92	.3925
183	374.18	379.19	848.86	1228.06	.3944
184	374.63	379.66	848.53	1228.20	.3962
185	375.08	380.13	848.20	1228.33	.3980
186	375.52	380.59	847.88	1228.47	.3999
187	375.97	381.05	847.55	1228.61	.4017
188	376.41	381.51	847.22	1228.74	.4035
189	376.85	381.97	846.90	1228.87	.4053
190	377.29	382.42	846.58	1229.01	.4072
191	377.72	382.88	846.26	1229.14	.4089
192	378.16	383.33	845.94	1229.27	.4107
193	378.59	383.78	845.62	1229.41	.4125

Properties of Saturated Steam—*Continued*.

Pressure above Zero.	Temperature.	Sensible Heat above Zero Fahr.	Latent Heat.	Total Heat above Zero Fahr.	Weight of One Cubic Foot.
Lbs. per sq. in.	Fahr. Deg.	B.T.U.	B.T.U.	B.T.U.	Lbs.
194	379.02	384.23	845.30	1229.54	.4143
195	379.45	384.67	844.99	1229.67	.4160
196	379.97	385.12	844.68	1229.80	.4178
197	380.30	385.56	844.36	1229.93	.4196
198	380.72	386.00	844.05	1230.06	.4214
199	381.15	386.44	843.74	1230.19	.4231
200	381.57	386.88	843.43	1230.31	.4249
201	381.99	387.32	843.12	1230.44	.4266
202	382.41	387.76	842.81	1230.57	.4283
203	382.82	388.19	842.50	1230.70	.4300
204	383.24	388.62	842.20	1230.82	.4318
205	383.65	389.05	841.89	1230.95	.4335
206	384.06	389.48	841.59	1231.07	.4352
207	384.47	389.91	841.29	1231.20	.4369
208	384.88	390.33	840.99	1231.32	.4386
209	385.28	390.75	840.69	1231.45	.4403
210	385.67	391.17	840.39	1231.57	.4421

QUANTITIES OF HEAT CONTAINED IN ONE POUND OF WATER AT VARIOUS TEMPERATURES, RECKONED FROM ZERO, FAHRENHEIT.

[From Charles T. Porter's treatise on *The Richards' Steam-Engine Indicator*.]

Temperature.	Heat contained above Zero.	Temperature.	Heat contained above Zero.	Temperature.	Heat contained above Zero.
Fahr. Deg.	B.T.U.	Fahr. Deg.	B.T.U.	Fahr. Deg.	B.T.U.
35	35.00	155	155.33	275	276.98
40	40.00	160	160.37	280	282.09
45	45.00	165	165.41	285	287.21
50	50.00	170	170.45	290	292.32
55	55.00	175	175.49	295	297.45
60	60.00	180	180.54	300	302.58
65	65.01	185	185.59	305	307.71
70	70.02	190	190.64	310	312.84
75	75.02	195	195.69	315	317.98
80	80.03	200	200.75	320	323.13
85	85.04	205	205.81	325	328.28
90	90.05	210	210.87	330	333.43
95	95.06	215	215.93	335	338.59
100	100.08	220	221.00	340	343.75
105	105.09	225	226.07	345	348.92
110	110.11	230	231.15	350	354.10
115	115.12	235	236.23	355	359.28
120	120.14	240	241.31	360	364.46
125	125.16	245	246.39	365	369.65
130	130.19	250	251.48	370	374.84
135	135.21	255	256.57	375	380.04
140	140.24	260	261.67	380	385.24
145	145.27	265	266.77	385	390.45
150	150.30	270	271.87	390	395.67

Table No. XII.

COMPOSITION OF VARIOUS FUELS OF THE UNITED STATES.

	C.	H.	O.	N.	S.	Moisture.	Ash.	Spec. Grav.
Pennsylvania Anthracite	78.6	2.5	1.7	0.8	0.4	1.2	14.8	1.45
Rhode Island "	85.8	10.5	....	....	3.7	....	....	1.85
Massachusetts "	92.0	6.0	....	....	2.0	....	....	1.78
North Carolina "	83.1	7.8	....	....	9.1	....	....	....
Welsh "	84.2	3.7	2.3	0.9	0.9	1.3	6.7	1.40
Maryland Semi-bituminous	80.5	4.5	2.7	1.1	1.2	1.7	8.3	1.33
Pennsylvania "	75.8	20.2	....	....	....	....	4.0	1.32
" "	59.4	38.8	....	....	....	....	1.8	1.30
Indiana "	70.0	28.0	....	....	....	....	2.0	1.24
" "	52.0	39.0	....	....	....	....	9.0	1.27
Illinois Bituminous	62.6	35.5	....	....	....	....	1.9	1.30
" (Block) Bituminous	58.2	37.1	....	....	....	....	4.7	....
Illinois and Indiana (Cannel) Bituminous	59.5	36.6	....	....	....	....	3.9	1.27
Kentucky (Cannel) Bituminous	48.4	48.8	....	....	....	....	2.8	1.25
Tennessee Bituminous	71.0	17.0	....	....	....	....	12.0	1.45
" "	41.5	56.5	....	....	....	....	2.5	....
Alabama "	54.0	42.6	....	....	1.0	1.2	1.2	....
Virginia "	55.0	41.0	....	....	....	....	4.0	....
" "	74.0	18.6	....	....	....	....	7.4	....
California and Oregon Lignite	50.1	3.9	13.7	0.9	1.5	16.7	13.2	1.32

			Theoretical Value.	
STATE.	KIND OF COAL.	Per Cent. of Ash.	In Heat Units.	In Pounds of Water Evaporated.
Pennsylvania	Anthracite	3.49	14,199	14.70
"	"	6.13	13,535	14.01
"	"	2.90	14,221	14.72
"	Cannel	15.02	13,143	13.60
"	Connelsville	6.50	13,368	13.84
"	Semi-bituminous	10.77	13,155	13.62
"	Stone's Gas	5.00	14,021	14.51
"	Youghiogheny	5.60	14,265	14.76
"	Brown	9.50	12,324	12.75
Kentucky	Caking	2.75	14,391	14.89
"	Cannel	2.00	15,198	16.76
"	"	14.80	13,360	13.84
"	Lignite	7.00	9,326	9.65
Illinois	Bureau County	5.20	13,025	13.48
"	Mercer County	5.60	13,123	13.58
"	Montauk	5.50	12,650	13.10
Indiana	Block	2.50	13,588	14.38
"	Caking	5.66	14,146	14.64
"	Cannel	6.00	13,097	13.56
Maryland	Cumberland	13.98	12,226	12.65
Arkansas	Lignite	5.00	9,215	9.54
Colorado	"	9.25	13,562	14.04
"	"	4.50	13,866	14.35
Texas	"	4.50	12,962	13.41
Washington	"	3.40	11,551	11.96
Pennsylvania	Petroleum	....	20,746	21.47

TABLE No. XII.—Continued.

DRY ANTHRACITE COAL—AVERAGE TABLE OF RESULTS.*

Mine.	Locality.	Volatile Matter.	Ash.	Fixed Carbon.	Specific Gravity.	Heat-units per Pound Dry Coal.	Lbs. of Water Evaporated from and at 212°.
L. V. Buckwheat	W.-Barre, Pa.	6.21	15.5	76.04	1.3	11,959	12.38
" "	Unknown	6.8	13	80.2		12,000	12.42
" "	"	5	14	81		11,800	12.22
D., L. & W.	"	5	11	84		12,400	13.13
Jermyn Stove	Schuy. Co., Pa.	6.08	11.02	82.90	1.425	12,316	13.05
Woodward	Scranton, Pa.	4.06	14.07	81.87	1.42	12,554	13
Cayuga	" "	6.50	9.12	84.38	1.49	12,413	12.77
Mt. Pleasant	" "	7.63	10.78	81.59	1.42	12,458	12.90
L. V. Pea	L. V. region	7.49	16.23	76.28	1.52	11,920	12.37
Forty-foot	Scranton, Pa.	5.07	10.01	84.92	1.41	13,045	13.49
Manville Shaft	" "	6.12	7.33	86.5	1.42	13,064	13.54
Continental	" "	5.78	10.03	84.19	1.615	13,107	13.57
Avondale	Avondale, Pa.	6	6.91	87.78	1.44	13,218	13.71
Oxford	Scranton, Pa.	6.49	2.24	91.27	1.415	13,433	13.91
Mammoth(Buckwh't)	Drifton, Pa. (Slate removed)	2.44	6.97	90.59	1.55	13,720	14.20
Buck Mountain	Cross Creek, Pa (Slate removed)	2.17	5.42	92.41	1.56	14,220	14.72

BITUMINOUS COAL—AVERAGE TABLE OF RESULTS.*

Mine.	Locality.	Volatile Matter.	Ash.	Fixed Carbon.	Specific Gravity.	Heat-units per Pound of Dry Coal.	Lbs. of Water Evaporated from and at 212°.	Percentage of Combustible Matter in Black Smoke.
Gillespie	Gillespie, Ill.	36.26	12.33	51.41	1.26	10,902	11.28	0.56
Auburn Screenings	Sugar Creek, Ill.	37.5	15.2	47.3		11,200	11.6	
LittlePittsburg, Va.	Morgantown, W. Va.	37.5	6.6	55.9		12,800	13.3	
Bernmont	Monongahela R., Pa.	32	8.04	59.96	1.275	13,424	13.9	1.04
Antrim	New Blossburg, Pa.	18.54	11.30	70.16	1.42	13,695	14.18	0.27
Eureka	Clearfield Co., Pa.	23.79	5.82	70.39	1.32	13,897	14.39	0.43
Turtle Creek	Monongahela R., Pa.	34.95	4.33	60.72	1.28	14,450	14.96	0.30
Nova Scotia	No. 2 Slope, U. S	32.38	4.11	63.61	1.31	15,324	15.86	0.27
Reynoldsville	Reynoldsville, Pa.	24.67	5.37	69.96	1.34	15,134	15.67	0.33
Leisenring	Connellsville, Pa.	29.26	6.25	64.49	1.34	15,285	15.82	0.92
Pocahontas	New River, Va.	17.84	3.72	78.45	1.255	15,283	15.82	0.2
Cooperstown	Nova Scotia.	30.75	4.09	65.16	1.345	15,435	15.98	0.5

* From experiments made by Flory and Gilbert at Sibley College, Cornell University. The heat-units are given per pound of dry coal. Coal in ordinary conditions contains from 3 to 10 per cent of moisture, and the results must be reduced accordingly. Seventy per cent of the theoretical heating value represents the average results obtained in practice.

TABLE No. XII.—*Continued.*

ANALYSES OF ASH.

	Specific Grav.	Color of Ash.	Silica.	Alumina.	Oxide Iron.	Lime.	Magnesia.	Loss.	Acids S.&P.
Pennsylvania Anthracite....	1.559	Reddish Buff.	45.6	42.75	9.43	1.41	0.33	0.48	....
" Bituminous....	1.372	Gray.	76.0	21.00	2.60	....	..	0.40	....
Welsh Anthracite............	1.32	....	40.0	44.8	....	12.0	trace	....	2.97
Scotch Bituminous...........	1.26	..	37.6	52.0	....	3.7	1.1	....	5.02
Lignite.....................	1.27	....	19.3	11.6	5.8	23.7	2.6	....	33.8

TABLE No. XIII.

FOR REDUCING BAROMETRIC OBSERVATIONS TO THE FREEZING-POINT.

Reading of Barometer. Inches.	Correction at 10° Fahr. Inches.	Correction at 40° Fahr. Inches.	Correction at 70° Fahr. Inches.	Correction at 90° Fahr. Inches.
	+	−	−	−
27	0.045	0.028	0.100	0.148
27.5	0.046	0.028	0.102	0.151
28.0	0.047	0.029	0.104	0.153
28.5	0.048	0.029	0.106	0.156
29	0.049	0.030	0.108	0.159
29.5	0.050	0.030	0.109	0.162
30.0	0.051	0.031	0.111	0.164
30.5	0.052	0.032	0.113	0.167
31.0	0.053	0.032	0.115	0.170

Table No. XIV.

Thermal Conductivities.

PER DEGREE DIFFERENCE OF THE SUBSTANCE.

Substances.	Thickness, one metre. Calories per sq. metre.	Thickness, one foot. B. T. U. per sq. ft. per hr.	Authority.
Copper	326	594	
Iron	57.5	104	
Zinc	56	102	
Lead	28	50.5	
Air, Oxygen, Nitrogen, Carbonic oxide,	0.0177	0.323	Clausius and Maxwell, according to kinetic theory.
Carbonic acid	0.0137	0.0249	Do. do. do.
Hydrogen	0.0125	0.0227	Do. do. do.
Glass	0.82	1.49	Péclet.
Porphyritic trachyte	2.12	3.86	Aryton & Perry, Phil. Mag., 1878, first half year, p. 241.
Marble	3.13	5.67	Péclet.
Underground strata	1.8	3.29	Forbes and Wm. Thomson.
Limestone	1.82	3.31	
Sandstone of Craigleith Quarry	3.84	7.0	Do. do. do.
Trap-rock of Calton Hill	1.5	2.73	Do. do. do.
Sand of experimental garden	0.94	1.72	Do. do. do.
Water	0.72	1.82	J. P. Bottomley.
Fir across fibres	0.093	0.169	Péclet in Everett's Units and Physical Constants.
" along fibres	0.169	0.308	Do. do. do.
Walnut across fibres	0.105	0.192	Do. do. do.
" along fibres	0.173	0.315	Do. do. do.
Oak across fibres	0.212	0.387	Do. do. do.
Cork	0.105	0.192	Do. do. do.
Hempen cloth, new	0.052	0.095	Do. do. do.
" " old	0.043	0.078	Do. do. do.
Writing paper, white	0.043	0.078	Do. do. do.
Gray paper, unsized	0.0337	0.0515	Do. do. do.
Calico, new, of all densities	0.05	0.91	Do. do. do.
Wool, carded, of all densities	0.044	0.08	Do. do. do.
Finely carded cotton-wool	0.04	0.073	Do. do. do.
Eider-down	0.039	0.017	Do. do. do.
Indian rubber	0.17	0.308	
Brick dust	0.15	0.272	
Wood ashes	0.06	0.109	
Coke	4.96	9.01	

TABLE No. XV.

WROUGHT-IRON WELDED STEAM-, GAS-, AND WATER-PIPE.

TABLE OF STANDARD DIMENSIONS.

Diameter.			Thickness.	Circumference.		Transverse Areas.			Length of Pipe per Square Foot of		Length of Pipe Containing One Cubic Foot.	Nominal Weight per Foot.	Number of Threads per Inch of Screw.
Nominal Internal.	Actual External.	Actual Internal.		External.	Internal.	External.	Internal.	Metal.	External Surface.	Internal Surface.			
Inches.	Inches.	Inches.	Inches.	Inches.	Inches.	Sq. In.	Sq. In.	Sq. In.	Feet.	Feet.	Feet.	Pounds.	
⅛	.405	.27	.068	1.272	.848	.129	.0573	.0717	9.44	14.15	2513.	.241	27
¼	.54	.364	.088	1.696	1.144	.229	.1041	.1249	7.075	10.49	1383.3	.42	18
⅜	.675	.494	.091	2.121	1.552	.358	.1917	.1663	5.657	7.73	751.2	.559	18
½	.84	.623	.109	2.639	1.957	.554	.3018	.2492	4.547	6.13	472.4	.837	14
¾	1.05	.824	.113	3.299	2.589	.866	.5333	.3337	3.637	4.635	270.0	1.115	14
1	1.315	1.048	.134	4.131	3.292	1.358	.8626	.4954	2.904	3.645	166.9	1.668	11½
1¼	1.66	1.38	.14	5.215	4.335	2.164	1.496	.668	2.301	2.768	96.25	2.244	11½
1½	1.9	1.611	.145	5.969	5.061	2.835	2.038	.797	2.01	2.371	70.66	2.678	11½
2	2.375	2.067	.154	7.461	6.494	4.43	3.356	1.074	1.608	1.848	42.91	3.609	11½
2½	2.875	2.468	.204	9.032	7.753	6.492	4.784	1.708	1.328	1.547	30.1	5.739	8
3	3.5	3.067	.217	10.996	9.636	9.621	7.388	2.243	1.091	1.245	19.5	7.536	8
3½	4.0	3.548	.226	12.566	11.146	12.566	9.887	2.679	.955	1.077	14.57	9.001	8
4	4.5	4.026	.237	14.137	12.648	15.904	12.73	3.174	.849	.955	11.31	10.665	8
4½	5.0	4.508	.246	15.708	14.162	19.635	15.061	3.674	.764	.848	9.02	12.34	8
5	5.563	5.045	.259	17.477	15.849	24.306	19.99	4.316	.687	.757	7.2	14.502	8
6	6.625	6.065	.28	20.813	19.054	34.472	28.888	5.584	.577	.66	4.98	18.762	8
7	7.023	7.023	.301	23.955	22.063	45.664	38.738	6.926	.501	.544	3.72	23.271	8
8	8.625	7.982	.322	27.096	25.076	58.426	50.04	8.386	.478	.478	2.88	28.177	8
9	9.625	8.937	.344	30.238	28.076	72.76	62.73	10.03	.397	.427	2.29	33.701	8
10	10.75	10.019	.366	33.772	31.477	90.763	78.839	11.924	.355	.382	1.82	40.065	8
11	11.75	11.0	.375	36.914	34.558	108.434	95.033	13.401	.335	.347	1.51	45.028	8
12	12.75	12.0	.375	40.055	37.7	127.677	113.098	14.579	.299	.319	1.27	48.085	8
13	13.25	13.25	.375	43.982	41.626	153.938	137.887	16.051	.273	.288	1.04	53.921	8
14	14.0	13.25	.375	47.124	44.768	176.715	159.485	17.23	.255	.268	.903	57.893	8
15	15.0	14.25	.375	50.26	48.48	201.06	187.04	14.02	.239	.248	.77	47.11	8
16	16.0	15.43	.284	53.41	51.52	226.93	211.24	15.74	.225	.233	.68	52.89	8
17	18.0	17.32	.34	56.55	54.11	254.47	235.61	18.86	.212	.241	.61	63.37	8

TABLE No. XVI.

WEIGHT OF WATER PER CUBIC FOOT FOR VARIOUS TEMPERATURES.*

Weight of Water per Cubic Foot, from 32° to 212° F., and Heat-units per Pound, Reckoned Above 32° F.

Temperature, Deg. F.	Weight, Lbs. per Cubic Foot.	Heat-units.	Temperature, Deg. F.	Weight, Lbs. per Cubic Foot.	Heat-units.	Temperature, Deg. F.	Weight, Lbs. per Cubic Foot.	Heat-units.	Temperature, Deg. F.	Weight, Lbs. per Cubic Foot.	Heat-units.
32	62.42	0.	78	62.25	46.03	123	61.68	91.16	168	60.81	136.44
33	62.42	1.	79	62.24	47.03	124	61.67	92.17	169	60.79	137.45
34	62.42	2.	80	62.23	48.04	125	61.65	93.17	170	60.77	138.45
35	62.42	3.	81	62.22	49.04	126	61.63	94.17	171	60.75	139.46
36	62.42	4.	82	62.21	50.04	127	61.61	95.18	172	60.73	140.47
37	62.42	5.	83	62.20	51.04	128	61.60	96.18	173	60.70	141.48
38	62.42	6.	84	62.19	52.04	129	61.58	97.19	174	60.68	142.49
39	62.42	7.	85	62.18	53.05	130	61.56	98.19	175	60.66	143.50
40	62.42	8.	86	62.17	54.05	131	61.54	99.20	176	60.64	144.51
41	62.42	9.	87	62.16	55.05	132	61.52	100.20	177	60.62	145.52
42	62.42	10.	88	62.15	56.05	133	61.51	101.21	178	60.59	146.52
43	62.42	11.	89	62.14	57.05	134	61.49	102.21	179	60.57	147.53
44	62.42	12.	90	62.13	58.06	135	61.47	103.22	180	60.55	148.54
45	62.42	13.	91	62.12	59.06	136	61.45	104.22	181	60.53	149.55
46	62.42	14.	92	62.11	60.06	137	61.43	105.23	182	60.50	150.56
47	62.42	15.	93	62.10	61.06	138	61.41	106.23	183	60.48	151.57
48	62.41	16.	94	62.09	62.06	139	61.39	107.24	184	60.46	152.58
49	62.41	17.	95	62.08	63.07	140	61.37	108.25	185	60.44	153.59
50	62.41	18.	96	62.07	64.07	141	61.36	109.25	186	60.41	154.60
51	62.41	19.	97	62.06	65.07	142	61.34	110.26	187	60.39	155.61
52	62.40	20.	98	62.05	66.07	143	61.32	111.26	188	60.37	156.62
53	62.40	21.01	99	62.03	67.08	144	61.30	112.27	189	60.34	157.63
54	62.40	22.01	100	62.02	68.08	145	61.28	113.28	190	60.32	158.64
55	62.39	23.01	101	62.01	69.08	146	61.26	114.28	191	60.29	159.65
56	62.39	24.01	102	62.00	70.09	147	61.24	115.29	192	60.27	160.67
57	62.39	25.01	103	61.99	71.09	148	61.22	116.29	193	60.25	161.68
58	62.38	26.01	104	61.97	72.09	149	61.20	117.30	194	60.22	162.69
59	62.38	27.01	105	61.96	73.10	150	61.18	118.31	195	60.20	163.70
60	62.37	28.01	106	61.95	74.10	151	61.16	119.31	196	60.17	164.71
61	62.37	29.01	107	61.93	75.10	152	61.14	120.32	197	60.15	165.72
62	62.36	30.01	108	61.92	76.10	153	61.12	121.33	198	60.12	166.73
63	62.36	31.01	109	61.91	77.11	154	61.10	122.33	199	60.10	167.74
64	62.35	32.01	110	61.89	78.11	155	61.08	123.34	200	60.07	168.75
65	62.34	33.01	111	61.88	79.11	156	61.04	124.35	201	60.05	169.77
66	62.34	34.02	112	61.86	80.12	157	61.06	125.35	202	60.02	170.78
67	62.33	35.02	113	61.85	81.12	158	61.02	126.36	203	60.00	171.79
68	62.33	36.02	114	61.83	82.13	159	61.00	127.37	204	59.97	172.80
69	62.32	37.02	115	61.82	83.13	160	60.98	128.37	205	59.95	173.81
70	62.31	38.02	116	61.80	84.13	161	60.96	129.38	206	59.92	174.83
71	62.31	39.02	117	61.78	85.14	162	60.94	130.39	207	59.89	175.84
72	62.30	40.02	118	61.77	86.14	163	60.92	131.40	208	59.87	176.85
73	62.29	41.02	119	61.75	87.15	164	60.90	132.41	209	59.84	177.86
74	62.28	42.03	120	61.74	88.15	165	60.87	133.41	210	59.82	178.87
75	62.28	43.03	121	61.72	89.15	166	60.85	134.42	211	59.79	179.89
76	62.27	44.03	122	61.70	90.16	167	60.83	135.43	212	59.76	180.90
77	62.26	45.03									

WEIGHT OF WATER AT TEMPERATURES ABOVE 212° F.

Porter (Richards' "Steam-engine Indicator," p. 52) says that nothing is known about the expansion of water above 212° F. Applying formulæ derived from experiments made at temperatures below 212° F., however, the weight and volume above 212° F. may be calculated, but in the absence of experimental data we are not certain that the formulæ hold good at higher temperatures.

---

* Kent's "Pocket-book for Mechanical Engineers."

TABLE NO. XVI.—Continued.

Thurston, in his "Engine and Boiler Trials," gives a table from which we take the following (neglecting the third decimal place given by him):

Temperature, Deg. F.	Weight, Lbs. per Cubic Foot.	Temperature, Deg. F.	Weight, Lbs. per Cubic Foot.	Temperature, Deg. F.	Weight, Lbs. per Cubic Foot.	Temperature, Deg. F.	Weight, Lbs. per Cubic Foot.	Temperature, Deg. F.	Weight, Lbs. per Cubic Foot.
212	59.71	280	57.90	350	55.52	420	52.86	490	50.03
220	59.64	290	57.59	360	55.16	430	52.47	500	49.61
230	59.37	300	57.26	370	54.79	440	52.07	510	49.20
240	59.10	310	56.93	380	54.41	450	51.06	520	48.78
250	58.81	320	56.58	390	54.03	460	51.26	530	48.36
260	58.52	330	56.24	400	53.64	470	50.85	540	47.94
270	58.21	340	55.88	410	53.26	480	50.44	550	47.52

Box on Heat gives the following:

Temperature F	212°	250°	300°	350°	400°	450°	500°	600°
Lbs. per cubic foot	59.82	58.85	57.42	55.94	54.34	52.70	51.02	47.64

TABLE NO. XVII.

PRESSURE OF WATER PER SQUARE INCH FOR DIFFERENT HEIGHTS IN FEET.*

At 60° F. 1 foot head = 0.433 lb. per square inch, .433 × 144 = 62.352 lbs. per cubic foot.

Head, Feet.	0	1	2	3	4	5	6	7	8	9
0		0.433	0.866	1.299	1.732	2.165	2.598	3.031	3.424	3.897
10	4.330	4.763	5.196	5.629	6.062	6.495	6.928	7.361	7.794	8.227
20	8.660	9.093	9.526	9.959	10.392	10.825	11.258	11.691	12.124	12.557
30	12.990	13.423	13.856	14.289	14.722	15.155	15.588	16.021	16.454	16.887
40	17.320	17.753	18.186	18.619	19.052	19.485	19.918	20.351	20.784	21.217
50	21.650	22.083	22.516	22.949	23.382	23.815	24.248	24.681	25.114	25.547
60	25.980	26.413	26.846	27.279	27.712	28.145	28.578	29.011	29.444	29.877
70	30.310	30.743	31.176	31.609	32.042	32.475	32.908	33.341	33.774	34.207
80	34.640	35.073	35.506	35.939	36.372	36.805	37.238	37.671	38.104	38.537
90	38.970	39.403	39.836	40.269	40.702	41.135	41.568	42.001	42.436	42.867

HEAD IN FEET OF WATER, CORRESPONDING TO PRESSURES IN POUNDS PER SQUARE INCH.

1 lb. per square inch = 2.30947 feet head, 1 atmosphere = 14.7 lbs. per square inch = 33.95 feet head.

Pressure.	0	1	2	3	4	5	6	7	8	9
0			2.309	4.619	6.928	9.238	11.547	13.857	16.166	18.476	20.785
10	23.0047	25.404	27.714	30.023	32.333	34.642	36.952	39.261	41.570	43.880
20	46.1801	48.499	50.808	53.118	55.427	57.737	60.046	62.356	64.665	66.975
30	69.2841	71.594	73.903	76.213	78.522	80.831	83.141	85.450	87.760	90.069
40	92.3788	94.688	96.998	99.307	101.62	103.93	106.24	108.55	110.85	113.16
50	115.4735	117.78	120.09	122.40	124.71	126.02	129.33	131.64	133.95	136.26
60	138.5682	140.88	143.19	145.50	147.81	150.12	152.42	154.73	157.04	159.35
70	161.6629	163.97	166.28	168.59	170.90	173.21	175.52	177.83	180.14	182.45
80	184.7576	187.07	189.38	191.69	194.00	196.31	198.61	200.92	203.23	205.54
90	207.8523	210.16	212.47	214.78	217.09	219.40	221.71	224.02	226.33	228.64

* Kent's "Pocket-book."

TABLE No. XVIII.

CONTENTS IN CUBIC FEET AND U. S. GALLONS OF PIPES AND CYLINDERS OF VARIOUS DIAMETERS AND 1 FOOT IN LENGTH.*

1 gallon = 231 cubic inches.   1 cubic foot = 7.4805 gallons.

Diameter in Inches.	For 1 Foot in Length.		Diameter in Inches.	For 1 Foot in Length.		Diameter in Inches.	For 1 Foot in Length.	
	Cu. Ft., also Area in Sq. Ft.	U. S. Gals., 231 Cu. In.		Cu. Ft., also Area in Sq. Ft.	U. S. Gals., 231 Cu. In.		Cu. Ft., also Area in Sq. Ft.	U. S. Gals., 231 Cu. In.
1/4	.0003	.0025	6¾	.2485	1.859	19	1.969	14.73
5/16	.0005	.004	7	.2673	1.999	19½	2.074	15.51
3/8	.0008	.0057	7¼	.2867	2.145	20	2.182	16.32
7/16	.001	.0078	7½	.3068	2.295	20½	2.292	17.15
1/2	.0014	.0102	7¾	.3276	2.45	21	2.405	17.99
9/16	.0017	.0129	8	.3491	2.611	21½	2.521	18.86
5/8	.0021	.0159	8¼	.3712	2.777	22	2.640	19.75
11/16	.0026	.0193	8½	.3941	2.948	22½	2.761	20.66
3/4	.0031	.0230	8¾	.4176	3.125	23	2.885	21.58
13/16	.0036	.0269	9	.4418	3.305	23½	3.012	22.53
7/8	.0042	.0312	9¼	.4667	3.491	24	3.142	23.50
15/16	.0048	.0359	9½	.4922	3.682	25	3.409	25.50
1	.0055	.0408	9¾	.5185	3.879	26	3.687	27.58
1¼	.0085	.0638	10	.5454	4.08	27	3.976	29.74
1½	.0123	.0918	10¼	.5730	4.286	28	4.276	31.99
1¾	.0167	.1249	10½	.6013	4.498	29	4.587	34.31
2	.0218	.1632	10¾	.6303	4.715	30	4.909	36.72
2¼	.0276	.2066	11	.66	4.937	31	5.241	39.21
2½	.0341	.2550	11¼	.6903	5.164	32	5.585	41.78
2¾	.0412	.3085	11½	.7213	5.396	33	5.940	44.43
3	.0491	.3672	11¾	.7530	5.633	34	6.305	47.16
3¼	.0576	.4309	12	.7854	5.875	35	6.681	49.98
3½	.0668	.4998	12½	.8522	6.375	36	7.069	52.88
3¾	.0767	.5738	13	.9218	6.895	37	7.467	55.86
4	.0873	.6528	13½	.994	7.436	38	7.876	58.92
4¼	.0985	.7369	14	1.069	7.997	39	8.296	62.06
4½	.1134	.8263	14½	1.147	8.578	40	8.727	65.28
4¾	.1231	.9206	15	1.227	9.180	41	9.168	68.58
5	.1364	1.020	15½	1.310	9.801	42	9.621	71.97
5¼	.1503	1.125	16	1.396	10.44	43	10.085	75.44
5½	.1650	1.234	16½	1.485	11.11	44	10.559	78.99
5¾	.1803	1.349	17	1.576	11.79	45	11.045	82.62
6	.1963	1.469	17½	1.670	12.49	46	11.541	86.33
6¼	.2131	1.594	18	1.768	13.22	47	12.048	90.13
6½	.2304	1.724	18½	1.867	13.96	48	12.566	94.00

To find the capacity of pipes greater than the largest given in the table look in the table for a pipe of one half the given size, and multiply its capacity by 4; or one of one third its size, and multiply its capacity by 9, etc.

To find the *weight* of water in any of the given sizes multiply the capacity in cubic feet by 62¼ or the gallons by 8⅓, or, if a closer approximation is required, by the weight of a cubic foot of water at the actual temperature in the pipe.

Given the dimensions of a cylinder in inches to find its capacity in U. S. gallons: square the diameter, multiply by the length and by .0034. If d = diam. l = length, gallons = $\frac{d^2 \times .7854 \times l}{231} = .0034\, d^2 l$.

---

* Kent's "Pocket-book."

TABLE No. XIX.

EQUALIZATION OF PIPE AREAS.*

Sizes of Pipe.	Number of small pipes required to make area equivalent to one larger pipe, with allowance for friction.													
	½ in.	¾ in.	1 in.	1¼ in.	1½ in.	2 in.	2½ in.	3 in.	3½ in.	4 in.	4½ in.	5 in.	6 in.	7 in.	8 in.
½ inch	1	2.0	3.7	7.6	11.3	19	37	55	80	108	146	188	290	427	595
¾ "	....	1	1.8	3.7	5.4	9.2	16.7	25.5	39	53	70	90	143	210	295
1 "	....	....	1	2.0	3.1	5.1	9.3	14.7	27	30	39	53	80	117	165
1¼ "	....	....	....	1	1.5	2.6	4.5	7.3	10.6	14.7	19.5	25	39	57	80
1½ "	....	....	....	....	1	1.7	3.1	4.7	7.1	9.8	13.4	16.8	26	38	54
2 "	....	....	....	....	....	1	1.83	2.9	4.1	5.8	7.8	9.9	16	23	32
2½ "	....	....	....	....	....	....	1	1.7	2.5	3.5	4.7	5.9	9.3	13.7	19
3 "	....	....	....	....	....	....	....	1	1.5	2.4	2.7	3.5	5.4	7.7	11
3½ "	....	....	....	....	....	....	....	....	1	1.4	1.8	2.5	3.7	5.8	7.6
4 "	....	....	....	....	....	....	....	....	....	1	1.3	1.7	2.7	4.1	5.5
4½ "	....	....	....	....	....	....	....	....	....	....	1	1.25	2	3.3	4.1
5 "	....	....	....	....	....	....	....	....	....	....	....	1	1.6	2.5	3.2
6 "	....	....	....	....	....	....	....	....	....	....	....	....	1	1.5	2
7 "	....	....	....	....	....	....	....	....	....	....	....	....	....	1	1.4
8 "	....	....	....	....	....	....	....	....	....	....	....	....	....	....	1

* Especially computed.

TABLE No. XX.

TEMPERATURES OF VARIOUS LOCALITIES.

COMPILED FROM OBSERVATIONS OF THE SIGNAL SERVICE, U. S. A., AND BLODGETT'S "CLIMATOLOGY OF THE UNITED STATES."

NOTE.—In the United States the comfortable temperature of the air in occupied rooms is generally 70 degrees when walls have the same temperature.

Station.	No. of months fire is required.	Mean temp. of cold months.	Av. No. of deg. temp. to be raised.	Max. No. deg. temp. to be raised.	Minimum temperature F°.
Albany, N. Y.	7	35	35	87	− 17
Baltimore, Md.	6	39	31	72	− 2
Boston, Mass.	7	37	33	81	− 11
Buffalo, N. Y.	8	35	35	83	− 13
Burlington, Vt.	7	32	38	90	− 20
Chicago, Ill.	7	35	35	90	− 20
Charleston, S. C.	3	52	18	47	+ 23
Cincinnati, O.	7	42	28	77	− 7
Cleveland, O.	7	38	32	83	− 13
Detroit, Mich.	7	35	35	90	− 20
Duluth, Minn.	8	28	42	108	− 38
Indianapolis, Ind.	7	41	21	88	− 18
Key West, Fla.	0	0	0	26	+ 44
Leavenworth, Kan.	6	37	33	90	− 20
Louisville, Ky.	6	42	28	80	− 10
Memphis, Tenn.	5	39	31	68	+ 2
Milwaukee, Wis.	8	37	32	95	− 25
New Orleans, La.	0	0	0	44	+ 26
New York, N. Y.	7	40	30	76	− 6
Philadelphia, Pa.	7	40	30	75	− 5
Pittsburg, Pa.	7	39	31	82	− 12
Portland, Me.	8	33	37	82	− 12
Portland, Ore.	6	43	27	67	+ 3
San Francisco, Cal.	4	53	17	34	+ 36
St. Louis, Mo.	5	37	33	86	− 16
St. Paul, Minn.	7	25	45	102	− 32
Washington, D. C.	5	40	30	73	+ 3
Wilmington, N. C.	4	50	20	55	+ 15

TABLE XXI.

PRICE-LIST FITTINGS. VALVES. ETC.

Size, inches		1/8	1/4	3/8	1/2	3/4	1	1¼	1½	2	2½	3	3½	4	4½	5	6
Elbows, cast-iron, R. H	Each	....	$0.04	$0.05	$0.06	$0.09	$0.13	$0.20	$0.25	$0.40	$0.75	$1.10	$1.35	$1.80	$2.50	$2.85	$3.90
" " R. & L	"	....	.05	.06	.07	.11	.16	.23	.29	.46	.85	1.25	....	....	....	....	....
" " reducing	"	....	....	....	.07	.11	.16	.23	.29	.46	.85	1.25	1.50	2.10	3.00	3.25	4.50
" " 45°	"	....	....	.10	.10	.15	.20	.26	.35	.50	1.30	1.60	1.90	2.50	3.50	4.50	5.50
" " No. 1, water	"	....	....	....	....	....	.25	.35	.45	.60	1.00	1.50	2.00	2.50	3.50	4.50	6.50
Tees, cast-iron	"	....	.06	.07	.09	.13	.20	.30	.38	.60	1.10	1.50	2.00	2.50	3.50	4.00	5.50
" " reducing	"	....	....	....	.11	.15	.23	.35	.44	.70	1.25	1.75	2.30	2.90	4.00	4.60	6.35
" " Nos. 2 and 3, water	"	....	....	....	....	....	.38	.52	.68	.90	1.50	2.25	3.00	3.75	5.25	6.75	9.75
Crosses, cast-iron	"	....	....	....	.12	.18	.28	.40	.50	.80	1.50	2.20	2.70	3.50	5.00	5.70	7.80
" " reducing	"	....	....	....	.14	.21	.32	.46	.58	.92	1.70	2.50	3.00	4.00	6.00	6.60	9.00
Plugs, cast-iron	"	....	.03	.03	.04	.05	.06	.10	.13	.20	.35	.50	.75	.85	1.35	1.75	2.40
Bushings	"	....	....	.05	.06	.07	.09	.13	.17	.27	.42	.60	.80	1.00	1.50	1.85	2.50
Caps	"	....	.03	.04	.05	.08	.12	.16	.24	.32	.45	.85	1.00	1.20	1.60	2.00	2.35
Lock-nuts	"	....	.02	.03	.04	.05	.07	.09	.11	.18	.40	.50	.70	.95	1.25	1.35	1.90
Reducers	"	....	.03	03	.05	.10	.16	.20	28	.45	.70	1.00	1.50	1.85	2.75	3.00	4.00
Unions	"	....	.15	.18	.20	.28	.34	.46	.60	.80	1.50	2.10	3.00	4.00	....	....	....
Flange-unions, cast-iron	"	....	....	....	....	.65	.70	.85	1.15	1.50	1.75	2.25	2.75	3.15	4.50	5.00	6.50
Return-bends, cast-iron, C. P	"	....	....	....	.10	.15	.22	.34	.45	.75	1.50	2.25	....	....	....	....	....
" " " O. P	"	....	....	....	....	.20	.30	.48	.68	1.15	1.75	2.75	....	....	....	....	....
Nipples, short	"	$0.05	.05	.06	.07	.09	.10	.14	.17	.25	.56	.75	1.00	1.25	1.75	2.00	2.75
" long	"	.07	.07	.09	.10	.11	.15	.20	.25	.35	.75	.95	1.25	1.60	2.25	2.60	3.60
Couplings, wrought-iron	"	.05	.05	.06	.07	.10	.13	.17	.21	.28	.40	.60	.80	1.00	1.50	1.65	2.40
Expansion pipe-hangers	"	....	....	....	....	.22	.25	.30	.38	.44	.55	.65	.90	1.15	1.40	1.50	2.25
Ceiling-plates	"	....	....	....	....	.16	.18	.20	.25	.30	.35	.50	.60	.70	....	....	....
Floor-plates	"	....	....	....	.06	.08	.10	.15	.18	.23	.30	.40	.50	.65	....	1.00	1.40
Globe and angle-valves, brass	"	.60	.60	.75	1.00	1.35	1.80	2.80	3.90	5.00	11.25	16.00	30.00	40.00	....	....	....
Globe and angle-valves, comp. discharge	"	....	1.10	1.25	1.60	2.20	2.80	4.00	5.50	8.00	15.75	22.00	....	....	....	....	....
Check-valves, brass	"	.50	.50	60	85	1.15	1.55	2.30	3.25	5.20	10.00	14.00	....	....	....	....	....
Safety-valves, brass	"	....	2.00	2.25	2.75	3.50	5.00	7.00	8.50	12.00	20.00	30.00	....	....	....	....	....
Steam-cocks, brass	"	.70	.70	.75	1.10	1.50	2.25	3.75	4.80	7.25	14.00	20.00	....	....	....	....	....
Gas-service cocks, brass	"	....	.55	.65	.75	1.00	1.40	2.20	3.00	5.00	....	....	....	....	....	....	....
Gas-meter cocks, brass	"	....	....	.70	.85	1.20	1.70	2.60	3.60	5.75	....	....	....	....	....	....	....
Gas-service cocks, brass, heavy, with stops	"	....	....	....	1.05	1.55	2.20	3.25	4.50	7.40	....	....	....	....	....	....	....
Globe and angle-valves, iron, screwed	"	....	....	....	....	....	2.00	2.50	3.50	5.00	7.50	10.50	....	....	....	....	....
" " " " " flanged	"	....	....	....	....	....	3.00	3.75	5.00	6.75	9.50	13.50	....	....	....	....	....
" " " " " yoke, screwed	"	....	....	....	....	....	....	....	....	8.00	10.50	14.50	18.00	21.00	28.00	32.00	44.00
" " " " " " flanged	"	....	....	....	....	....	....	....	....	9.75	12.50	17.50	21.50	25.00	32.00	36.00	49.00
Check-valves, iron, screwed	"	....	....	....	....	....	1.50	2.25	2.75	3.75	6.25	9.75	12.75	15.00	20.00	24.00	33.00
" " flanged	"	....	....	....	....	....	2.50	3.50	4.25	5.50	8.25	12.75	16.25	19.00	24.00	28.00	38.00
Safety-valves, iron, screwed	"	....	....	....	....	....	3.50	5.00	6.00	8.00	13.00	18.00	24.00	30.00	36.00	44.00	60.00
" " flanged	"	....	....	....	....	....	....	....	....	10.50	16.00	22.50	29.25	36.00	42.00	50.00	67.50
All-iron cocks	"	....	....	....	.80	.90	1.25	1.50	2.00	2.60	4.50	6.50	12.00	16.00	....	33.00	45.00
Iron cocks, brass plug	"	....	....	....	1.20	1.60	2.00	2.75	4.00	5.00	9.50	13.50	30.00	40.00	....	70.00	95.00

TABLE XXI.—*Continued.*

PRICE-LIST OF PIPE, ADOPTED JAN. 29, 1895.

Diameter, inches		1/8	1/4	3/8	1/2	3/4	1	1¼
Butt-weld, black	Each	$0 05½	$0.05½	$0.05½	$0.07	$0.08½	$0.11½	$0.15½
" galvanized	"	.08	.07½	.07½	.09½	.11½	.16	.22

Diameter, inches		1½	2	2½	3	3½	4	4½	5	6
Lap-weld, black	Each	$0.26	$0.35	$0.52	$0.68	$0 81	$0.95	$1.25	$1.42	$1.85
" galvanized	"	.31	.42	.62	.80	.98	1.16	1.50	1.75	2.20

For selected pipe, or pipe cut to specified lengths, the discount will be five (5) per cent less in the gross (i.e., 5 per cent higher in gross list discount) than on regular pipe.

On pipe lighter than standards, or without threads or sockets, no extra allowance will be made.

INDEX

A

	PAGE
Absolute pressure, defined	120
zero	6
Air, analysis of	27
, change of, in a room	51
delivered in pipes, table	286
, discharge, different temperatures, table of	45
, discharge, different pressures	42
, flues, indirect heating	232
, force required for moving	35
, humidity of	29
, inlet, location of	46
, measurement of velocity	37
, microbe organisms in	23
, properties of, table	381
, relation between velocity and force, table	45
required for ventilation	31, 202
-supply for furnace	272
-trap	180
-valves	102
, velocity of, how computed	39
, weight of	22
Anemometer, description of	37
Angle-valves	100
Area of main pipe	192
of pipes, hot-water heating	229
of safety-valve	150
of steam-pipes	222, 226
of ventilating-flues	52
Areas and circumferences of circles	376
Argon	27
Atmosphere, composition and pressure of	21

B

Bacteria in air	23
Bailey, L. H., tests	245
Barometer	21
Blower, capacity of	292

	PAGE
Blow-off cocks and valves	157
Boiler explosions	172
Boiler horse-power, standard established	122
Boiler-setting, depth of foundation for	145
Boiler, size of, hot-blast heating, practical construction	292–4
Boiler specifications	326
Boilers, appliances for	147
, brick settings for	143
, fire-tube	128
, heating, classes of	130
, forms of	129
, for soft coal	142
, setting of	147
, horizontal tubular	130
, locomotive and marine	131
, portable settings for	147
, power	128
, sectional	140
, steam-heating, care of	169
, steam, requisites of	127
, tubular	138
, types of	128
, water-tube	133, 138
Boiling-points, gases	9
Books, list of, on heating	353
Bourdon pressure-gauge	154
Branch tees	97
Breeching	145
Brick settings for boilers	143
Bucket traps	165
Buildings, loss of heat from	54

C

Calorie, defined	4
Capacity of boiler	122
Capital invested in manufacture	2
Carbonic acid, CO_2, or carbon dioxide	24
Carbonic oxide, CO	26
Ceiling and floor plates	98
Check-valves	102
Chimneys, form of	160
, size of	161
Chimney-tops	162
Coal, soft, kind of heater for	142
Cocks and valves	98
Cocks, blow-off	157
, try	152

	PAGE
Combination heaters	188
Conduction of heat	188
Connection to radiators in hot-water heating systems	195
Contents of pipe in gallons	396
Convected heat	60
Convection	19
, formula for	63
Cooling of rooms	300
Couplings, right and left	93
, union	93
Coverings for pipes	198

D

Damper-regulators	156
Dead-weight safety-valve	149
Density and weight per cubic foot of water	394
Diagram of heat from radiating surfaces	204
Diathermancy, defined	16
Diffusion, amount of	35
of gases	24
of radiant heat	17
Dimensions of steam, gas, and water pipe, table	399
Direct and indirect heating	60
Drip-pipes	228
Drop-tubes for boilers	139

E

Elbows and bends	94
Electrical and heat equivalents	301
heaters	306
heating	301
expense	303
units, value of	5
Equalization of pipe-areas, use of table	286
table for air	287
for steam	397
Equalizing valve	168
Exhaust-steam heating	247
, table of dimensions	249
Expansion of pipes	260
Expansion-joints	106
-tank	158
-traps	166
Explosions, boiler	172
Explosions of hot-water heaters	176
Extended-surface heaters	139

F

	PAGE
Factory and workshop heating	245
Fans and blowers	289
Field tube	139
Fittings, miscellaneous	96
pipe	92
Flange-union, joint	94
Float traps	165
Flues, dimensions of	52
for forced-blast system	284
, indirect heating, table	233
, ventilation, size of	49
Forced-blast systems of heating	283
Foot-pound, defined	3
Foundation, depth of, for boiler-setting	145
Fuels of the United States, table	389
Furnace, form of	270
, heating, formula for dimensions	274
, proportions of	272
Furnaces, directions for operating	118

G

Gases and air, flow of	40
, diffusion of	24
Gate-valve	100
Gauge-pressure	153
Gauges, Bourdon	154
, U-shaped, water	38
, vacuum	155
Globe valve	99
Governor for pump	255
Grates, kind of	163
Gravity circulating system	178
Green-house heating	236

H

Heat, bodily sensation of	19
, conduction of	18
, demand for	1
, flow through metals	61
, latent	15, 121
, loss of, from buildings	54
, measurement of, in test	71
, mechanical equivalent	3
, nature of	2
, radiant	15, 61

	PAGE
Heat, radiant, diffusion of	17
——, transmitted, table of	17
——, relation to electricity and work	2
—— removed by convection	17, 63
—— required for ventilation	58
——, specific	14
—— supplied by radiating surfaces	60
——, total, emitted from radiating surfaces, diagram	203
—— transformation	4
—— transmission, table of	69
——————, test of	264
——-unit	14
Heaters, extended surface	136
————, hot-water	135
————————, care of	171
————————, explosions of	176
————————, setting of	148
————, indirect, setting of	118
Heating-boilers, classification of	130
——————, for soft coal	142
——————, setting of	147
—————— with magazines	141
Heating, indirect, amount of surface required for	209
——-surfaces, indirect, tests of	79
————, systems of	20
————, with fan	283
High-pressure system	178, 254
Hot air and steam, combination of	190
——————— heating	268
——————————— formula for	273
Hot-blast heating, radiating surface required	291
———— system, air heated	293
————————, heating-surface	293
————————, size of blower	294
Hot-water and steam heating, tests	242
———————— heaters	135
————————————, care of	171
————————————, explosions of	176
————————————, setting of	148
———————— heating, general table, proportions	237
————————————, rule for pipes	232
————————————, table of data	229
————————————— of pipes	231
———————— radiators	112
Howard regulator	316
Humidity of air	29, 363
Hygrometer, description of	29

I

	PAGE
Indirect heating, air-flues	232
, dimensions of registers	235
, factors for flues	234
, surface required, table for	84
, table of proportions	238
, tests of surfaces	79
radiators	116
, efficiency of	84
, experiments on	81
Industry, magnitude of	1
Insulating substance, best known	198

J

Johnson system of heat regulation	318
Joint, lead, how formed	87
, rust, how made	88
Joints, flange-union	94

L

Lap-welding, process of	89
Latent heat	15
Lawler regulator	312
Lead joint, how formed	87
Leader-pipe	276
Lever safety-valve	149
Literature and references	353
Logarithms, how to use	357
of numbers, table of	377
Loss in transmitting steam	260

M

Main pipes, exhaust-steam	249
, hot-water heating	231
, steam-heating table	226
, steam and hot-water	237
Manometer	153
Marine boilers	132
Mason reducing valve	260
Maynard (S. T.) greenhouse test	242
Melting-points, table of	12
Mill's experiments on steam-heated surfaces	77
system of piping	181

N

	PAGE
Nipples, hooks, etc	97
Nitrogen	27

O

| Oxygen | 25 |
| Ozone | 25 |

P

Papers devoted to heating	355
Paul system	254
Pettersson's apparatus for determining CO_2	28
Petticoat-pipe	131
Pipe-boilers	138
-connections, hot-water heating systems	193
, steam-heating systems	191
-fittings	92
, radiating surface of	107
, return	179
, steel	91
Pipe systems, comparisons of	197
, table of dimensions	399
, wrought-iron	89
, thickness and size of	89
Pipes, method of computing area	222
Piping for indirect heaters	196
, method of, in greenhouses	239
, in hot-water heating	185
, systems of	180
Pitch, defined	179
Pitot's tube, description of	38
Plain-surface boilers	136
Plates, ceiling and floor	98
Pop-valve	150
Portable setting for boilers	147
Power's regulator	313
Pressure-gauge	153
, Bourdon	154
Pressure, methods of measuring	153
systems of hot-water heating	159
Properties of air, table of	381
of steam, table of	384
Proportions, hot-air heating	275
Protection of main pipe from loss of heat	197
of pipes	261
Pump-governor	255
Strength of materials, table	379

	PAGE
Pyrometers	11
—, calorimetric	12

R

Radiant heat, defined	60
—, diffusion of	17
—, emissive power, table of	16
—, transmission of	17
Radiating surface, exhaust-steam heating	249
for greenhouses	241
—, hot-blast heating	291
—, measurement of	73
of pipe	107
—, proportioning of	201
—, results of tests	75
—, rules for	215
surfaces, effect of painting	74
Radiation	15
—, amount of	61
—, direct	60
—, indirect	60
Radiators, contents of, how determined	74
—, direct indirect	116
—, effect of grouping surfaces	67
—, extended surface	79
—, flue	112
—, heat from	60
—, hot-water	112
—, testing	72
—, indirect	116
—, efficiency of	84
—, experiments on	81
—, material of	67
—, method of testing	69
—, proportion of parts of	119
—, sectional	110
—, tests	78–83
—, valves	101
—, vertical-pipe	109
Reducing-valves	258
Reflection and transmission of radiant heat	16
power, table of	16
Refrigerating machines, heating with	299
Registers, area of, hot-air heating	274
—, dimensions of	52, 235
for forced-blast system	200
—, table of	280
—, table of dimensions	275

	PAGE
Regulator for temperature	310
Regulators, damper	156
Relations of units for measuring pressures	155
Relay, term defined	179
Relief- or drip-pipe	179
Reliefs	228
Return pipes	179
,, table	227
steam-traps	167
Risers	179
Rules, approximate, for estimating radiating surface	215
, hot-water mains	232
, steam-mains	224
Rust-joint, how made	88

S

Safety-valve	149
, area of	150
Setting of heating-boilers	147
of hot-water heaters	148
of indirect heaters	118
Settings, brick, for boilers	143
Single-pipe system for hot-water heating	188
Siphon, term defined	179
-trap	164
Specific heat	14
Specifications, heating apparatus	322
, tubular boiler	341
Stacks, table for	278
Standard forms, hot water and steam specifications	323
Steam and water, flow of	217
, circulation, comparisons of	82
Steam-boiler, requisites of	121–127
-fitter's tools	349
-heating, general table of proportions	237
-loop	257
radiators, cast-iron	110
, vertical-pipe	109
tables, explanation of	120
-thermometer	13
transmission	260
-traps	164, 180
Steel pipe	91

T

Tables, see list on page	350
Taft, L. R., tests	244

	PAGE
Tank, expansion	158
Tees, Y's, pipe-junction, etc.	95
Temperature, boiling, table of	22
in various localities of the United States, table	398
, melting-points	12
, measured by color	12
produced by given amount of surface	85
regulators	310
, saving due to	320
required	1
Test of loss in steam-transmission	265
Thermal conductivity, table of	392
Thermometer, air and mercurial	10
-cup	13
, Fahrenheit and centigrade	7
, maxima and minima	12
, steam	13, 156
, use of	13
Thermostat	310
Tools, steam-fitter's	349
Transmission of steam	260
Traps, bucket	165
, counterweighted	165
, expansion	166
, float	165
, gravitating-return	169
, siphon	164
, steam	164–180
, steam-return	167
Tredgold's experiments, summary of	76
Try-cocks	152
Tubular boilers	138
, horizontal	130
, specifications for	341
Two-pipe system of steam-heating	184
Types of boilers	128

U

Underground pipe systems	261
Unit of heat	4

V

Vacuum-gauges	155
Valves, air	102
, angle	100
, check	102

	PAGE
Valves, corner and cross	101
, equalizing	168
, gate	100
, globe	99
, pop	150
, position of, in pipes	195
, radiator	101
, safety	149
Velocity of air due to heat	43
of water and steam	219
of water, hot-water heating	220
Ventilation, air required	31
by heat	35
by suction	42
ducts	269
-flues, size of	49
, table of	238
, influence of size of room	34
inlet for air	44
, mechanical	36
, principles of	21
, relation to heating	21
space for each person	52
, summary of problems	50
, systems of	298
Vertical boilers	132

W

	PAGE
Warming, systems of	20
Water and steam circulation, comparison of	82
, flow of	217
Water-columns	153
-hammer	180
-surface, steam and water space	126
-tube boilers	133, 138
Watts	5
Welding-lap, process of	89
Willame's system	253
Windows, loss of heat from	54
Wolff's rule for steam-mains	225
Workshop and factory heating	245
Wrought-iron pipe	89

www.ingramcontent.com/pod-product-compliance
Lightning Source LLC
Chambersburg PA
CBHW051736300426
44115CB00007B/581